AF386027

Thermodynamics: A Modern Approach

Thermodynamics:
A Modern Approach

HAL TASAKI
Professor, Department of Physics, Gakushuin University

GLENN PAQUETTE

OXFORD
UNIVERSITY PRESS

Great Clarendon Street, Oxford, OX2 6DP,
United Kingdom

Oxford University Press is a department of the University of Oxford.
It furthers the University's objective of excellence in research, scholarship,
and education by publishing worldwide. Oxford is a registered trade mark of
Oxford University Press in the UK and in certain other countries

Published in the United States of America by Oxford University Press
198 Madison Avenue, New York, NY 10016, United States of America

British Library Cataloguing in Publication Data
Data available

Library of Congress Control Number: 2025942884

ISBN 9780198842033

DOI: 10.1093/9780191878091.001.0001

Printed and bound by
CPI Group (UK) Ltd, Croydon, CR0 4YY

The manufacturer's authorized representative in the EU for product safety is
Oxford University Press España S.A. of Parque Empresarial San Fernando de Henares,
Avenida de Castilla, 2 – 28830 Madrid (www.oup.es/en or product.safety@oup.com).
OUP España S.A. also acts as importer into Spain of products made by the manufacturer.

Cover illustration: Mari Okazaki

To our families.

Contents

Preface

This book employs a modern point of view to present a novel framework for the theory of thermodynamics.[1] Its content should be accessible to sufficiently motivated undergraduate students in science and engineering, and it provides anyone with a background in calculus and classical mechanics the material necessary to teach themselves the fundamentals of thermodynamics, along with basic applications to physics and chemistry. In addition, owing to its novel point of view, this book allows readers who already have a working knowledge of thermodynamics to gain a new appreciation for the depth and beauty of its theoretical structure. It is designed as a textbook for a single-semester college course.[2] While it covers a great deal of material, that presented in the last few chapters is not necessary for a basic knowledge of the field. The theoretical framework of thermodynamics is constructed in the first six chapters. From these chapters, the reader can gain a deep understanding of the fundamental nature of thermodynamics and the concept of entropy. The final four chapters treat important applications of this theoretical framework.

Our goal in writing this book was to present the theory of thermodynamics in a clear and logical form that brings its concepts to life. In addition, we have aimed to present content and arguments that form a complete understanding, with the stance that the reader should not be left feeling that something has been glossed over or left vague. This emphasis on clarity and logic begins with our approach to constructing the overall theory, and it extends to the manner in which we lay out and connect the individual arguments used in this construction. In addition, we attempt to elucidate the relationship between the theory discussed here and the larger frameworks of physics and natural science in general, with the intention of providing the reader with a clear understanding of the position that thermodynamics occupies within the general scientific framework as well as the peculiar point of view that it offers.

With the above-stated aim, we do not concern ourselves with repeating the explanations found in "classic" textbooks. This book does not present a "standard" treatment of thermodynamics, and it is not organized as a handy reference book of thermodynamic relations. Rather, it represents our attempt to rethink the fundamental form of thermodynamics, from its starting point to its overall theoretical structure. We believe that the reformulation of thermodynamics that we offer here provides the reader with an intuitive understanding of this field in its clearest and most logical form.

Of course, the reformulation of thermodynamics carried out in this book would not have been possible if it were not for the groundbreaking work of the scientists who laid the conceptional foundation on which it is built. In particular, the 1999

---

[1] This book is based largely on a book published in Japanese by one of the authors (H.T.) in 2000. That book has become one of the most popular thermodynamics textbooks in Japan. However, this English version has been extensively rewritten, with many significant improvements made throughout. Numerous parts (including all of Chapter 2) are entirely new. Supplemental information will be posted at https://haltasaki.github.io/tasaki_paquette/TD/.

[2] We present an outline for such a course at the end of Section 1.3.

paper by Elliott Lieb and Jakob Yngvason [9] on the axiomatic formulation of thermodynamics was greatly influential in our attempt to understand thermodynamics. That work constructs the framework of thermodynamics in a mathematically elegant and rigorous manner, while providing a clear physical interpretation of entropy and its role in the theory.[3] From a more philosophical point of view, the understanding that we learned from Yoshi Oono—that macroscopic thermodynamics should be regarded as the empirical foundation for microscopic statistical mechanics—was also essential.

Sampling some of the countless thermodynamics textbooks, one comes to realize that almost all present the fundamental structure of the theory in the logical form developed by Rudolf Clausius more than one hundred years ago.[4] However, we feel that the formulation of Clausius is not ideal, first because it is not particularly easy to understand, and second because it encounters some subtle difficulties in the treatment of phase transitions.

In this book, we present the theory of thermodynamics through a novel reconstruction. While the content of this theory itself is, of course, identical to that of the conventional theory, the point of view that we take in its construction differs from that taken in conventional textbooks. The fundamental manner of thinking employed in this book is the following:

We consider thermodynamics from an operational point of view in which the most important concept is *work*. With the focus on work, after individually studying isothermal operations (and the second law) and adiabatic operations (and the first law), we seek a unified framework within which both can be understood. In this way, the overall structure of thermodynamics naturally takes form.

We employ a "semi-axiomatic" approach, in which the results needed to construct the theory are derived from postulates and definitions that are based on empirical observation. For the sake of simplicity and clarity, however, we have intentionally avoided a presentation that is strictly rigorous in a mathematical sense.

Our purpose in presenting an alternative to the conventional construction is not simply to display our eccentric views. Rather, like many physicists, we found the conventional approach to thermodynamics difficult to understand, and, while studying a number of modern works, we searched for a clearer approach. The formulation presented here is the result of that search.

We do not wish to claim that what we present here is the clearest possible formulation of thermodynamics. Indeed, we await the criticism and advice of readers. Our hope is that this book will provide a stimulus that leads to the formulation of new approaches to the understanding and teaching of thermodynamics by scientists with various points of view. More generally, we hope that the point of view

---

[3] While our formulation of thermodynamics owes much to that of Lieb and Yngvason, it is important to note that the two formulations are conceptually different. Also, our formulation is mathematically somewhat less rigorous than theirs.

[4] There are a few, however, that are based on the axiomatic systemization of Constantin Carathéodory or the formulation of Josiah Willard Gibbs. Of course, there are also some that attempt neither to establish a foundation for the theory nor to derive it in a systematic manner, simply presenting the necessary formulas in piecemeal treatments that provide an incomplete understanding. Indeed, perhaps the majority of textbooks fall into this category.

expressed in this book with regard to both thermodynamics and natural science as a whole will be of benefit to future generations of scientists as they attempt to expand the frontiers of our understanding.

Our attempt to find a new way of understanding thermodynamics, which led to the reorganization of the theory presented here, was in many ways a profound and rewarding experience. But our greatest reward would be the joy of knowing that this book allows the reader to experience even a small part of the exhilaration that we have felt in this pursuit.

Numerous people have made contributions to this book. In particular, we are indebted to Yoshi Oono, Shin-ichi Sasa, Shinji Takesue, Takuma Tanaka and Ichiji Tasaki, who all contributed both directly, through useful comments and suggestions on the manuscript at various stages, and indirectly, through valuable discussions of science, which helped form our understanding of thermodynamics. We would also like to acknowledge Joel Lebowitz and Elliott Lieb, who have greatly influenced our way of thinking about the physics of macroscopic systems. Finally, we would like to thank Mari Okazaki for the cover artwork. We believe that this illustration beautifully captures the spirit of this book in Mari's distinctive manga style.

January 2025

Hal Tasaki and Glenn Paquette

Notation

The following notation is employed for the case of the single-component systems treated in Chapters 1–8. A simple generalization of this notation is used in Chapter 9, where we treat multi-component systems. In Chapter 10, where we treat magnetic systems, the roles played in the previous chapters by the volume, V, and pressure, p, are instead played by the magnetization, M, and external magnetic field, H, respectively.

- **Extensive variables:** V (volume), N (amount of substance[5]). [Section 2.2]

- **Collective extensive variable:** We often use X (and occasionally X', Y, Z, etc.) to collectively represent the extensive variables characterizing a system (most often, V and N). [Section 2.2]

- **Temperature:** T. [Sections 2.4.4, 2.4.6, 5.2.1, A.3]

- **Equilibrium state:** We represent an equilibrium state by $(T; V, N)$ or $(T; X)$, $(T; X, Y)$, etc. [Section 2.4]

- **Isothermal transition:** $(T; X_1) \xrightarrow{\text{i}} (T; X_2)$ represents a transition from the equilibrium state $(T; X_1)$ to the equilibrium state $(T; X_2)$ resulting from an operation carried out on a system under isothermal conditions, i.e., a system in thermal contact with an environment at constant temperature. [Section 3.1]

- **Quasi-static isothermal transition:** $(T; X_1) \xrightarrow{\text{qi}} (T; X_2)$ represents an isothermal transition resulting from an operation carried out quasi-statically. [Sections 3.1, A.2]

- **Adiabatic transition:** $(T_1; X_1) \xrightarrow{\text{a}} (T_2; X_2)$ represents a transition from the equilibrium state $(T_1; X_1)$ to the equilibrium state $(T_2; X_2)$ resulting from an operation carried out on a system under adiabatic conditions, i.e., a system separated from the external world by adiabatic walls.[6] [Section 4.1]

- **Quasi-static adiabatic transition:** $(T_1; X_1) \xrightarrow{\text{qa}} (T_2; X_2)$ represents an adiabatic transition resulting from an operation carried out quasi-statically. [Sections 4.1, A.2]

- **Generalized isothermal transition:** $(T_1; X_1) \xrightarrow{\text{i}'} (T_2; X_2)$ represents a transition in which the system begins in an equilibrium state under adiabatic conditions with temperature T_1 and ends in an equilibrium state under isothermal conditions with temperature T_2. [Section 4.3.1, Problem 5.2]

- **Maximum work:** $W_{\max}(T; X_1 \rightarrow X_2)$ represents the maximum amount of work that can be performed by a system on the external world in an isothermal

---

[5] Sometimes we simply use *amount* in place of *amount of substance* when the meaning is clear.

[6] In mechanics (both classical and quantum), the term *adiabatic* is commonly used in reference to a slow variation of system parameters. In thermodynamics, however, it is used to describe a situation in which a system is perfectly thermally insulated, exchanging no heat with its environment.

operation under which the system undergoes a transition from X_1 to X_2 at temperature T. [Section 3.5]

- **Work per cycle:** W_{cyc} represents the amount of work performed by a system on the external world in a "cycle" (i.e., an operation under which the initial and final states of the system are identical). [Sections 3.2, 5.2.2, 5.3]

- **Adiabatic work:** $W_{\mathrm{ad}}((T_1; X_1) \to (T_2; X_2))$ represents the amount of work performed by a system on the external world in an adiabatic operation giving rise to the transition $(T_1; X_1) \stackrel{\mathrm{a}}{\longrightarrow} (T_2; X_2)$. [Section 4.2.2]

- **Maximum heat:** $Q_{\mathrm{max}}(T; X_1 \to X_2)$ represents the maximum amount of heat that can be absorbed by a system from its environment in an isothermal operation under which the system undergoes a transition from X_1 to X_2 at temperature T. [Section 5.1.3]

- **Helmholtz free energy:** $F[T; V, N]$. [Sections 3.6, 7.1]

- **Pressure:** $p(T; V, N)$. [Section 3.7]

- **Energy:** $U(T; V, N)$. [Section 4.3]

- **Entropy:** $S(T; V, N)$. [Section 6.1]

- **Chemical potential:** $\mu(T; V, N)$ and $\mu(T, p)$. [Sections 7.1.3, 8.2.1]

- **Enthalpy:** $H(T; V, N)$. [Section 7.5.4]

- **Gibbs free energy:** $G[T, p; N]$. [Section 8.1]

- **Volume (as a function of T, p and N):** $V(T, p; N)$. [Section 8.1.3]

- **Saturated vapor pressure:** $p_{\mathrm{v}}(T)$. [Section 7.5.2]

- **Boiling point:** $T_{\mathrm{b}}(p)$. [Section 7.5.4]

- **Critical point:** $(T_{\mathrm{c}}, p_{\mathrm{c}})$. [Section 7.5.4]

- **Enthalpy of vaporization:** $H_{\mathrm{vap}}(T; N) = h_{\mathrm{vap}}(T)N$ and $H_{\mathrm{vap}}(p; N) = h_{\mathrm{vap}}(p)N$. [Sections 7.5.4, 8.4.2]

1

What Is Thermodynamics?

In this chapter, as an introduction, we present an elementary treatment of a thermodynamic system employing the equation of state of an ideal gas. Then, focusing on the concept of universality, we discuss the status of thermodynamics within science as a whole, with particular emphasis on elucidating the relationship between thermodynamics and statistical mechanics. We next explain the distinguishing features of our formulation of thermodynamics. Finally, we present a brief outline of the organization of this book.

1.1 From the thermodynamics of gases to a general theory of thermodynamics

The reader may already have some experience with the field of thermodynamics, and perhaps this includes the treatment of an ideal gas. In this chapter, we study the equation of state for an ideal gas and briefly describe how the concepts and behavior illustrated by an ideal gas appear throughout the theory of thermodynamics developed in this book. We hope that through this simple investigation, the reader will be able to gauge how their present understanding of thermodynamics compares with the actual theory. In this chapter, we purposely avoid formality and refrain from identifying the theoretical status of each component of the arguments—i.e., as a definition, an empirical result, a conjecture, and so on—in order to lay out the basic ideas in an intuitively understandable manner. This style differs from the more formal style used in the rest of the book.

1.1.1 The equation of state and work

Here, we consider the system depicted in Figure 1.1, consisting of a closed cylindrical vessel filled with a gas. This vessel has circular ends of area A and an adjustable length, which is controlled by a moveable piston, acting as one of its ends. Suppose that the entire apparatus is situated in an environment of constant temperature, T.[1] Let us assume that the system and the environment are in *thermal contact* and that the system has existed in the environment for a sufficiently long time that it is in an equilibrium state with the same temperature. Further, suppose that the vessel contains N moles of gas and that its volume is fixed at some value V. The reader may already have in mind the well-known equation

$$pV = NRT \,, \tag{1.1}$$

---

[1] This is the *absolute temperature*, measured in kelvins (see Section A.3).

Thermodynamics: A Modern Approach. Hal Tasaki and Glenn Paquette, Oxford University Press.
© Hal Tasaki and Glenn Paquette (2026). DOI: 10.1093/9780191878091.003.0001

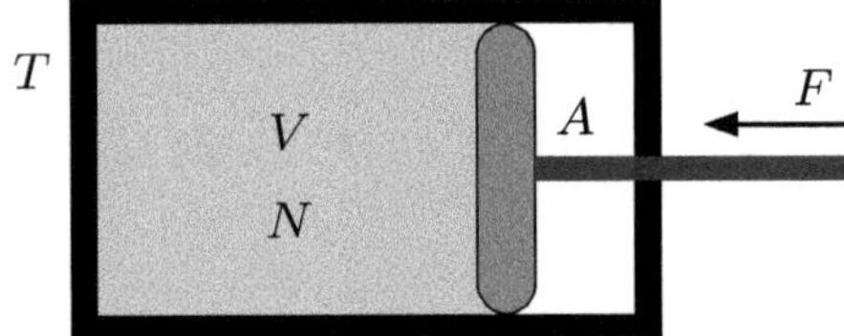

Figure 1.1 An amount of gas N (measured in units of moles) is contained within a closed vessel whose volume, V, is controlled by the position of a piston. In order to maintain a constant volume, it is necessary to apply a force F on the piston from the right. The entire system is situated in an environment (a surrounding gas) at temperature T. With this familiar system as our starting point, in this chapter we elucidate in basic terms the class of phenomena studied and the type of description sought within the field of thermodynamics.

where p is the pressure of the gas and R is the *gas constant*.[2] Below, we study the significance of this equation.

We all have an intuitive understanding of the pressure of a gas as a measure of how strongly it presses outward on its container. In the presently considered situation, the pressure is easily measured. Here, in order to keep the piston fixed, it is necessary to apply some force to it from the right. Let us call this force F. Then, the pressure is given simply by $p = F/A$.[3] Of course, rearranging (1.1), we have another expression for the pressure,

$$p = \frac{NRT}{V} \,.$$
(1.2)

We thus see how (1.1) allows us to determine from quantities characterizing the state of the gas—its amount, its temperature, and its volume—how this gas interacts mechanically with the external world. A hypothetical gas for which (1.2) holds exactly is referred to as an *ideal gas*.

In general, an equation like (1.2), expressing the pressure as a function of N, V and T, is known as an *equation of state*. Despite its simplicity, (1.2) provides a very accurate approximation even for real gases under conditions of sufficiently low concentration and sufficiently high temperature. However, in general, the actual pressure and that given by this equation can differ significantly. In addition, in a real system, as T or V becomes small, we can observe *phase transitions*, through which the gas changes to a liquid or a solid. Here, however, let us ignore such more complicated situations and limit our consideration to only those simple cases in which (1.2) holds to the desired precision. At this point, we simply note that systems exhibiting more complicated behavior do exist and that these too are studied within the context of thermodynamics.[4]

---

[2] The gas constant is a universal constant with the *exact* value $R = 8.31446261815324\,\mathrm{J\,K^{-1}mol^{-1}}$.

[3] In a typical experimental situation, there will be a gas surrounding the system that exerts a pressure on the piston, and this must be taken into account when determining the pressure of the gas inside the container. However, here, in order to simplify the discussion, let us assume that the external region to the right of the piston is a vacuum and thus that there is no such external pressure.

[4] Indeed, categorizing the types of phase transitions that can take place in thermodynamic systems with respect to their structures and elucidating the universal relations that hold among the various physical quantities characterizing such systems during phase transitions are important areas of research in this field. We treat phase transitions in Sections 7.5 and 10.2–10.4.

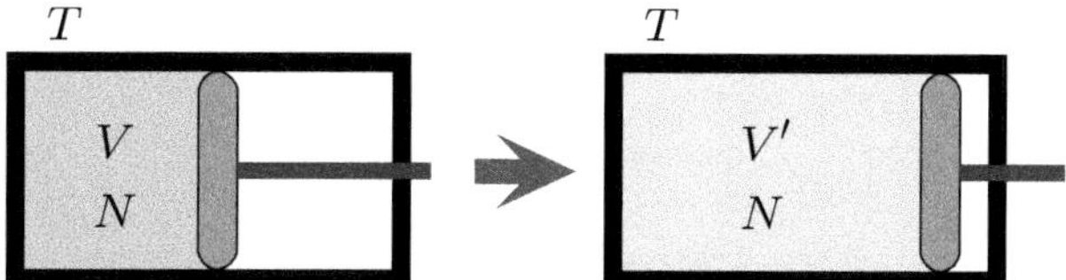

Figure 1.2 A process in which the volume of a vessel containing an amount N of an ideal gas in an environment at temperature T is allowed to change from V to V'.

Now, suppose that with the situation depicted in Figure 1.1, the piston is allowed to move a small distance Δx to the right. Assuming that the gas continues to push on the piston to the right with the force F introduced above,[5] under this process the gas will perform an amount of work $\Delta W = F\Delta x$ on the external world (i.e., on the external system pushing to the left on the piston). Next, noting that the amount by which the volume occupied by the gas changes during this process is $\Delta V = A\Delta x$, and recalling the equality $p = F/A$, we find that we can also write $\Delta W = p\Delta V$. Hence, from the equation of state (1.2), we obtain the relation

$$\Delta W = p\Delta V = \frac{NRT}{V}\Delta V \tag{1.3}$$

for the amount of work performed by the gas on the external world as the piston moves the distance Δx. With these very elementary considerations, we thus see how the equation of state (1.2) is directly related to work. This is a realization of fundamental importance. Indeed, the simple example considered here demonstrates the main approach employed in this book—namely, to characterize the properties of thermodynamic systems in terms of mechanical work.

Next, let us consider the situation depicted in Figure 1.2, where the piston moves some large amount to the right, resulting in a change of volume from V to V'. The amount of work, W, performed in such a process can be determined by dividing this process into many short subprocesses and adding up the amounts of work performed in all of them. Proceeding in this manner and using the form (1.3) for each subprocess, we obtain

$$W = \int_V^{V'} \frac{NRT}{\widetilde{V}}d\widetilde{V} = NRT\log\frac{V'}{V}\,. \tag{1.4}$$

However, it turns out that this treatment is too simple, and, in fact, this relation holds only if the operation is carried out sufficiently slowly. If an experiment like that described by Figure 1.2 is carried out rapidly,[6] the amount of work actually done by the gas on the external world will be *less* than the amount W given in (1.4). More generally, the total amount of work performed by the system in an operation changing its volume will increase as the speed of the operation decreases. The value W appearing in (1.4) is the amount of work performed by the gas in the limiting case in which the piston moves infinitesimally slowly throughout the operation (see Figure 1.3).[7]

---

[5] In other words, we assume that this operation itself does not affect the force exerted on the piston by the gas.

[6] A process of the kind described here (carried out at any speed) is termed an *isothermal operation*.

[7] This type of process is referred to as a *quasi-static isothermal operation*.

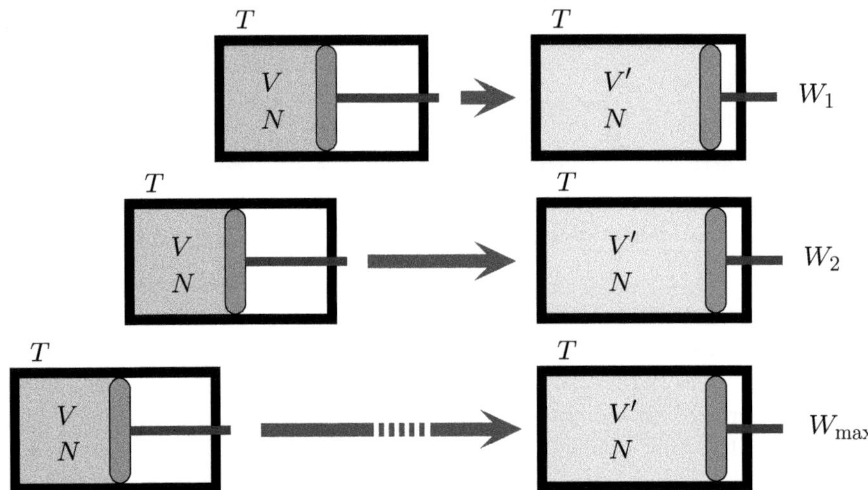

Figure 1.3 The work performed on the external world in a process through which gas in thermal contact with an environment at constant temperature is allowed to expand. The speed with which the operation is carried out decreases from top to bottom. The amounts of work performed in the three cases are related as $W_1 < W_2 < W_{\mathrm{max}}$. In the case that the operation is carried out very rapidly, because the gas cannot "keep up with" the piston, the amount of work it performs is small. As a general rule, the amount of work increases as the speed of the piston decreases, and in the limit of vanishing speed, the maximum amount of work, W_{max}, is realized.

Because the quantity given in (1.4) is always greater than or equal to the amount of work performed by the gas in a process through which its volume changes from V to V', we refer to it as the *maximum work* for such processes. In subsequent chapters, we express quantities of this kind as $W_{\mathrm{max}}(T; (V, N) \to (V', N))$. The maximum work represents something of a reference quantity for all possible processes with the same beginning and ending states (in the present case, those corresponding to the volumes V and V'), and for this reason, it provides a way of relating the amounts of work performed in these processes. Establishing relations among the quantities of work done in different processes is one of the fundamental investigative methods employed in thermodynamics. Indeed, this method is of essential importance, because (at least within this book) it is *work* (not *heat*) that plays the central role in thermodynamics. This is a fundamental idea lying at the foundation of the formulation presented in this book, as is seen throughout our formal treatment, which begins in the next chapter. The maximum work is closely related to the *Helmholtz free energy*, which is an extremely important quantity in thermodynamics. The precise connection between these two quantities is elucidated in Chapter 3.

1.1.2 Adiabatic operations and reversibility

To this point, we have studied an experimental system consisting of an ideal gas that is in thermal contact with an environment at a constant temperature. Now, let us consider the case in which the vessel containing the gas is surrounded by walls

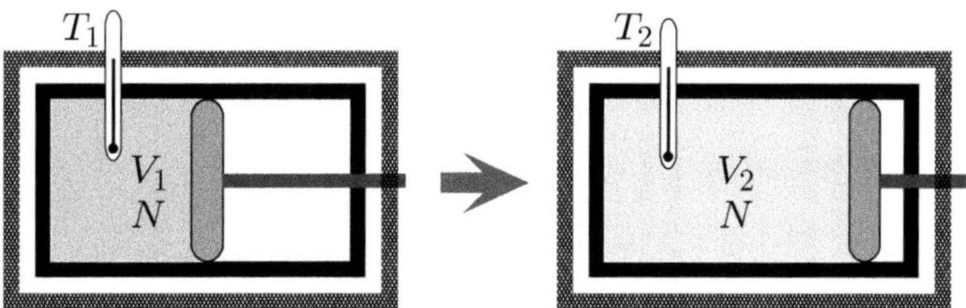

Figure 1.4 An adiabatic operation. If we increase the volume of a gas-filled container enclosed within adiabatic walls from V_1 to V_2, the gas will experience a decrease in temperature (unless the operation is carried out extremely rapidly). In the case that the operation is carried out very slowly, the temperature will return to its original value if the operation is subsequently reversed, and thus the system will return to its original state.

consisting of some material that prevents thermal contact between the system and its environment (for example, foam polystyrene). Walls of this kind are referred to as *adiabatic walls*.

First, suppose that the system has been prepared with volume V_1 and amount of substance N. Then, suppose that after this system has been placed in thermal contact with a surrounding gas at temperature T_1 and left undisturbed for a long time, it is enclosed within adiabatic walls. Clearly, the operation of simply placing these walls around the container should not change the temperature of the gas inside. However, if the volume of the container is subsequently altered to some new value V_2, in general, the temperature of the gas will change (see Figure 1.4). In particular, in the case that the volume changes infinitesimally slowly[8] and the system remains an ideal gas throughout the operation, the initial and final temperatures, T_1 and T_2, will be related to V_1 and V_2 in accordance with the following Poisson relation:[9]

$$T_1^{3/2}V_1 = T_2^{3/2}V_2 \,. \tag{1.5}$$

This relation can be derived from the general principle of energy conservation for a thermodynamic system. Of course, the pressure of the system will also change under this operation, as determined by (1.2).

Next, let us consider the case in which the operation described by Figure 1.4 is carried out in reverse after the completion of the original operation; i.e., with the adiabatic walls remaining in place, the volume is decreased infinitesimally slowly from V_2 to V_1. The relation (1.5) holds for this process too, and thus after reaching the volume V_1, the gas will possess its original temperature, T_1. Because N is fixed, the fact that the volume and temperature have returned to their original values implies that the system has returned to its original state. An operation that can be completely "undone" in this way is termed *reversible*. It is important to note that the situation is different if these operations are not carried out slowly. In general, if the volume of this insulated system is changed rapidly from V_1 to V_2 and then back to V_1, the state of the gas after it returns to the original volume will differ from

---

[8] An operation of the type considered here (carried out at any speed) is termed an *adiabatic operation*. An adiabatic operation carried out infinitesimally slowly is referred to as a *quasi-static adiabatic operation*.

[9] The exponent 3/2 appearing here is correct only in the case of a monatomic ideal gas.

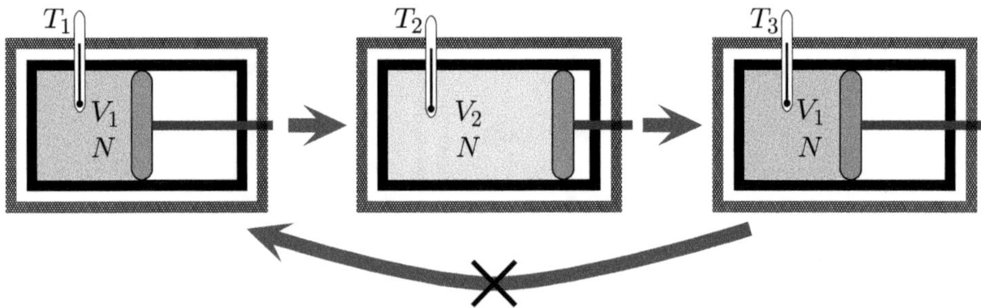

Figure 1.5 An operation under which the gas contained in a vessel surrounded by adiabatic walls undergoes rapid expansion followed by rapid compression. In such a situation, the gas does not return to its original state at the completion of the compression (see Problem 1.1). This is a typical example of an irreversible thermodynamic process.

the original state. In this case, as shown in Figure 1.5, at the end of the second operation, although V has been returned to its initial value, the temperature will have some value T_3 that is necessarily higher than the original temperature, T_1. An actual example of a pair of operations of this kind that can be treated exactly is given in Problem 1.1.

Although a gas undergoing rapid adiabatic expansion and subsequent compression will not return to its original state, it is interesting to consider whether there is some additional operation that can be applied to the gas after these changes that will result in a decrease of temperature to the original value, and hence cause the gas to again realize its original state. Obviously, this could be done by simply removing the adiabatic walls and allowing the gas within the container to re-establish thermal contact with the external gas. However, we wish to know if there is some other way of returning the system to its original state. Specifically, is it possible—with the adiabatic walls still in place—to move the piston in such a way that the gas will return to its original volume and temperature? Recalling that the temperature increased when the volume of the gas was first increased and then decreased, it may seem logical that the temperature would decrease if the volume were first decreased and then increased. However, in fact this is not the case, and the result of such operations can only be to increase the temperature further. We are thus led to the conclusion that, without removing the adiabatic walls, there is no possible manipulation of the piston's position that would cause the gas to return to its original state. But, looking beyond this simple type of operation, it is natural to ask if the temperature increase could be undone by utilizing some more complicated operation carried out on the system. Whether or not it is possible to realize such a result is in its essence the fundamental thermodynamic problem of reversibility/irreversibility in adiabatic operations. In this book, we consider a general solution to this problem. This is done through the introduction of the very important physical quantity called *entropy* (see Chapter 6). One of the main goals of the first half of this book (namely, Chapters 2 through 6) is to acquire an understanding of entropy from a macroscopic, operational point of view.

1.1.3 A simple heat engine and the goal of thermodynamics

In Section 1.1.1, we saw how an ideal gas system performs work when an operation altering its volume is carried out. We now consider a simple situation demonstrating the application of this behavior. Figure 1.6 describes how a gas can be used as the working substance of a heat engine that lifts a weight by converting "heat energy" into mechanical work. The source of this heat energy is a high-temperature heat source. The ratio of the amount of work produced by the engine to the amount of heat that it absorbs from the heat source is called the *efficiency*. (The amount of heat absorbed from the heat source can be determined by measuring the amount of fuel needed to keep its temperature constant while the engine performs work.)

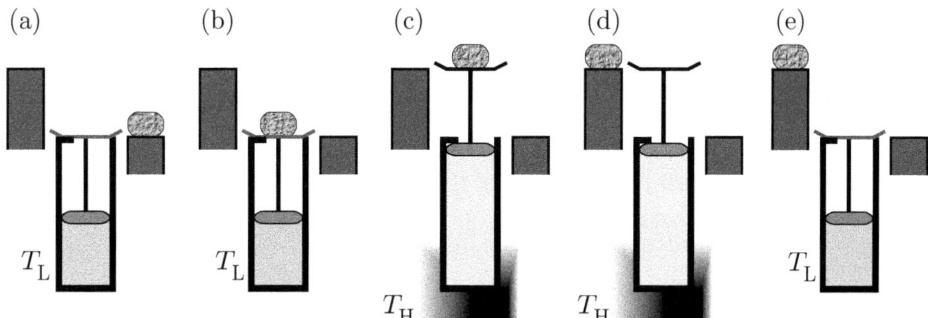

Figure 1.6 A heat engine that raises a weight step by step. The engine contains a gas that acts as the working substance. The entire apparatus is surrounded by a separate gas of temperature T_L and pressure p_0. (For simplicity, we assume that the masses of the piston and the plate on which the weight rests are sufficiently small that they can be ignored.) The process of lifting the weight is carried out in the following steps. (a) The gas in the container is in equilibrium with the external gas at temperature T_L. Here, the two gases have the same pressure, p_0. (b) The weight is moved from the lower shelf to the plate situated atop the piston. (We assume that the plate is supported by some external device that prevents it from moving downward as a result.) (c) The gas is placed in contact with a heat source at a temperature T_H. If T_H is sufficiently higher than T_L, the gas in the container will expand and lift the weight to the height of the higher shelf. We can understand this process as a conversion of one kind of energy into another: Energy in the form of "heat" flows into the gas from the heat source, and some of this energy is converted into the mechanical work that raises the weight. (d) Next, the weight is moved onto the higher shelf. (Removing the weight from the plate would result in further expansion of the gas, causing the plate to move higher, but, again, we assume that this is prevented by some external device.) (e) Finally, the high-temperature heat source is removed, and hence the gas begins to cool, with its temperature gradually approaching that of the ambient gas, T_L. This drop in temperature occurs as heat flows from the internal gas to the external gas. The steps (a) through (e) constitute one cycle of this heat engine. At the conclusion of step (e), the piston-gas system has returned to its original state, and thus this entire process can be repeated indefinitely. (See Section 5.3 and Problems 1.2 and 5.4 for further consideration of heat engines.)

Because increased efficiency allows for a reduction in fuel consumption, designing ever more efficient heat engines is a problem of great importance in modern society.

The efficiency of an engine like that depicted in Figure 1.6 would certainly not be high. (Calculating this efficiency is the topic of Problem 5.4.) Developing high-efficiency heat engines by combining well-suited materials and carefully designed mechanisms is one important application of thermodynamics. Although the problem of designing actual efficient heat engines is beyond the scope of this book, using only the fundamental concepts of thermodynamics, we are able to elucidate the basic nature of heat engines and obtain an understanding of the concept of efficiency. Indeed, with very simple considerations employing only fundamental principles, it can be shown that the efficiency of a heat engine can never exceed a certain value determined uniquely by the temperatures of the heat source and the environment and that this limiting value is necessarily less than 1. This absolute limitation on the performance of heat engines cannot be overcome—no matter how cleverly designed the mechanism nor how exquisitely fabricated the material. If the efficiency were equal to 1, then all of the energy absorbed from the heat source as heat would be converted into mechanical work. In other words, no energy would be wasted. The fact that there is a limiting value less than 1 for the efficiency implies that in our real world, such a perfectly efficient heat engine cannot exist. This remarkably powerful result is one of the numerous groundbreaking discoveries reported in Nicolas Léonard Sadi Carnot's short book *Réflexions sur la puissance motrice du feu et sur les machines propres à développer cette puissance* (*Reflections on the Motive Power of Fire and on Machines Fitted to Develop that Power*), published in 1824. This revolutionary book can be considered the starting point for the field of thermodynamics.

The reader may be surprised that the topic of discussion here has jumped suddenly from gas in a simple container to cleverly designed mechanisms and exquisitely fabricated materials. The fact that physical systems of such disparate types can be treated within the framework of thermodynamics hints at the extremely broad applicability of this field. Indeed, it can be said that the great range of its applicability is the most distinctive characteristic of the field of thermodynamics. The main goal of thermodynamics is to establish a theoretical framework describing macroscopic systems that provides universal,[10] exact relations characterizing equilibrium states and the realizable transitions among them, along with the exchanges of energy that take place in such transitions. These relations are applicable to physical systems and phenomena studied in a broad range of fields, including physics, chemistry, engineering, meteorology and physiology. No inconsistency has been found between the predictions of thermodynamic theory and experimental results, and in many contexts, thermodynamics is regarded as a theory of absolute validity.

While there is almost no limit to the number of applications of thermodynamics, in this book we focus mainly on describing the fundamental thermodynamic theory of gases and liquids and give brief treatments of chemical reactions and ferromagnets.

---

[10] A *universal relation* is a relation that applies uniformly to all physical systems within a broad class.

1.2 Thermodynamics and universality

1.2.1 Introduction

As emphasized in the previous section, in studying the fundamentals of thermodynamics, our goal is not to describe the specific features of individual thermodynamic systems but, rather, to understand the universal properties shared by all thermodynamic systems, properties that are independent of the details of the actual physical systems that possess them.[11] Of course, the proposition that we could find some kind of quantifiable universal structure characterizing the behavior of all macroscopic systems is in no way trivial. Indeed, we regard the discovery that such a structure does exist as something truly remarkable.[12] That we can discern and understand such structure is an empirical fact that has come to be understood during the course of human history, through our experience in observing the world, as well as through formal scientific investigation, both experimental and theoretical.[13]

1.2.2 Science and universality

The reader may wonder how thermodynamics is related to the mechanical descriptions of the world provided by such microscopic theories as classical mechanics and quantum mechanics or, even, quantum field theory, which forms the foundation of the current standard model of particle physics. Are thermodynamic laws merely approximate relations that can be derived from microscopic, mechanical laws? Or, do thermodynamic laws constitute some different type of scientific framework that is independent of microscopic laws?

---

[11] In this section, we present an understanding of science from our point of view as scientists. However, it should be noted that the theory constructed in the remainder of the book does not depend on the validity of this point of view, and thus the reader's acceptance of the arguments given here is not a prerequisite for understanding this theory. In fact, the reader could skip this entire section without missing any of the content of the theory that we present.

[12] Obtaining an answer in terms of a microscopic description to the question of why we observe such strong universality in macroscopic phenomena is an important topic in modern physics. We are still seeking such an answer.

[13] A fine treatment of the history of thermodynamics is given in Ref. [15] (in Japanese) by Yoshitaka Yamamoto. Most accounts of the history of thermodynamics written by physicists seem to be based on the premise that progress was made through the inevitable process of modern, "correct" ideas coming to replace outdated, "incorrect" ideas, as the absurdity of these older ideas became revealed. (Such accounts express a clearly Whiggish view of history, in a negative sense.) By contrast, the historical treatment given in Ref. [15] presents previous interpretations and ideas in the context in which they arose, citing the experimental and theoretical evidence that supported them, and then proceeds to describe the developments that eventually led to their revision or abandonment. For example, the interpretation that heat is an attribute of "heat particles" (the so-called *caloric theory*) is often presented as a foolish idea, lacking scientific merit. Of course, the validity of the caloric theory was eventually refuted, but obtaining the evidence necessary to do this was by no means simple. Note that the analogous interpretation of electric phenomena, based on the idea that electric charge is an attribute possessed by particles, remains an integral part of present-day physics. Clearly, if we simply compare the types of macroscopic behavior exhibited by systems to which these two theories are meant to apply (heat current and electric current, frictional heating and triboelectric charging, etc.), it is certainly not easy to recognize the fundamental difference between heat and electricity.

We believe that, in fact, thermodynamic laws are neither completely derivable from mechanical laws nor completely independent of them.[14] It is our understanding that quantum mechanics, classical mechanics and thermodynamics each forms a universal structure that provides a description of some aspects of the natural world. As we look around us at the limitless variety of phenomena that this world displays, it is difficult to imagine how we, with our tiny reasoning power, could hope to obtain any kind of systematic understanding of it. However, we have found that if we properly limit our inquiry, and focus on only certain aspects of the phenomena that we observe, it is possible to construct concise theoretical frameworks that indeed do provide descriptions of these particular aspects. This is a truly miraculous empirical fact that the pioneers of science discovered through the long course of scientific inquiry. In the ideal case, the theoretical framework that comes to be constructed through such a process possesses a *universal structure* that provides a description of some large class of phenomena in a universal form.

Before continuing with the present discussion, allow us to clarify the meaning with which we use the term *universal structure*. As a rough guide, we present the following as properties necessary for a theoretical construct to be deemed universal structure.

- The structure provides quantitative predictions that are consistent with experimental results for some class of physical phenomena. In particular, we can imagine some limiting, ideal physical situations in which the agreement between quantifiable aspects of the phenomena under consideration and their description provided by this structure becomes exact.
- The structure describes general aspects of behavior that are common to physical systems of a broad class, and it is insensitive to the causes (both known and unknown) determining the specific properties and conditions that vary from system to system within this class.[15]
- The structure possesses a mathematically closed form.[16]

It is important to understand that universal structure, as described above, is not something that is *in that very form* intrinsic to the real world. Thus, the task we face in attempting to discover these types of structures is not simply one of observing nature and picking them out. However, it is also not the case that humans can impose these structures upon nature, at their convenience or in accordance with their own purposes and tastes.[17] Rather, universal structure is an

---

[14] What we present here is our own understanding and interpretation of science, but we believe that there are many scientists today who possess similar understandings. Also, we wish to point out that the discussion given in this section is not meant to be a systematic treatment of the philosophy of science. Rather, our goal is to simply sketch the form of science as it is perceived by many practicing scientists.

[15] For example, the relation $F = ma$ itself applies uniformly to a vast range of physical systems, while the explicit form that it takes in any individual case will depend on the properties and conditions characterizing the individual system under consideration. In general, the causes determining these properties and conditions may be quite complicated and even unknown to us, but this does not affect the validity of $F = ma$.

[16] We are not yet sure how important this last property is.

[17] Within the philosophy of science, the interpretation that scientific theories are not *objective truths* but merely products of sociological processes occurring within scientific communities is propounded by some extreme sociological constructivists. This represents a misunderstanding that results from a simplistic view of science obtained from a very superficial observation of scientific activity and history. (For related discussion, see the insightful book *Fashionable Nonsense: Postmodern Intellectuals' Abuse of Science*, by A. Sokal and J. Bricmont, Picador, 1999, and particularly Chapter 4 therein.) The "rules"

abstraction that is extracted from nature through a long process consisting of both thorough observation employing painstakingly designed experiments and careful theoretical investigation, which are repeated back and forth, again and again. In this sense, universal structure is neither "picked from" nor "imposed upon" nature but instead lies hidden within nature and comes to be known to humans by virtue of their investigative and reasoning powers.

Through the long history of scientific inquiry, humans have managed to extract many types of universal structures from the natural world and to describe these structures in theoretical, mathematical forms. For example, focusing on the particular class of phenomena constituted by the motion of macroscopic objects, from the scale of everyday objects to the scale of celestial bodies, the universal structure that we know as classical mechanics was formulated. Then, beginning with the study of atomic structure, the observation of microscopic phenomena led to the construction of quantum mechanics, another type of universal structure. Similarly, through the investigation of the countless types of phenomena involving heat that are exhibited by macroscopic matter, the universal structure of thermodynamics came to be developed.

At the time that the theory of thermodynamics was established—the middle of the 19th century—quantum mechanics had not yet been developed, and the idea that matter is composed of tiny particles was still not universally accepted. In this context, on the basis of a great variety of observational and experimental results, the exact science of thermodynamics was formulated. Subsequently, the field of statistical mechanics was conceived, atomic theory was established, and quantum mechanics was formulated as the theory describing the microscopic structure and behavior of matter. Remarkably, while the development of each of these theories can truly be regarded as a "revolution," as each had a profound effect on our understanding of nature, none of them necessitated a revision of thermodynamics, and the content of this theory survived, completely unaffected. It has been found that in its form developed in the 19th century, without assuming even the existence of molecules, thermodynamics is rigorously consistent with, for example, quantum solid-state physics and quantum field theory.[18] This is a clear manifestation of the fact that the theory of thermodynamics possesses the robustness of universal structure, which is truly unaffected by the details of phenomena that are beyond the realm of consideration.

In this section, we have put forth the idea that a great variety of individual types of universal structures lie hidden in nature and that each of these exists in some sense independently of the others. However, despite this interpretation, we do not wish to assert that science is merely a collection of disconnected universal structures existing in isolation. On the contrary, we believe that the aim of science is to obtain ever more general descriptions of nature by finding connections among universal structures, with the ultimate goal of realizing a single, unified

of science are not determined by simple social consensus. The process by which they are constructed is much more difficult (and therefore much more interesting). The speciousness of this simplistic sociological interpretation is clearly evident from the consistency and utility of natural science (here understood in a broad sense to include biology, medicine, engineering and other fields), but in order to avoid this and other simple misunderstandings, we believe that it is also important to present an understanding of science that emphasizes the concept of universality.

[18] It is noteworthy that the investigation of the thermodynamics of electromagnetic fields carried out by Max Karl Ernst Ludwig Planck, Albert Einstein, and others led to the birth of quantum physics.

understanding. Indeed, we have repeatedly seen this kind of progress as connections have been discovered among universality classes and we have found how they fit within larger frameworks. For example, in the proper macroscopic limit, the universal structure of quantum mechanics reduces to the universal structure of classical mechanics. As another example, it is believed that the universal structure of fluid mechanics is obtained if we apply an appropriate limiting procedure to the classical mechanical description of the particles that make up a fluid.[19]

As the above-mentioned examples illustrate, we know that many types of universal structures are closely connected on a purely mathematical level. Now, there is an interesting philosophical point regarding the nature of the relationships among various types of universal structures. The interpretation that such theories as thermodynamics and fluid mechanics, which describe behavior on macroscopic scales (so-called *phenomenological theories*), should be interpreted as mere approximations of more microscopic, "fundamental" theories is the essence of the reductionist point of view. While it is true that reductionism provides a coherent interpretation of science,[20] we believe that "good" phenomenological theories (i.e., those which describe universal structures) should not be understood as mere approximations of microscopic theories. Rather, they should be regarded as existing independently of microscopic theories, just as the universal structures that they describe are independent of microscopic universal structures.[21]

It is important to note that universal structures are not limited to theories with very broad scopes, such as classical mechanics or quantum field theory. Indeed, universal structures of myriad forms lie hidden in natural phenomena on innumerable levels. For example, within the realm of classical mechanics, in addition to the fundamental equation describing the motion of particles, there is an equation describing the motion of rigid bodies. The form of this equation is simple and universal, being independent of the characteristics of the particles that form any particular rigid body—their number, their configuration, the nature of the forces that bind them, and so on. The equation of motion for rigid bodies itself possesses a closed mathematical form and represents a typical example of universal structure. Furthermore, this equation of motion can be derived exactly from the equation of motion for the particles forming a rigid body, and hence there is a perfect theoretical connection between this universal structure and the general universal structure of classical mechanics. As a more interesting example, it is known that the critical phenomena seen in a wide variety of physical contexts, including magnetic systems and fluid systems, are almost completely independent of variations in the specific properties characterizing the individual systems that exhibit them. Hence, the theory describing these phenomena too constitutes a kind of universal

---

[19] However, rigorous demonstration of this point is quite difficult, and a definitive result has not yet been obtained. However, there is abundant evidence that such a connection does indeed exist.

[20] For an interpretation of science from the reductionist point of view, see S. Weinberg, *Dreams of a Final Theory: The Scientist's Search for the Ultimate Laws of Nature*, Vintage, 1994.

[21] This is clearly reflected by the fact that, in general, the class of behavior described by a single phenomenological theory is exhibited by many types of physical systems whose microscopic descriptions require many different microscopic theories. In this sense, it is often the case that a single phenomenological theory corresponds to many microscopic theories. (For example, the Navier-Stokes equation used in fluid mechanics can describe the macroscopic behavior of a collection of molecules of air, water, liquid metal, etc.) The idea that a single universal structure becomes "incarnated" in various concrete forms in different physical systems provides an intuitively appealing understanding of this kind of correspondence. (See Chapter 1 of Ref. [13] for further discussion.)

structure.[22] Moreover, it has been found, rather unexpectedly, that this universal structure is mathematically equivalent to that describing quantum field theories.

We believe that science fundamentally consists of the unity of the multitudinous universal structures that we have discovered in nature. Progress in fundamental science is made through discoveries of both previously unknown universal structures and previously unknown relations among universal structures. Indeed, progress usually results from developments that involve an interplay between both types of discoveries. This point of view leads us to interpret fundamental science as an intricate network formed by universal structures existing on many levels in nature that are interconnected in an almost organic manner through theoretical relations. The fact that we humans have been able to acquire such an understanding of this world is, we believe, nothing short of miraculous.

The network of universal structures that we call "science," which we have gradually woven through our reasoning power to make sense of countless empirical discoveries, has allowed us to understand many types of phenomena that were previously beyond our comprehension. Thus, with the advance of science, there has also been an advance in our power to comprehend, which, in turn, has allowed us to weave an ever richer network providing an ever more profound understanding. We believe that this interpretation of science is much more attractive (and realistic) than the reductionist interpretation, according to which there is a single, "ultimate" microscopic theory, with all other theories being simply approximations of it that apply in certain limited situations.[23]

1.2.3 Thermodynamics as a strictly macroscopic theory

This book presents thermodynamics as a universal structure providing a description of the macroscopic world. We wish to construct a closed theory in terms of only quantities that can be measured through macroscopic means without reference to microscopic theories. Because thermodynamics describes macroscopic phenomena, we believe that such a purely macroscopic formulation, if possible, is most natural.

Although taking a purely macroscopic approach to thermodynamics may seem unnecessarily restricted and, in some sense, austere, this was obviously the approach used to establish the field. But since that time, with the development of molecular theory and statistical mechanics, it has become common to use such theories of the microscopic world in the formulation of thermodynamics with the intention of "simplifying" the theory.[24] However, with a formulation of thermodynamics that begins from microscopic theories, the understanding described in

---

[22] A simple treatment of critical phenomena is given in Chapter 10.

[23] Although we have emphasized the difference between these two interpretations of science—one that posits a single ultimate theory and one that recognizes as essential the many universal structures and relations among them—in fact the difference between the meanings that they attribute to actual theoretical relations is extremely subtle. Indeed, this difference is less one of theory than of the tastes and intuitive understandings of individual scientists. In the course of actual research, this difference in point of view rarely leads to substantially different conclusions, while in the course of debate, it rarely leads to misunderstanding or a failure to communicate.

[24] It is commonly thought that thermodynamics formulated strictly in terms of macroscopic quantities and concepts is "difficult." But are formulations employing microscopic concepts actually easier to understand? For example, does the microscopic expression for the entropy due to Boltzmann, $S = k_{\mathrm{B}} \log \Omega$, provide a simple understanding of the thermodynamic entropy? Of course, this equation is concise and easy to memorize, but we must not confuse the memorization of a formula with the

the previous section that thermodynamics constitutes a universal structure with an independent existence becomes lost. Basing thermodynamics on statistical mechanics results in a theoretical framework that is merely a derivative of molecular theory. While it may seem that this is simply an alternative, valid approach to formulating thermodynamics, in fact the resulting framework is flawed, possessing at least the three weaknesses described below. These considerations provide compelling evidence for the validity of our claim that thermodynamics exists independently as a universal structure.

The first weakness of a microscopically-based approach is that, from the point of view of empirical science, to regard thermodynamics as being derived from molecular theory through statistical mechanics is to some extent invalid. Several points closely related to this are discussed in detail in subsequent chapters, but here let us simply point out that there are many assumptions used in the foundation and construction of both molecular theory and statistical mechanics whose validity cannot be directly verified. The reason that we believe in the validity of these microscopic theories themselves despite this fact is that this validity is indirectly (but extremely convincingly) supported by a great body of macroscopic empirical evidence (see Problem 1.3). Furthermore, most of this empirical evidence takes the form of phenomena within the realm of thermodynamics. Thus, from the point of view of empirical science, it is not microscopic statistical mechanics that provides the foundation for thermodynamics but, rather, macroscopic thermodynamics that provides the foundation for statistical mechanics.[25]

The second weakness concerns the present state of the field of (equilibrium) statistical mechanics. Statistical mechanics is actually a small part of a much broader field called *statistical physics*. In simple terms, statistical physics is a framework for treating physical systems possessing many degrees of freedom, and its goal is to derive theoretical laws describing macroscopic phenomena from microscopic theories. The scope of statistical physics is not restricted to any particular type of methodology or class of phenomena. Rather, this field employs insightful theoretical ideas to search for the subtle and often wondrous relations among the (universal) structures characterizing the micro and macro worlds, and it is intended to cover a vast range of physical systems. Although deriving such relations is in many cases very difficult, the results that can be obtained within this theoretical framework are of profound significance in the broader scope of science. Given the very far-reaching goal of statistical physics, it is not surprising that as a theory, it is yet far from complete. At this time, only the statistical theory of equilibrium phenomena (which constitute a very limited class of macroscopic behavior) can be regarded as established in a general sense, and this is, of course, the theory of equilibrium statistical mechanics. Within this context, however, the formalism of statistical mechanics is quite powerful. In fact it can be used to derive the thermodynamic functions characterizing thermal equilibrium states for any mechanical system with well-defined energy.[26] Considering the great generality of

---

actual understanding of a concept. The aim of this book is to present a macroscopic formulation of thermodynamics that is truly easy to understand.

[25] This important point was brought to our attention by Yoshi Oono.

[26] However, in general, the formulas representing these thermodynamic functions are quite complicated (involving, for example, very high-dimensional integrals), and therefore, although formal expressions for the desired quantities can be derived quite readily, obtaining useful relations that provide an intuitive understanding of physical phenomena is in general very difficult.

this statement, it may seem that with such relations, the entire theory of thermodynamics could be derived. But this is not true. The reason for this is that thermodynamics describes not only the properties of equilibrium states, but also transitions among equilibrium states (and the accompanying exchanges of energy) engendered by externally applied operations of any (physically realizable) kind. The important point here is that if the initial and final states of the system are both equilibrium states, then no matter how far from equilibrium the system is forced during the evolution between these states, the theory of thermodynamics is applicable and yields rigorous quantitative results concerning the relationship between them. By contrast, in its present form, the theory of statistical mechanics cannot be applied to phenomena of this type if the system deviates too far from equilibrium.[27] It is thus seen that only a small part of the theory of thermodynamics can actually be derived from statistical mechanics.

The third weakness regards practical matters of experimental investigation. For example, let us consider the energy of a system. In statistical mechanics, the energy consists of the sum of the kinetic and potential energies of all of the particles composing the system. However, it is not possible to carry out direct, exhaustive measurements of this set of microscopic quantities for a macroscopic system. The energy of a macroscopic system can only be determined through measurement of the work and heat exchanged by the system in thermodynamic processes. The situation is similar for essentially all other thermodynamic quantities. For this reason, it is fair to say that we always need thermodynamic interpretations in order to apply the results of statistical mechanical analyses to experiments on macroscopic systems.

Before ending the present section, let us briefly comment on the status of the foundation of statistical mechanics. Equilibrium statistical mechanics is based on the premise (called the *principle of equal weights*) that thermal equilibrium can be described by a probabilistic model in which every microscopic state of the system consistent with the macroscopic conditions is realized with equal likelihood. From the point of view of reductionism, this fundamental premise used in connecting the microscopic and macroscopic worlds should itself be based on a more microscopic law, that is, some kind of mechanical law. This problem of the *foundations of statistical mechanics* has been investigated by many people, beginning with Ludwig Eduard Boltzmann, the founder of the field, and great progress in our understanding of this problem has been made through these investigations.[28] However, in the more than one hundred years that have passed since Boltzmann's work, we have not yet obtained a definitive argument justifying this premise. Indeed, many researchers (including the present authors) now believe that demonstrating the validity of the fundamental premise of statistical mechanics requires some additional information that cannot be derived from mechanical laws. Nevertheless, the validity of the formalism of equilibrium statistical mechanics is generally not regarded as a matter

---

[27] There are, however, various (partially successful) attempts to derive, for example, the second law of thermodynamics on the basis of equilibrium statistical mechanics and quantum or classical mechanics (see Footnote 8 on p. 48).

[28] For recent progress, see, for example, L. D'Alessio, Y. Kafri, A. Polkovnikov, and M. Rigol, "From Quantum Chaos and Eigenstate Thermalization to Statistical Mechanics and Thermodynamics," Advances in Physics, vol. 65, iss. 3, 2016, 239–362 (https://doi.org/10.1080/00018732.2016.1198134); H. Tasaki, "Typicality of Thermal Equilibrium and Thermalization in Isolated Macroscopic Quantum Systems," Journal of Statistical Physics, vol. 163, 2016, 937–997 (https://doi.org/10.1007/s10955-016-1511-2).

of debate. Our strong belief in the validity of this theory is based primarily on an overwhelming body of empirical evidence—the quantitative consistency of the predictions provided by this theory and experimental results involving countless types of systems—while it is also supported by partially successful theoretical studies elucidating the logical connections between the underlying microscopic mechanics and the laws of equilibrium statistical mechanics. Thus, it can be said that the theory of equilibrium statistical mechanics is founded partially on macroscopic evidence and partially on microscopic theories.

1.3 About this book

1.3.1 Our formulation of thermodynamics

This book contains a detailed presentation of the fundamental theory of thermodynamics, along with its basic applications to physics and chemistry. It assumes an understanding of classical mechanics and, in particular, the concept of work, as well as an elementary knowledge of calculus. The explanations have been laid out in as logical a manner as possible, with an emphasis on clarity. Rather than presenting the theory of thermodynamics at once in its completed form, we start from a set of selected empirical facts and construct the theory in a logical, stepwise manner. This book provides readers studying thermodynamics for the first time with a sound and complete understanding of the fundamentals of the field. For readers who have already studied thermodynamics, we hope that our treatment leads to a new appreciation for the power, depth and beauty of its theoretical framework.

While there exist many textbooks on thermodynamics, this book differs from most with regard to several important points.[29] Most importantly, in contrast to the approach used in nearly all textbooks, where the concepts of entropy and free energy are introduced through arguments based on the concept of heat,[30] in the development of the theory given here, operationally defined work plays the central role. Of course the content of the resulting theory is equivalent to that of the original,[31] but we believe that the formulation presented here is clearer and more concise.

While the formulation of thermodynamics presented in this book is novel, it owes a great deal to the work of many scientists. In particular, our approach was strongly influenced by the axiomatic formulation of thermodynamics developed by Lieb and Yngvason [9] and its simplified reformulation carried out by Sasa (private communication).

Below, we summarize the most important characteristics of thermodynamics, as this theory is constructed in the present book.

---

[29] Where useful, we compare and contrast the definitions and arguments given in this book with those used in other textbooks.

[30] This heat-based approach was introduced in 1865 by Clausius in his work *Ueber verschiedene für die Anwendung bequeme Formen der Hauptgleichungen der mechanischen Wärmetheorie* (*On Several Convenient Forms of the Fundamental Equations of the Mechanical Theory of Heat*).

[31] However, it should be noted that some of the conventional definitions used in the standard formulation lead to subtle problems when the theory is applied to phase transitions. For this reason, the definitions presented in this book have broader application.

- Although the concept of heat is very important within thermodynamics, it need not be defined in a general manner for the purpose of formulating the established thermodynamic theory describing equilibrium states and transitions among them. In older treatments of thermodynamics, heat is usually introduced as an undefined concept, with an intuitive understanding used in place of a formal definition. In this book, by contrast, the concept of heat is used only within a limited context, and within this context, it is defined formally. Thus, none of the theoretical treatment given here employs heat as a concept that relies on an intuitive understanding alone. In the formulation of thermodynamics presented in this book, it is not heat but, rather, operationally defined work that plays the central role. In fact, heat itself does not appear explicitly in the final form of the theoretical framework constructed here. Instead, it exists somehow behind this framework, exercising its control as a kind of hidden "guiding hand."

- The concepts of entropy and free energy are introduced differently in this book than in conventional approaches. Here, we first define the Helmholtz free energy and the energy of a system through consideration of the work that it performs under isothermal and adiabatic conditions, respectively. Then, examining the difference between the free energy and the energy in light of Carnot's theorem, we show that the concept of entropy emerges naturally. With this approach, entropy can be defined explicitly through the concept of work. Also, no problems arise even at phase transitions, where various state functions become non-analytic. (The formulations presented in most thermodynamic textbooks break down in such situations.) Finally, this approach affords a very intuitive understanding of the free energy, because it is defined in a simple manner, directly in terms of mechanical work. This is important for practical purposes, because the free energy plays a central role in most applications of thermodynamic theory.

- Complete thermodynamic functions are of fundamental importance in thermodynamics.[32] In this book, we introduce the concept of complete thermodynamic functions and elucidate their role within the theory. Because of their special status, we distinguish them notationally from other types of functions.

- Convex functions and the Legendre transformation are mathematical concepts that play important roles in thermodynamic theory. We stress how these concepts emerge naturally from physical considerations. In particular, we explain in detail how the fact that the stability of equilibrium states results from the convexity of the free energy can be understood in terms of a variational principle. Detailed mathematical treatments of convex functions and the Legendre transformation are given in the appendices.

- This book is written in a partially mathematical style, with a semi-axiomatic presentation. Those components of the theory referred to as *postulates*, *results* and *derivations* correspond to axioms, theorems and proofs in a mathematical work.[33] We use the symbol ∎ to indicate that a derivation has been completed. The term *postulate* is used in reference to universal properties

---

[32] Complete thermodynamic functions correspond to the *fundamental equations* of Gibbs.

[33] Keeping in mind that this is a physics textbook, we have purposely avoided a strictly rigorous presentation. Indeed, the assumptions used in the construction of the theory are not restricted entirely to the "postulates," and for the purpose of conciseness and clarity, we occasionally assume certain technical points without explicit mention.

of thermodynamic systems abstracted from empirical observations.[34] These postulates are assertions concerning the properties of physical systems that cannot be derived within the thermodynamic framework itself. Postulates and results that play particularly fundamental roles in the theory of thermodynamics are referred to as *principles*. (In this regard, we mainly follow the established convention.)

1.3.2 Organization of this book

Below we briefly outline the organization of the book.

The core of the book consists of Chapters 2–6. In these chapters, we construct the theoretical framework of thermodynamics. It is the treatment given in this part of the book that most distinguishes it from previous textbooks. This treatment begins in Chapter 2 with the introduction of the concept of equilibrium states and their quantitative description. In Chapter 3, through consideration of mechanical (externally applied) operations carried out on systems at constant temperature (i.e., isothermal systems), we introduce the Helmholtz free energy. At this point in our construction of the theory, however, the temperature dependence of the Helmholtz free energy is not yet established. In Chapter 4, the concept of energy is introduced through the study of mechanical operations applied to thermally closed systems (i.e., adiabatic systems). Next, in Chapter 5, by comparing the amounts of work performed in isothermal and adiabatic operations, we are naturally led to the concept of heat as a quantity exchanged by an isothermal system and its environment. Then, through Carnot's theorem, we demonstrate the very universal nature of a quantity that we refer to as the *maximum heat*. In Chapter 6, we introduce and investigate the concept of entropy. It is seen there that this concept emerges quite naturally through consideration of Carnot's theorem when we study the heat exchanged in isothermal operations. Then, employing a natural definition of entropy in this context, we are able to establish the temperature dependence of the Helmholtz free energy. Next, we carry out a detailed analysis of the intimate relationship between entropy and adiabatic operations. With the completion of this step, all of the postulates needed to formulate the theory of thermodynamics are in place.

In the next two chapters, building on the framework completed in Chapter 6, we give detailed treatments of the Helmholtz free energy (Chapter 7) and the Gibbs free energy (Chapter 8), which are of great theoretical and practical importance, and discuss their basic application. With these treatments, our construction of the fundamental theory of thermodynamics is complete.

In the final two chapters, we present important applications. In Chapter 9, we extend the theory developed in the preceding chapters to multi-component systems and treat the topics of phase equilibrium, chemical equilibrium, and equilibrium electrochemistry. In Chapter 10, phase transitions and critical phenomena in ferromagnetic systems are investigated. While these applications themselves may not be

---

[34] In general, for both physical and mathematical systems, there is some degree of arbitrariness with respect to what is regarded as a postulate and what is regarded as a result. Indeed, in some cases, with a change in point of view, the roles of postulates and results can be interchanged. But, of course, a system is unaffected by the manner in which we attempt to understand it, and hence changing the identifications of postulates and results by changing our point of view does not alter the system.

of primary interest to some readers, even those with a casual interest will acquire a deeper understanding of the fundamental theory by studying how it is applied in these contexts.

In the appendices, we present supplemental material, consider several advanced topics, and treat some important mathematical concepts.

We present several problems at the end of most chapters. These problems are quite diverse in both their purpose and their level of difficulty. Some of the problems are used to further elucidate topics that were not treated completely in the main text, while others are used to raise important points about the theory. At the end of the book, detailed guidance is given for all problems whose solutions require non-trivial considerations.

1.3.3 Note to the instructor

Here we give several points of advice to instructors who wish to use this book as a course textbook.

In a physics-oriented course, the main focus should be on elucidating the theoretical structure of thermodynamics. Particular emphasis should be placed on acquiring a deep understanding of entropy and obtaining a familiarity with the Helmholtz free energy and its uses. A thorough treatment of the main topics in Chapters 2–7 is sufficient for this purpose.[35] Covering this material should not require a great deal of time, and therefore such a treatment can easily be incorporated into a course plan that includes other materials. Then, a brief treatment of Chapters 8–10 will provide a basic foundation in chemical physics as well the background needed for studying the statistical mechanics of spin systems.[36]

For a course whose focus is on the application of thermodynamics to chemistry, we recommend instruction that emphasizes how the fundamental concepts introduced in Chapters 2–8 form the foundation of the treatment presented in Chapter 9. However, for a deep understanding of chemical reactions and equilibrium electrochemistry, the treatment given in this book is insufficient, and it should be supplemented with other materials containing more detailed studies. Our treatment of chemical applications is focused on elucidating the relationship between the fundamental concepts of thermodynamics and chemical phenomena.

1.4 Mathematical notation

Below we summarize the mathematical notation and terminology used in this book.

- Consider a function $f(x)$ of a single independent variable x and suppose that the relation $f(x_1) < f(x_2)$ holds for every x_1 and x_2 in the domain of f satisfying $x_1 < x_2$. Then $f(x)$ is said to be an *increasing function*.

---

[35] The applications that we treat in Chapter 7 have been chosen with this type of course in mind.

[36] Since 2000, many instructors have taught undergraduate courses using the Japanese version of this book as the primary textbook. In a single semester (consisting of roughly 20 hours of lecture), most have been able to cover the material through Chapter 8, and some have also treated material from Chapters 9 and 10.

If $f(x_1) \leq f(x_2)$ holds for every x_1 and x_2 satisfying $x_1 < x_2$, then $f(x)$ is said to be a *non-decreasing function*. The terms *decreasing function* and *non-increasing function* are defined similarly.

- We assume that the reader is familiar with the definition of a continuous function. We say that a function is *n times differentiable* with respect to a given variable if the nth derivative of that function with respect to that variable exists at every point in the domain of the function. When the nth derivative of a function is continuous, we say that the original function is *n times continuously differentiable*. We often use the shorthand terms *differentiable* and *continuously differentiable* to mean *once differentiable* and *once continuously differentiable*, respectively.

- Unless otherwise noted, the logarithm represented by the notation "log" is understood to be the natural logarithm (i.e., that with base e).

- As a general rule, the independent variables of functions are expressed explicitly. Thus, for example, a function $f(x, y, z)$ of three independent variables will generally not be written simply as f (although there are some exceptional cases).[37]

A partial derivative $\dfrac{\partial f(x, y, z)}{\partial x}$ is defined as follows:

$$\frac{\partial f(x, y, z)}{\partial x} := \lim_{\varepsilon \to 0} \frac{f(x + \varepsilon, y, z) - f(x, y, z)}{\varepsilon} . \tag{1.6}$$

Here, it is understood that x is varied with y and z held fixed. In the context of thermodynamics, this quantity is often expressed as

$$\left(\frac{\partial f}{\partial x} \right)_{y, z} , \tag{1.7}$$

but we do not employ such notation.

In the case that it is necessary to specify the value of an independent variable in a partial derivative, we use notation illustrated by the following:

$$\left. \frac{\partial f(x, y, z)}{\partial x} \right|_{x=a} := \lim_{\varepsilon \to 0} \frac{f(a + \varepsilon, y, z) - f(a, y, z)}{\varepsilon} . \tag{1.8}$$

- The expression $x \searrow x_0$ means that x approaches x_0 from above, while $x \nearrow x_0$ means that x approaches x_0 from below. We use $x \to x_0$ in the situation that the same result is obtained in either case, and hence there is no need to distinguish between them.

Although this is not a mathematics textbook, it is important to keep the distinction between these two types of limiting procedures clear. As a physical example illustrating this point, consider a sample of H_2O at a pressure of 1 atm and suppose that we change the temperature in some manner, eventually ending up at the freezing point, 273.15 K. In this situation, whether this value is approached as $T \searrow 273.15$ K or $T \nearrow 273.15$ K has significant

---

[37] Although including all independent variables is sometimes tedious, we believe that it is better to forgo notational conciseness than to risk the confusion that can result if the complete set of independent variables is not explicitly expressed.

implications, as in the former case we will have water, while in the latter case we will have ice.

- Let $h(\varepsilon)$ be a quantity that converges to 0 as $\varepsilon \searrow 0$. The expression $h(\varepsilon) = O(\varepsilon^n)$ for a positive integer n means that there exist constants $A > 0$ and $\varepsilon_0 > 0$ such that $|h(\varepsilon)| \leq A\,\varepsilon^n$ holds for every ε satisfying $0 \leq \varepsilon \leq \varepsilon_0$. For example, we can write $\sin x = O(x)$ and $\sin x = x + O(x^3)$.

- We use the relational symbol $\simeq$ to express the meaning that the quantities related are "nearly equal" in the sense that, for the present purposes, the difference between them is negligible or irrelevant. In several cases, it is used to present a truncated value of a quantity known to higher precision, as in the expression for the gas constant $R \simeq 8.31\,\mathrm{J\,K^{-1}mol^{-1}}$, appearing on p. 61.

- We use the relational symbol $\approx$ in expressions that present currently accepted experimental values. Thus, as a general rule, such a value should be regarded has having an uncertainty on the order of the smallest significant figure given. For example, the expression $E_b \approx 0.513\,\mathrm{K\,kg\,mol^{-1}}$ for the ebullioscopic constant of water (p. 211) is understood to imply an experimental uncertainty of order $10^{-3}\,\mathrm{K\,kg\,mol^{-1}}$.

- In physics, it is a general convention that a symbol representing a physical quantity includes physical dimensions. We follow this convention. In other words, suppose that the symbol ℓ is used to represent a length. Then ℓ is regarded as a dimensional quantity, with the dimension of length, not a pure number. Hence, for example, if we are using it to represent a length of 1 meter, then we can write $\ell = 1$ m, $\ell = 100$ cm, etc. Similarly, for a velocity $v = 10$ m/s and time $t = 1$ h, their product can be written as, for example,

$$vt = 10 \text{ m/s} \times 1 \text{ h} = 10 \text{ m/s} \times 3600 \text{ s} = 36 \text{ km}. \tag{1.9}$$

Problems 1

1.1 (Section 1.1.2) Let us investigate the operation depicted in Figure 1.5, through which the temperature of an adiabatic system is increased. We suppose that the system consists of an ideal gas of amount N contained in a vessel that is surrounded by adiabatic walls. Let V_1 and T_1 represent the initial volume and initial temperature. Then, suppose that in the first part of the operation, the piston is moved to the right so rapidly that the gas molecules cannot keep up with it. (We acknowledge that, because of the very large velocities of molecules in a gas under typical conditions, such a situation is somewhat unrealistic.) Let V_2 represent the volume of the system realized at the end of this expansion. As can be understood from the explanation given in Section 4.4, the temperature of the gas will not change under such a rapid adiabatic expansion. Thus, as a result of this operation, an equilibrium state with volume V_2 and temperature T_1 will be obtained. (And thus, in Figure 1.5, we have $T_2 = T_1$.) Next, suppose that the piston is pushed very slowly inward, compressing the gas until the original volume, V_1, is again realized. During such an operation, the temperature and volume satisfy the relation $T^{3/2}V = $ [constant], which is known as a Poisson relation (see Section 4.4.2). Find the temperature of the final equilibrium state, T_3. Confirm the relation $T_3 > T_1$.

1.2 (Section 1.1.3) Let us consider a heat engine like that schematically described by Figure 1.6. We assume that the working substance of the engine is an ideal gas of amount N. Let A and m denote the cross-sectional area of the piston and the mass of the weight. Regarding the combined mass of the platform and piston to be negligible, determine the heights of the piston in (a) and (c). Using realistic values for the size of the system and the temperatures T_L and T_H, determine the amount of mechanical work that can be obtained from this heat engine. Is it a practically useful device?

1.3 (Section 1.2.2) It is probably safe to assume that the reader believes in the existence of atoms and molecules. However, given that we can neither see nor handle such tiny particles, why do we believe in their existence? (The fact that within modern-day science the existence of atoms and molecules is generally accepted without question does not in itself constitute evidence.) Formulate an argument supporting the existence of atoms and molecules.[38]

---

[38] More generally, how would you change the view of a person who claims that modern science is simply a magnificent story that has, for one reason or another, become the consensus view of the scientific community, or who claims that to believe in scientific theories, for which there exists no absolute proof, is no different from believing in mythology and fables? There indeed are people who make such claims (see Footnote 17 on p. 10).

2
Thermodynamic Systems, Operations and Equilibrium States

In this foundational chapter, we formally introduce thermodynamic systems and the elements of the external world with which they interact. These elements are distinguished by the manners in which they interact with the thermodynamic system—mechanically and non-mechanically. Next, we give detailed discussion establishing the mechanical operations used as the primary means of investigating thermodynamic systems. Through investigations employing these operations, we then define equilibrium states and the conditions under which they are realized. Finally, we identify the quantities in terms of which equilibrium states are parameterized. Throughout this treatment, the work performed in mechanical operations plays a central role, while temperature is introduced as the quantity controlling non-mechanical interactions.

2.1 Overview

In this section, we give a brief overview of the theory of thermodynamics that we construct in this book and describe the basic point of view that we employ in this construction. In the discussion given here, a number of the fundamental concepts are used in an informal, intuitive manner, without formal definition. All of these concepts are defined in subsequent sections of this chapter.

2.1.1 The basic elements of thermodynamics

In the treatment of thermodynamics presented in this book, we employ the premise—or idealization—that when investigating thermodynamic phenomena, we can always unambiguously separate the world into three systems playing entirely distinct roles: the thermodynamic system under investigation and two systems external to it, termed the *mechanical world* and the *environment* (see Figure 2.1). We refer to the combined system consisting of both the mechanical world and the environment as the *external world*. Below, we briefly describe each of these basic elements.

A *thermodynamic system* can be regarded simply as any macroscopic system, as the theory of thermodynamics applies universally to essentially all macroscopic systems of practical interest. In this book, we mainly study fluid (i.e., liquid or gas) systems, with minimal explicit consideration of the solid phase (as discussed in Section 2.4.5). The simplest such example is a single-component fluid contained in a single, unpartitioned vessel, but we also consider more complicated systems of

Thermodynamics: A Modern Approach. Hal Tasaki and Glenn Paquette, Oxford University Press.
© Hal Tasaki and Glenn Paquette (2026). DOI: 10.1093/9780191878091.003.0002

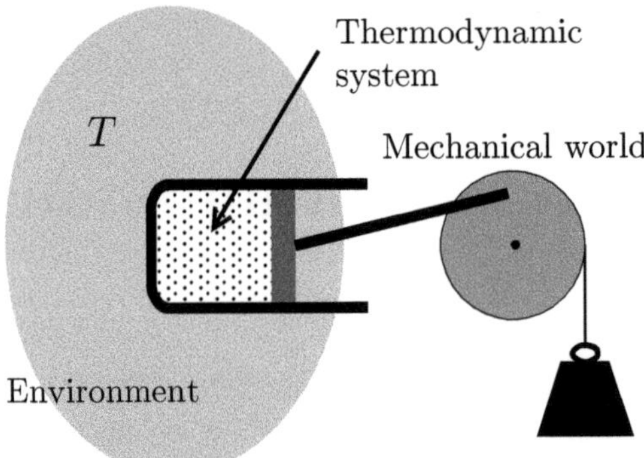

Figure 2.1 The basic elements of thermodynamics. The central element is the thermodynamic system whose behavior we intend to describe. In the situation depicted in the figure, this system consists of the gas contained in a vessel with a movable piston. External to this thermodynamic system are the mechanical world and the environment. The mechanical world, which is represented by the pulley and weight in the figure, operates on the system by moving the piston. The environment is a large external system that surrounds the thermodynamic system of interest and provides a constant ambient temperature. Depending on the insulating properties of the vessel, the system may or may not be in thermal contact with the environment.

various kinds. (We will always assume that the system has some "ordinary" three-dimensional shape, excluding, e.g., very thin, quasi-two-dimensional geometries.) Among these systems, for example, some act as heat engines, some exhibit phase transitions, and some undergo chemical reactions. In addition to fluid systems, we also study magnetic systems that exhibit ferromagnetic phase transitions.

The *mechanical world* is that part of the external world that interacts strictly mechanically with the thermodynamic system under consideration. Within the theory, this system is regarded not as a proper thermodynamic system, but as a purely mechanical system, consisting only of macroscopic objects whose behavior is described entirely by the laws of classical mechanics. Such a treatment is possible because the actual physical system to which the mechanical world corresponds is chosen or constructed in such a manner that its non-mechanical properties are irrelevant for the purpose of describing the behavior of the thermodynamic system. Of particular note, the mechanical world includes an *external agent* that carries out operations on the thermodynamic system. As we discuss in Section 2.1.2, these operations play a fundamental role in our theory.

The *environment* is that part of the external world that interacts purely non-mechanically with the thermodynamic system under consideration. The environment is itself a thermodynamic system, and as such is capable of exhibiting a wide range of behavior, but within the theory of thermodynamics, its role is limited to that of providing a constant ambient temperature for the system under investigation.[1] Generally, in order to function in this way while the system is

---

manipulated by the external agent, the environment must be much larger than the system. In some cases, the environment consists simply of the ambient air in a laboratory, while in other cases, it may be a carefully prepared system consisting of, for example, water or liquid helium.

The mechanical world and the environment together form the *external world*. Our investigation of thermodynamic systems is based on consideration of controlled mechanical operations that are carried out by the external agent. All other influences of the external world on a system under investigation are exerted by the *external conditions*. The external conditions are determined first by whether or not the system interacts with the environment, and if so, then also by the properties of the environment.

A thermodynamic system that is either in continuous thermal contact with a fixed-temperature environment (*isothermal conditions*) or thermally shielded from the external world (*adiabatic conditions*) will eventually reach an *equilibrium state* if it experiences no mechanical perturbations for a sufficiently long time. In such a state, all macroscopically observable properties of the thermodynamic system are independent of time. In the theory of thermodynamics, it is a premise of fundamental importance that an equilibrium state is uniquely specified by a small number of macroscopic quantities. In the case of a single-component fluid, the equilibrium state is generally specified by the volume, V, amount of substance, N, and temperature, T. Because V can be directly controlled by the external agent, it plays a special role in the investigation of thermodynamic behavior.

In the following sections, we carry out a systematic treatment, formalizing all of the concepts introduced above. Before proceeding, however, we point out that because of the very broad scope and general nature of thermodynamics, a number of its important concepts are inherently difficult to define in explicit, universally applicable manners. Fortunately, however, such definitions are in many cases unnecessary in the construction and application of thermodynamics. In this regard, we employ an operational, phenomenological approach. With this approach, rather than defining *what* certain things are, we define *how* they behave and interact with each other. For example, in defining the concept of *adiabatic walls* (i.e., walls that prevent the passage of heat), a direct, explicit definition would seem to first require a definition of the concept of *heat*. However, in our theory, we do not explicitly define heat. Instead, we define adiabatic walls by specifying how thermodynamic systems behave when surrounded by them.

2.1.2 The operational point of view in thermodynamics

In this book, we make the important assumption that a thermodynamic system can be manipulated in a controlled manner through mechanical operations

system. Theoretically, however, it is possible to formally separate the mechanical and non-mechanical influences of the external fluid by modeling its pressure effect with a mechanical device, and thus maintain the separation between the mechanical world and the environment. In this book, we employ such a modeling technique. It should be noted that, with regard to thermodynamic behavior, models obtained in this manner are entirely faithful, as long as we can assume the ideal situation of constant ambient temperature and pressure. (Systems of this type are studied in detail in Chapters 8 and 9.)

carried out by an agent existing in the mechanical world. However, a thermodynamic system cannot be described in terms of a (macroscopic) mechanical theory alone. As we know from experience, thermodynamic systems exhibit many types of phenomena that cannot be observed in systems described completely by macroscopic mechanical laws, including relaxation to equilibrium, exchange of heat, variation of temperature, phase transitions, and chemical reactions. Thus, a thermodynamic system and the mechanical world with which it interacts are of very different types.

Although a thermodynamic system is not purely mechanical in nature, it interacts with the mechanical world in a purely mechanical manner. Through mechanical operations applied to a thermodynamic system, the external agent controls the extensive variables (e.g., volume) that characterize this system. These variables play the role of an intermediary through which the thermodynamic system and the mechanical world are linked. From the point of view of the external agent, a system's extensive variables appear as part of the mechanical world, but their actual behavior reflects the inherently non-mechanical properties of the thermodynamic system.

We are able to alter a thermodynamic system by manipulating its extensive variables, and when doing so, we are able to feel the "reaction" of the thermodynamic system. More precisely, when an external agent carries out an operation through which the extensive variables are altered, the amount of work performed in this operation can be measured.[2] The measurement of the system's reaction, as represented by this work, provides quantitative information concerning the processes taking place within the thermodynamic system. In this book, we regard this type of thinking as fundamental to thermodynamics. In fact, we regard the information that can be obtained through mechanical operations either measuring or manipulating the extensive variables to be all the information that is possible to acquire about a thermodynamic system.[3]

With the type of thinking described above, a thermodynamic system is viewed as a "black box" that we, the experimenters, probe through variation of the extensive variables, adjusting them as desired—perhaps rapidly, perhaps slowly. In this way, we can exercise some degree of control over the thermodynamic system, and by measuring the reaction that we feel as we do so, we can determine the state of the system. Within the framework of thermodynamic theory, we neither observe nor imagine the "inner workings" of the system. Theoretically, our fundamental stance is that we understand the thermodynamic system only through our manipulation of the extensive variables and our measurement of the system's reaction to this manipulation.

---

[2] Among thermodynamics textbooks, some define the amount of work performed in an operation as that done by the system on the mechanical world and some define it as that done by the mechanical world on the system. In this book, we define the work to be that done by the system on the mechanical world (unless explicitly stated otherwise), and the heat (see Chapter 5) to be that absorbed by the system from the mechanical world. This is the traditional point of view, having its inception in the study of heat engines.

[3] Of course, information can also be obtained through measurements that do not involve mechanical operations—for example, temperature measurements using a thermometer. However, the same information can be obtained through measurements involving only mechanical operations.

2.2 Thermodynamic systems and extensive variables

2.2.1 Thermodynamic systems and walls

Thermodynamic systems

We define an *individual thermodynamic system* (or, simply, *individual system*) to consist of all of the substance existing within a macroscopic, contiguous region of space bounded entirely by macroscopically observable boundaries (i.e., persistent interfaces across which—according to macroscopic measurements—material properties change discontinuously and substance does not cross), while itself containing no such boundaries. We impose no restriction on the nature of the substance contained in a thermodynamic system, but we do stipulate that it exist in a sufficiently large amount to be macroscopically observable and mechanically controllable. Depending on the situation under consideration, this substance may be in a gas, liquid or solid phase, or some combination thereof. Then, a *thermodynamic system* is any system composed of one or more individual thermodynamic systems. We refer to a system containing more than one individual thermodynamic system as a *composite thermodynamic system* (or *composite system*).[4]

Throughout the book, for the sake of clarity, we generally study the simplest types of systems that exhibit the behavior under consideration. In particular, in the first several chapters, in which we establish the basic framework of thermodynamics, we focus mainly on single-component, single-phase fluid systems.[5] In the latter part of the book, we consider more complicated systems, for example, systems containing multiple phases (Section 7.5), systems containing multiple chemical species (Chapter 9), including systems undergoing chemical reactions (Sections 9.4–9.6), and systems composed of magnetic materials (Chapter 10).

Walls

Throughout this book, we regard the boundary of a system to be provided by *walls*. Intuitively, we can imagine walls as quasi-two-dimensional, impenetrable barriers, but rather than defining them in this way, here we define them by describing the roles that they play and the manners in which they interact with the system in these roles.

In addition to bounding systems, walls serve two distinct roles, both of which are fundamental in the theory of thermodynamics. In these roles, walls participate in the mechanical and non-mechanical interactions between the system and the external world, respectively.

First, for systems investigated through variation of the volume, this variation is carried out by the external agent through the motion of the walls. We assume that the external agent always has complete control over this motion, with the ability to move any wall precisely in accordance with any predetermined protocol. In this

---

[4] In most situations, either the distinction between individual systems and composite systems is unimportant or the type of system under investigation is clear from the context. For this reason, usually we simply use the term "system" in reference to both individual and composite systems. However, in situations that the distinction between individual and composite systems is important, the type will be specified.

[5] Even in the case that a fluid consists of multiple components (i.e., multiple chemical species), if it remains spatially homogeneous, and hence its multi-component nature is not apparent from macroscopic observation, then it can be rigorously treated as a single-component fluid. This is a fundamental point within an operationally formulated theory of thermodynamics.

role, walls act as part of the mechanical world, being the structure through which the system performs work.

Second, walls determine the nature of the non-mechanical interaction of the system with the environment. We assume the existence of two types of walls, adiabatic and diathermal, which, respectively, prevent and allow interaction of the system with the environment. These types of walls are defined precisely in Section 2.4.2. In all cases, we assume that the walls bounding a system can be freely replaced with walls of the other type.

We also assume that walls can be inserted into a system, partitioning it into subsystems in any desired manner, and, conversely, that any partitioning wall can be removed. Further, we assume that walls are of negligible thickness, so that the change in total volume resulting from their insertion into or removal from a system can be ignored.

Finally, we assume that the properties of walls are such that in each of their roles, the exchanges of energy between the external world and the system are unaffected by the walls themselves. In other words, we assume that walls are of negligible mass, so that the work required to manipulate them can be ignored, and of negligible heat capacity (see Sections 4.3.2 and 8.3), so that the internal energy (see Section 4.3.1) that they possess can be ignored.

2.2.2 Extensive variables

As discussed in Section A.1.1, the theory of thermodynamics is cast in terms of intensive and extensive quantities. Let us begin this discussion by defining these.

Suppose that we combine an individual system of volume V and amount of substance N with an identical system to form a new individual system.[6] Obviously, the volume and amount of substance of this new system will be $2V$ and $2N$. A property that behaves in such a manner, increasing by a factor of n when n identical systems are combined, is said to be *extensive*.[7] A property that is unchanged when identical systems are combined is said to be *intensive*.

We can imagine carrying out processes in which a system is combined with arbitrary numbers of identical systems, as well as the inverses of such processes, in which a system is divided into arbitrary numbers of identical systems. In this way, we can create "scaled copies" of the original system of various sizes. In this

---

[6] This process is understood to proceed as follows. The two systems are originally separated by a wall. This wall is then removed through an operation in which no work is performed on the systems.

[7] We point out that this definition of *extensivity* is not the only one found in the literature, and, in fact, it is probably not the most common. There are two other types of definitions that often appear, one in which an extensive quantity is defined as being proportional to (or, simply, "depending on") the number of particles or the "size" of a system, and one in which it is defined as being additive for subsystems. The key point missed in these definitions, however, is that for the unambiguous identification of the proper thermodynamic quantities, the definition of extensivity must be made through comparison of only *balanced* systems (see Section 7.4).

We also note here that in most works, *extensivity* is not distinguished from *additivity*. However, we find that, at least in the construction of thermodynamics carried out in this book, it is useful to distinguish these as separate attributes. We define *additivity* as follows: A quantity is *additive* if separate contributions of this quantity can be combined through addition to yield a physically meaningful total.

All extensive properties of a system are also additive, but the converse is not true. The distinction between extensive and additive properties is important because additive, non-extensive properties (for example, those depending on its shape—length, surface area, etc.) never play a role in (conventional) thermodynamics.

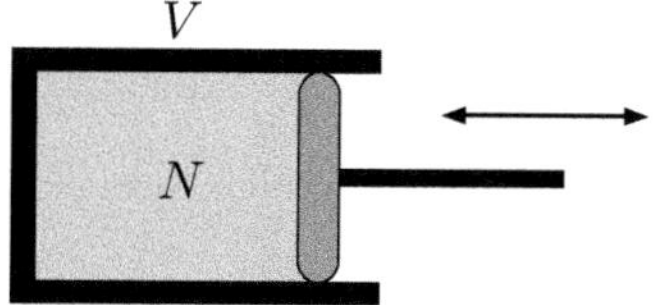

Figure 2.2 A vessel of volume V containing an amount N (in units of moles) of a single-component fluid. We consider N to be fixed throughout our investigation, while V can be controlled through mechanical operations applied by the external agent. The extensive variables of this system are V and N, and the collective extensive variable is $X = (V, N)$.

book, we consider the ideal situation and assume that for any given system and any positive real number λ, we can construct a scaled copy for which all extensive properties are changed by a factor of λ and all intensive properties are unchanged. This is used freely when investigating the behavior of various important quantities.

Among the extensive quantities that we use to describe thermodynamic systems, there is a special set that we term the *extensive variables*.[8] As we discuss below, some of the extensive variables can be directly controlled by the external agent through mechanical operations, and it is through the variation of these variables that the thermodynamic system is investigated.

Let us consider a single-component fluid existing in a closed container, as depicted in Figure 2.2. In most situations,[9] the extensive variables for such a system are its volume, V, and amount of substance, N. (In this book, we regard the amount of substance to be measured in moles.[10]) For convenience, we often write the volume and amount of substance together as (V, N) and refer to this combination as the *collective extensive variable* of the system. This notation is convenient, because we can formally treat a collective extensive variable using vector notation, writing, for example,

$$\lambda\,(V, N) = (\lambda V, \lambda N) \quad \text{or} \quad (V, N) = (V', N') + (V'', N'')\,. \tag{2.1}$$

Although the volume and amount of substance are both extensive variables, there is a fundamental difference between them. To see this, note that although N is a controllable quantity, it cannot be changed once the system has been prepared and enclosed within a container. By contrast, the volume can be varied by simply moving a wall, as illustrated in Figure 2.2. In this book, extensive quantities like the volume, which are uniquely determined by (and thus directly controlled through variation of) the macroscopic mechanical degrees of freedom characterizing the system, are referred to as *mechanically controllable extensive quantities*. Mechanical operations carried out through the variation of these types of quantities play an essential role in the construction of thermodynamics presented in this book.

---

[8] The special role played by the extensive variables is explained in Sections 2.4.5 and A.1.3. Simply stated, these variables, together with the temperature, allow for a "complete" description of thermodynamic behavior.

[9] The most general case is discussed in Footnote 11 on p. 30.

[10] This is not the only possible choice, however. For example, we could use the total mass, or even the total number of molecules. But, as we argue just after introducing the equation of state of an ideal gas in (3.34), the thermodynamic treatment is clearest if we use moles.

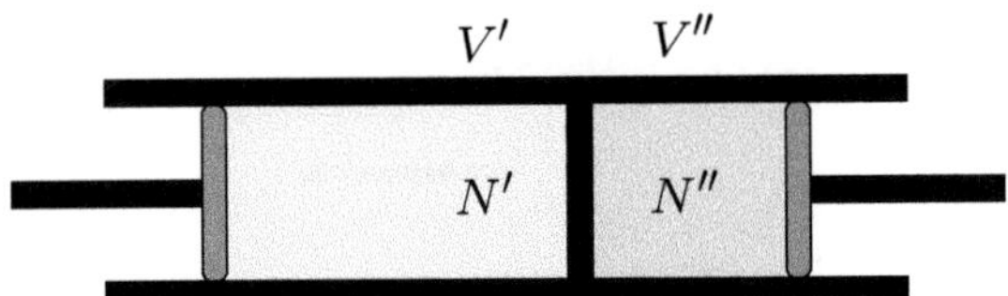

Figure 2.3 A composite system obtained by placing together two systems whose collective extensive variables are $X' = (V', N')$ and $X'' = (V'', N'')$. For this composite system, we have the collective extensive variable $X = \{X', X''\} = \{(V', N'), (V'', N'')\}$. Note that the quantities V' and V'' can be controlled independently by moving the two pistons.

The actual extensive variables characterizing a physical system will depend on the type of system in question, but the general form of thermodynamic theory does not. For this reason, in order to allow for a unified treatment, we often refer to the collective extensive variable in a generic manner, writing it as X (and sometimes X', X'',...,Y, etc.). For example, X represents (V, N) in the simplest example of a single-component individual system,[11] while it represents $\{(V', N'), (V'', N'')\}$ for a composite consisting of two such systems, and it represents $(V, N_1, N_2, \ldots, N_m)$ for a multi-component individual system containing m constituent species (see Chapter 9). In the case of a magnetic system, X represents (M, N), where M is the total magnetization of the system (see Chapter 10). The important point here is that in our development of the theoretical framework, it is unnecessary to specify any particular collection of extensive variables. However, in order to acquire an intuitive understanding from this formal presentation, the reader who is studying thermodynamics for the first time is encouraged to keep in mind the simple case in which X represents (V, N).

From this point, we will freely use such notation as λX and $\{X, Y\}$, representing a constant multiple of a collective extensive variable and a composite of two collective extensive variables (see Figure 2.3). Straightforward generalizations are used in more complicated situations.

2.3 Operations

2.3.1 Introduction

As discussed in Section 2.1.2, mechanical operations play a fundamental role in the formulation of thermodynamics constructed in this book. In general, an operation can be regarded as a controlled mechanical procedure through which the mechanical world interacts with the system under investigation. In this section

---

[11] There is an additional extensive variable that must be included in the case of investigations involving the triple point (see Appendix D). In such a situation, in order to obtain a complete thermodynamic description, as characterized in Section A.1.3, one extensive quantity providing information about the relative abundances of the three phases is necessary. For this purpose, the volume or amount of substance of any of the three phases can be used. We arbitrarily choose the amount of substance in the solid phase, written N_S. Thus, most generally, the extensive variables for a single-component individual system of this type are V, N and N_S.

we present the basic specifications for the class of mechanical manipulations regarded as operations and discuss the types of operations considered in this book.

As explained in Section 2.1.1, the mechanical world interacts with the system strictly mechanically, and it is described exactly by classical mechanics. We assume that for a given realization of a system, the mechanical configuration of the system is determined uniquely by that of the mechanical world. Also, we regard an operation to be defined by the time evolution of the mechanical world: For a given realization of a system, two operations are considered identical if and only if the mechanical world exhibits the same time dependence in each. Further, we regard every operation to begin and end with the system in equilibrium. In other words, all manipulations carried out between two consecutive realizations of equilibrium are regarded as forming a single operation. Let us also point out here that we use the term *operation* flexibly, with the understanding that the entire mechanical manipulation of the system carried out between the realizations of two equilibrium states can always be regarded as a single operation, whether or not the system realizes other equilibrium states between these two. Thus, any sequence of operations—punctuated by equilibrium states—can be regarded both as multiple operations and as a single operation.

We can imagine the mechanical world to consist of standard mechanical equipment—weights, springs, pulleys, motors, batteries, etc. Generally, there is no need to consider the precise nature of the mechanical world. It is only necessary to assume that in it there exists some mechanical apparatus with which all operations of interest can be carried out.

The mechanical world includes an agent that controls the operations carried out on the system. In the case of a system consisting of fluid in a container, a typical example of an agent is some apparatus that can alter the volume of the container by moving its walls, measure the force exerted by the gas, and record the history of the motion and forces experienced by the system (see Figure 2.1). The agent may even be (or include) a human experimenter who literally operates on the gas by pushing or pulling a wall.[12]

Next we introduce the types of operations studied in this book. In actual applications, most of the operations that we consider consist of just one type, although some consist of a combination of multiple types. In the latter case, we generally do not consider situations in which operations of different types are carried out simultaneously, in other words, without an equilibrium state being realized between.[13]

---

[12] Varying the position of a wall is the prototypical example of a mechanical operation carried out by an external agent. Other examples, however, cannot be interpreted in such an intuitively simple manner. In the case of concentration cells, treated in Section 9.6, the agent carries electric charge from one electrode to the other, while in the case of magnetic materials, treated in Chapter 10, the agent controls the total magnetization of the system.

[13] The only exceptional cases involve operations in which the collective extensive variable is varied while external walls are simultaneously added or removed. We consider two explicit examples of this kind, namely, the generalized isothermal operations studied in Problems 5.2 and 6.3. Also, as discussed at the beginning of Section 3.1.1, the general class of isothermal operations is sufficiently broad to include some combined operations of this type, but we do not consider explicit examples of such isothermal operations in this book.

2.3.2 Continuous extensive operations

Operations of the type described above, under which the volume of the system is varied continuously, exemplify a fundamentally important class of operations. We define this class of operations generally as follows.

Definition 2.1 (Continuous extensive operation) *A continuous extensive operation is an operation under which the mechanically controllable extensive quantities are varied in a controlled, continuous manner and the total work performed by the system is transmitted to the mechanical world entirely through the system degrees of freedom that exert this control.*

Continuous extensive operations play a central role in our construction of thermodynamics, being the main tool through which we investigate thermodynamic systems. As a necessary condition for these operations to provide definite information, we assume that the amount of work performed by the system on the mechanical world can always be measured by the agent. Hence, the work performed in any realizable continuous extensive operation is a well-defined, uniquely determined quantity.[14] Also, we regard these operations to be perfectly controlled in the sense that they exactly follow protocols that the agent has determined in advance. Haphazard manipulations that result in unpredictable, uncontrolled changes to the system are not considered.

It should be noted that although for any given continuous extensive operation there corresponds a unique change of the extensive variables, the converse is not true. For example, consider the system depicted in Figure 2.4. There, a fluid is contained in a vessel equipped with two movable pistons. In this case, the volume of the vessel can be varied by moving the left piston, the right piston, or both, and hence a given change in volume can be realized through various types of operations.

2.3.3 Other types of operations

In addition to continuous extensive operations, we make use of four other types of operations. These are considered only in the study of fluid systems.

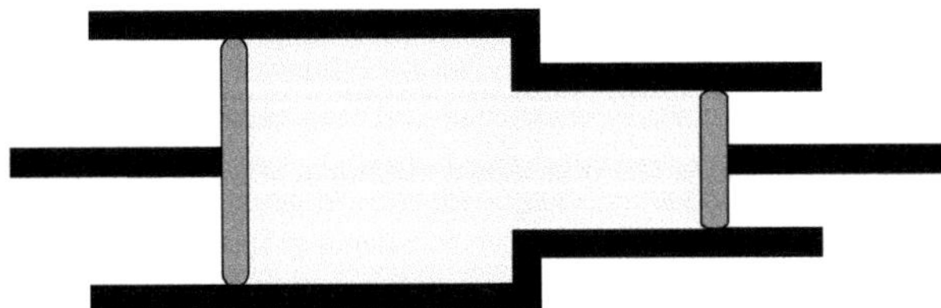

Figure 2.4 Fluid filling a vessel with two movable pistons. The volume of the container can be changed by moving either piston.

---

[14] We emphasize that this does not assume that the system is in equilibrium during the operation. Rather, it follows simply from the stipulation that for a thermodynamic system, independently of its state, the force required to vary any macroscopic degree of freedom in any particular manner can be determined through a mechanical measurement.

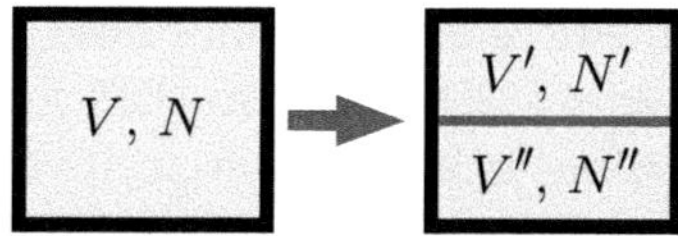

Figure 2.5 A wall is inserted into an individual system with volume V and amount of substance N. As a result, we obtain a composite system (or two separate systems) with volumes V' and $V'' = V - V'$ and amounts of substance N' and $N'' = N - N'$.

First, there are operations through which work is performed on the system while the extensive variables are held fixed. A typical example is the stirring of fluids. Operations of this type are considered in Sections 4.1.2 and 4.3.2. While such operations played an important role historically in the development of thermodynamics, they do not play an essential role in the formulation of thermodynamics presented in this book.

Second, we consider operations in which walls are inserted into or extracted from systems, resulting in the partitioning and merging of systems, respectively.[15] In the situation described by Figure 2.5, a wall is inserted into an individual system with volume V and amount of substance N. The container is thereby partitioned into two parts containing two separate individual systems, with volumes V' and $V - V'$ and amounts of substance N' and $N - N'$. We assume that operations of this type and their inverses are always possible.

Third, we often use operations in which a system is surrounded by or stripped of external adiabatic walls. Operations of this type change the external conditions of the system, as discussed in Section 2.4.2.

In all cases, we assume that the insertion or extraction of internal walls and the addition or removal of external walls is done without the performance of work on the system.

Fourth, we make limited use of operations consisting of the removal of a constraint on an internal wall, allowing it to move freely. We stipulate that these operations are carried out in such a manner that no work is performed by the system. Operations of this type are considered in Section 7.3, where we derive variational principles.

2.4 Equilibrium states

2.4.1 Introduction

Equilibrium states are the central objects of study in thermodynamics. In this section, we present a fundamental characterization of equilibrium states—defining them, elucidating the types of external conditions[16] under which they are realized, and specifying the quantities in terms of which they are parameterized.

---

[15] This includes operations in which an internal diathermal wall is replaced with an adiabatic wall and vice versa (see Section 2.4.2). We assume that operations of this kind are always realized through processes in which the diathermal wall is kept fixed and an adiabatic wall is inserted or extracted flush against it.

[16] Recall that, as stated in Section 2.1.1, the external conditions of a system are those conditions that determine all non-mechanical influences on it.

We all have an intuitive understanding of equilibrium states as time-independent states that result when a system is left undisturbed for a sufficiently long time. We can imagine that if a system is insulated from its surroundings or under the influence of a large system at constant temperature (situations that correspond to *adiabatic* and *isothermal* conditions, respectively), then if there is no mechanical perturbation, the system will eventually reach equilibrium. In this section, we formalize this understanding. Along the way, we define not only the concept of equilibrium, but also adiabatic and isothermal conditions, and temperature as a physical quantity.

2.4.2 Definition of equilibrium

In the following, we present fundamental definitions and postulates establishing the concept of equilibrium.[17]

Steady states

We begin by defining a class of states within which equilibrium states form a special subclass: A *steady state* of a thermodynamic system is a state in which the system exhibits no macroscopically measurable time dependence. Of course, the straightforward application of this definition to unequivocally identify a steady state in an actual system is unfeasible, as the condition that there be "no macroscopically measurable time dependence" is inherently open-ended. However, it can be assumed that in any given situation of experimental interest, the experimenter is able to identify a practical set of measurements sufficient to allow a sound judgment in this regard.

Adiabatic conditions

Now, consider a system in a steady state, and suppose that this steady state cannot be disturbed unless work is performed on the system. We characterize such a situation by saying that the system is enclosed within or surrounded by *adiabatic walls*, regardless of the actual material nature of the system's boundary. Then, we can turn this thinking around and define *adiabatic walls* to be any boundary that admits and maintains steady states that can be affected only by mechanical work. Also, we say that a system enclosed within adiabatic walls is under *adiabatic conditions* and refer to such a system as an *adiabatic system*.

We assume that every system is either itself an adiabatic system or a subsystem of some adiabatic system.[18] Then, for convenience, we stipulate the convention that no adiabatic system contains an adiabatic system as a subsystem. In other words, two systems that are separated by adiabatic walls are regarded as separate systems (or subsystems of separate systems), never as subsystems of the same system. This is simply a matter of terminology, and it in no way affects the content of the theory. We refer to a wall separating subsystems of an adiabatic system as a *diathermal wall*. Also, any two subsystems of an adiabatic system are said to be in *thermal contact*. We assume that we can always prepare a system with either type of walls,

---

[17] The concept of equilibrium can be formulated in various manners, but the resulting theory is not affected by the specific nature of this formulation. In fact, the Japanese version of this book adopts a different (less elegant) formulation.

[18] This simply amounts to the assumption that for any system of interest, we need not consider an unlimitedly large external world.

as desired, and that the two types of walls can be interchanged freely through controlled operations without the performance of work. In fact, we stipulate that an interchange of walls can *only* be realized through such an operation with that sole purpose, and thus that the state of thermal contact between any two systems never changes autonomously, unintentionally, or as a "side effect" of some other type of operation.

In rough terms, adiabatic and diathermal walls can be regarded as perfectly and non-perfectly insulating barriers, respectively. In other words, diathermal walls allow the passage of heat, while adiabatic walls do not.[19] Then, we intuitively understand that adiabatic conditions are realized by enclosing a system within insulating walls, and isothermal conditions are realized by enclosing a system within non-insulating walls and placing it in an environment with fixed temperature.[20]

This intuitive understanding of the roles of walls is made precise in the following sections, where isothermal conditions and temperature are defined formally.

Equilibrium

With the above preparation, we can now define an equilibrium state.

Definition 2.2 (Equilibrium state) *An equilibrium state of a thermodynamic system is a steady state that can be realized and maintained under adiabatic conditions.*

Experimentally, we can confirm that a given steady state is an equilibrium state by observing its behavior during and after an operation under which the system is enclosed within adiabatic walls (which, obviously, is trivial if the system is already under adiabatic conditions).[21] If such an operation causes no macroscopically measurable change, then we can conclude that the system is in an equilibrium state.[22] Note that from the above definition, the definitions of a steady state and adiabatic conditions, and the assumption that we can enclose any system within adiabatic walls without the performance of work, it follows that any subsystem of a system in equilibrium is also in equilibrium.

Equilibrium states are the only states treated explicitly within the theory of thermodynamics. More precisely, thermodynamics treats equilibrium states and

---

[19] Because no actual walls can be perfectly insulating, the distinction between adiabatic and diathermal is a matter of time scale; for any given set of walls, if we observe the system they enclose for a sufficiently long time, they will behave diathermally. Thus, in all investigations that we consider in this book, we implicitly assume some time scale of observation, and the distinction between adiabatic and diathermal behavior is made with respect to this time scale.

[20] To assume the existence of fixed-temperature environments and isothermal conditions is quite advantageous for developing the theory of thermodynamics in an intuitive manner. On the other hand, we could (as in Ref. [9]) construct the theory on the basis of adiabatic operations, without employing the concepts of temperature and isothermal conditions at all. With the theory developed in that manner, temperature and isothermal conditions can be derived within the completed theoretical framework. However, we wish to ground our understanding of the theoretical structure of thermodynamics in experience, and for this purpose, we believe it is best to explicitly employ the concept of isothermal conditions in our treatment.

[21] A steady state that is not an equilibrium state is called a *non-equilibrium steady state*. A typical example is a state in which there exists a steady heat flux. Such a state can be realized by placing a thermodynamic system between two environments with different temperatures.

[22] We assume that if a system is in equilibrium, an experiment of this kind always confirms this fact, i.e., that surrounding the system with adiabatic walls without the performance of work on the system never disturbs the system. Also, let us point out here that we regard an equilibrium state to be specified by the properties intrinsic to the system alone, and thus, a system in equilibrium that is subject to such an operation is regarded as being in the same state before and after.

transitions among them. During such transitions, a system may be arbitrarily far from equilibrium, but the transitions themselves are treated strictly in terms of the equilibrium states acting as their endpoints and the work performed by the system in the operations giving rise to them. The non-equilibrium states realized by the system during a transition are never considered explicitly. For this reason, throughout the book, we often refer to equilibrium states simply as "states."

Fixed external conditions

In thermodynamics, we study how systems change under mechanical operations of various kinds. In this study, it is assumed that any systems in the external world with which the system under investigation is in thermal contact are themselves not subject to mechanical operations.[23] In other words, the external conditions of a system under investigation are subject to no mechanical perturbation. We characterize this situation by saying that the system is under *fixed external conditions*.

The validity of the following postulate, which we can understand intuitively from our everyday experience, has been established through exhaustive experimental investigation.

Postulate 2.3 (Realization of equilibrium) *Consider a thermodynamic system under fixed external conditions. If this system is subject to no mechanical operation for a sufficiently long time, it will eventually reach an equilibrium state.*

The theory of thermodynamics is conventionally formulated in terms of two types of fixed external conditions. Above we defined the first type, adiabatic conditions. In the following section, we define the second type, isothermal conditions.

2.4.3 Isothermal conditions

First, let us introduce the concept of a *quasi-static operation*, which plays a fundamental role in both the construction and application of thermodynamics. Conceptually, it is sufficient to imagine a quasi-static operation to be an operation that is carried out very slowly—so slowly that the system can be regarded as being in equilibrium during its entire performance. With this simple characterization, we can proceed to our treatment of isothermal conditions (although a deeper understanding is provided by the formal definition of a quasi-static operation, given in Section A.2).

Consider a system under fixed external conditions, and suppose that it has been left unperturbed for sufficiently long that it is in an equilibrium state. Then, consider the following procedure. First, we carry out a quasi-static continuous

---

[23] Of course, we could consider the situation in which we apply an operation to some system and then ask how some other system in thermal contact with this system changes. However, in such a situation, the "system under investigation" would be regarded as the composite system formed by both of these systems, and the theory of thermodynamics would be applied to describing how this composite system changes under the operation. More generally, when we have a composite system that is subject to one or more mechanical operations, the system whose behavior we describe always includes all of the individual systems to which these operations are applied. Note, however, that this does not limit the scope of the theory of thermodynamics. It simply clarifies the form in which we seek thermodynamic descriptions.

extensive operation, the "test" operation, and measure the work performed. Next, we carry out a "return" operation that (at arbitrary speed and along an arbitrary path) eventually reverses the mechanical manipulation of the test operation. Finally, after the system has again been left unperturbed for a sufficiently long time, we repeat the test operation and again measure the work.

Definition 2.4 (Isothermal conditions) *Considering the procedure described above, if the amounts of work performed in the pair of test operations are equal for any return operation (no matter how fast or indirect), and this holds for any (practically realizable) such pair, then the system is said to be under isothermal conditions.*

A system under isothermal conditions is referred to as an *isothermal system*.[24]

Intuitively, we can understand the meaning of this definition to be that for a system under isothermal conditions, the combined effect of the test operation followed by the return operation on the resulting equilibrium state—no matter how wildly disturbing the return operation—is undone by leaving the system unperturbed for a sufficiently long time. We assume that the isothermal conditions of a system result from its thermal contact with another system, and we further assume that the system providing these conditions can be unambiguously identified. Formally, this system is defined as the *environment*. We can imagine that the environment is so large that it is unaffected by the operations applied to the system, and hence its influence on the equilibrium states realized by the system is also unaffected by these operations. Thus, the isothermal conditions provided by the environment are maintained as long as it remains in thermal contact with the system.

2.4.4 Temperature

Definition

Consider a thermodynamic system, and suppose that we prepare numerous environments in a variety of manners that can each individually be used to create isothermal conditions for this system. Empirically, it is known that the behavior of the system will in general depend on which environment is used. Here, we elucidate how the influence of the environment is accounted for.

Consider two environments. Then, consider a set of experiments consisting of various quasi-static continuous extensive operations carried out on various systems for which each environment provides isothermal conditions. The results of these experiments are regarded as the amounts of work performed in the operations. This set of experiments is carried out twice, once using each environment, and the results are compared. We regard the two environments to be equivalent if for any such set of experiments, they yield identical results (to the desired precision). We term this binary relation that of *equivalence* between environments.[25]

---

[24] It follows immediately from this definition that any subsystem of an isothermal system is also an isothermal system.

[25] This terminology reflects the mathematical situation here. If we consider a collection of environments as a mathematical set, then the relation under consideration is a type of *equivalence relation* over this set. This equivalence relation segregates environments into *equivalence classes*, which satisfy the conditions that each environment is in exactly one equivalence class, every pair of environments

The validity of the following postulate has been confirmed through countless experiments, while intuitively we can understand it through our everyday experience.

Postulate 2.5 (Temperature) *There is an experimentally observable quantity $T \in \mathbb{R}_+$ characterizing individual environments that uniquely parameterizes the relation of equivalence. In other words, any two equivalent environments will have the same value of T, while any two non-equivalent environments will have different values. We refer to T as the* temperature.

The above postulate formalizes the intuitive idea that, with regard to the equilibrium behavior exhibited by a system, temperature is the only relevant property of its environment. An environment may consist of any substance or combination of substances in gas, liquid and/or solid phases forming any kind of shape and structure, and it may be coupled to the system in many different ways. Nonetheless, while the precise nature of the non-equilibrium behavior exhibited by a system under isothermal conditions when mechanically perturbed will in general depend on numerous properties of the environment and the manner in which it is coupled to the system,[26] it is only the temperature of the environment that influences the system's equilibrium behavior.

Postulate 2.5 uniquely identifies temperature as a physical quantity. However, as with any physical quantity, there are unlimitedly many possible methods of quantification, corresponding to different scales of measure. In a sense, Postulate 2.5 merely identifies temperature as a "label" that categorizes environments. In order to develop a meaningful theory of thermodynamics, it is necessary to choose a temperature scale that quantifies this label in a manner consistent with the property of "hotness," as understood intuitively. This point is addressed in Section A.3, where we define the temperature scale used in this book, the Kelvin scale.[27] In practical terms, the Kelvin scale has two important properties. First, it is based on universal thermodynamic behavior, and its specification does not require reference to any particular systems or substances. Second, it is always positive, with a value of 0 corresponding to the unattainable state of vanishing internal energy. (Thus, it is an *absolute temperature scale*.) The unit of the Kelvin scale is the kelvin, abbreviated K. Temperature in degrees Celsius is obtained from temperature in kelvins by subtracting 273.15. The physical meaning of this choice for the temperature scale is discussed in Section 3.7, and its deeper implication will become clear in Section 5.2.1. As we will see, this is indeed the (essentially) unique "natural" choice within the logical structure of thermodynamics. However, at the present stage in our formal construction of the theory of thermodynamics, we need not be concerned with the actual temperature scale used.

in a given equivalence class are equivalent, and every pair of equivalent environments are in the same equivalence class.

[26] Although the general situation regarding the influence of an environment on non-equilibrium behavior is extremely complicated, one important factor that is intuitively easy to understand is the rate of "heat conduction" between the system and the environment. As we know from our experience, this has a direct effect on the time scales characterizing non-equilibrium processes.

[27] The Kelvin scale is sometimes also referred to as the *thermodynamic temperature scale*.

Temperature of isothermal and adiabatic systems

Postulate 2.5 defines temperature as a property of an environment that provides isothermal conditions for a thermodynamic system. We now extend this consideration, defining temperature also as a property of isothermal and adiabatic systems.

The case of isothermal systems is trivial: We define the temperature of a system under isothermal conditions to be the temperature of its environment. (Clearly, this implies that every subsystem of an isothermal system has the same temperature, because each subsystem of an isothermal system is also an isothermal system with the same environment.)

We define the temperature of an adiabatic system through a natural extension of that for an isothermal system. First, we assume that given any system in an equilibrium state under adiabatic conditions, there exists a unique temperature such that any environment characterized by that temperature can maintain this state. In other words, if we remove the adiabatic walls and place the system in thermal contact with one of these environments, the system will remain in equilibrium. We define the temperature of the adiabatic system—as well as that of any subsystems that it contains—to be the temperature of such environments.

With the above definitions and the assumption discussed in Footnote 22 on p. 35, we infer that the sets of equilibrium states realizable under isothermal and adiabatic conditions are identical: Any equilibrium state realizable under either type of external conditions can be obtained directly from an equilibrium state under the other type by either adding or removing adiabatic walls, and these identical states under different external conditions have the same temperature. We refer to this reciprocal situation as the *correspondence between states under isothermal and adiabatic conditions*.[28]

Intensive nature of temperature

From the above definitions, it is seen that, unlike the volume or amount of substance, the temperature of a system does not change under a size scaling, and hence it is an intensive quantity (see Section 2.2.2).

Practical measurement of temperature

Applying the formal experimental procedures described above and in Section A.3, in principle, the temperature can be measured in any environment or any system under isothermal or adiabatic conditions. However, obviously, these methods are impractical. As a slight diversion from our theoretical treatment, let us now briefly discuss the practical measurement of temperature.

We know that there are various types of measuring devices, i.e., thermometers, that can be used to measure temperature. And with advances in technology, these devices become increasingly precise. Furthermore, modern methods of standardization ensure a high degree of uniformity in measurement. However, the precision and standardization of such devices in no way eliminates the need for the formal experimental procedure used to define the temperature scale. The existence

---

[28] Recall that the idea of the maintenance of a given state under the change of external conditions also plays a central role in the definition of equilibrium.

of reliable devices to measure temperature requires careful calibration that is ultimately based on the formal definition. This is a fundamental point of science that transcends any kinds of technological advances.[29]

However, while problems involved in the process of advancing from the formal definition of temperature to the actual fabrication of practical, standard devices to measure it are indeed very important, we do not consider them beyond this point. In the remainder of the book, we assume that simple devices allow temperature to be readily measured in any equilibrium situation and that the results of these measurements are always identical to those that would be measured in accordance with our definitions. Empirically, we know that this is a valid assumption.

2.4.5 Parameterization of equilibrium

In Section 2.4.2, we defined the type of state that is regarded as equilibrium, but we have not yet explained how individual equilibrium states are distinguished. Here we do this.

Suppose that we are considering two separate systems or two separate realizations of the same system. How do we judge whether or not these systems or realizations are in the same equilibrium state? This is an important question, because the manner in which the general condition of equilibrium defined above is partitioned into individual states will have a significant effect on the content and applicability of the theory.

Of course, equilibrium states are characterized by their macroscopically observable properties, but in order for two states to be regarded as identical, we do not require that their respective systems be macroscopically indistinguishable. As discussed in Sections 1.2 and A.1, in science, developing a successful theory requires that we identify the important quantities and behavior and ignore the irrelevant details of the systems we study. This allows us to uncover the underlying universality that could otherwise be missed. For thermodynamic systems, focusing on just a few essential properties, we are able to construct a theoretical framework describing the universality of equilibrium states that applies uniformly to a very broad range of physical systems. As explained in Section A.1, the properties that we select for this purpose are a minimal set that allow us to account for the work performed in quasi-static operations. Empirically, it is known that the collective extensive variable, X, and the temperature, T, are one such set: All properties of a thermodynamic system in equilibrium are either uniquely determined by X and T or do not affect the work performed in quasi-static operations. We thus parameterize

---

[29] Here we describe the idealized construction of a thermometer. First, we prepare a thermodynamic system whose equilibrium states are uniquely characterized by a readily measurable property. (For example, for a mercury thermometer, this property is the height of the column of mercury.) Then, the system is calibrated by establishing a one-to-one correspondence between measured values of this property and a set of chosen temperatures by placing it into thermal contact with a much larger reference system (more precisely, a reference system with much larger heat capacity—see Section 4.3) prepared at each of these temperatures using one of the formal methods described in Section A.3 (or some other formal method). After the thermometer has been calibrated in this way, it can be used to measure the temperature in any system that is, like the reference system, sufficiently large.

equilibrium states with these variables, expressing an equilibrium state as $(T; X)$.[30]

The simplicity of the above parameterization is quite remarkable. Empirically, we know that this parameterization in terms of just a handful of readily measurable macroscopic quantities can account for behavior of a vastly universal nature. Elucidating the reason that there exists such a strong universality characterizing macroscopic systems is a problem addressed in the foundations of statistical mechanics. Within thermodynamics, this universal behavior is simply accepted as a necessary characteristic of equilibrium states, and we attempt to determine what it implies about macroscopic phenomena in general.

Because the states of a thermodynamic system are uniquely specified by the collective extensive variable and temperature, any state can be prepared by following a very simple prescription. First, let us assume that we are able to prepare an environment with any desired (nonzero) temperature. Then, given a thermodynamic system, we fix its collective extensive variable to the desired value, X, and place the system in an environment at the desired temperature, T. Doing so, the equilibrium state $(T; X)$ will be realized in the large time limit. If we wish to study the system under adiabatic conditions, we then simply enclose it within adiabatic walls.

For a given thermodynamic system, the set of all possible equilibrium states defines the *state space* of the system. We refer to the quantities parameterizing the state space as the *state variables* or *thermodynamic variables*, and all other quantities that are uniquely determined by the equilibrium state as *state functions* or *thermodynamic functions*. In the following chapters, we consider a number of state functions, for example, the pressure, $p(T; X)$, and the Helmholtz free energy, $F[T; X]$.

In Chapters 1–8, we almost exclusively study systems whose equilibrium states (for an individual system) are parameterized by T, V and N.[31] In the broadest context, these systems may be in the gas, liquid or solid phase (or some combination thereof).[32] But because the types of operations that we consider in this book are generally extremely impractical for solid systems, throughout the first nine chapters, we focus mainly on fluid systems. It should be kept in mind, however, that the general conclusions that we reach and the theoretical framework that we construct apply to the solid phase as well. In Chapter 9, where we investigate multi-component systems, the $(T; V, N)$ parameterization is generalized to

---

[30] The semicolon here is used to separate the intensive and extensive variables. We use this notation to emphasize that the distinction between these two types of quantities is of fundamental importance in the theory of thermodynamics. This type of notation is used throughout the book, with variables representing intensive quantities placed to the left of the semicolon and variables representing extensive quantities placed to the right.

[31] It is important to understand, however, that there are other possible choices for the state variables of these systems. For example, parameterizations in terms of U, V and N (where U is the internal energy) and T, p and N (where p is the pressure) are often found in the literature. These and other parameterizations are discussed in Section 8.1 and Appendices D and F.

[32] As pointed out in Footnote 11 on p. 30 and discussed in detail in Appendix D, in the case of investigations that include the triple point, along with T, V and N, one additional state variable is needed for a full specification of equilibrium states. For this purpose, we use the amount of substance in the solid phase, N_S.

a multi-component counterpart.[33] In Chapter 10, where we investigate magnetic systems, equilibrium states are parameterized by T, M and N.

2.4.6 The "zeroth law" and related properties of temperature

In the following, we give further discussion of temperature and its behavior. The results presented here are used repeatedly throughout the book.

Systems in thermal contact have the same temperature

Result 2.6 ("Zeroth law of thermodynamics") *Consider two individual thermodynamic systems that are prepared in arbitrary manners and then placed in thermal contact. Next, suppose that the composite system that they form is enclosed within adiabatic walls and then left unperturbed. After a sufficiently long time, the two individual systems will be in equilibrium states with the same temperature.*

Derivation *From Postulate 2.3 (p. 36), the entire composite system must reach an equilibrium state. The result then follows from the definition of the temperature of an adiabatic system.* ∎

As the name "zeroth law" suggests, in some formulations of thermodynamics, Result 2.6 is treated as a fundamental premise. Here, instead, we have derived it from basic postulates and definitions. However, this does not mean that we have made a new discovery elucidating the behavior described by Result 2.6 at a more fundamental level. Rather, we have merely presented sufficiently many assumptions and definitions that the zeroth law has become derivable.

Insertion of a partitioning wall does not change the temperature

Consider an individual system under adiabatic conditions with temperature T. Suppose that we insert a wall (diathermal or adiabatic) into this system, partitioning it into two individual systems.[34] Here we show that the temperature of each resulting individual system is also T.[35]

Suppose that the initial state of the system is $(T; X)$. From the correspondence between states under isothermal and adiabatic conditions discussed in Section 2.4.4, we know that the system in this state under adiabatic conditions can be realized by first preparing the system in this state under isothermal conditions and then surrounding it with adiabatic walls. The final state is then realized

---

[33] In Sections 9.1–9.3, N is replaced with $(N_1, \ldots, N_m)$, where N_i is the (fixed) amount of the ith substance (of an m-component system), while in Sections 9.4–9.6, where we study systems exhibiting chemical reactions, it is replaced with $(\widetilde{N}_1, \ldots, \widetilde{N}_m)$, where $\widetilde{N}_i$ is the amount of the ith substance in equilibrium.

[34] Throughout the book, it is important to keep in mind that, as stated in Section 2.3.3, any operation consisting of the insertion or extraction of an internal wall or the addition or removal of external adiabatic walls is understood to be done without the performance of work by the system, and, as stated in Section A.2, any operation consisting of the insertion of an internal wall or the addition of external adiabatic walls is understood to be carried out quasi-statically.

[35] The same result in the case of a composite system is obtained with a simple generalization of the derivation given here.

by inserting the partitioning wall. Let X' and X'' denote the values of the collective extensive variables in this state.

Next, let us consider a second treatment of the same system. Again, we begin in the state $(T; X)$ under isothermal conditions. Then, maintaining the isothermal conditions, we insert the partitioning wall, again realizing the values X' and X'' for the new collective extensive variables. (Let us for the time being suppose that the partitioning wall is adiabatic.) This yields two systems in the states $(T; X')$ and $(T; X'')$, assuming that the isothermal conditions are maintained for each system. Finally, these systems are individually surrounded by adiabatic walls. Again, from the correspondence between states under isothermal and adiabatic conditions, it follows that this yields two systems in the states $(T; X')$ and $(T; X'')$ under adiabatic conditions. (Clearly, the situation in the case of a diathermal partitioning wall is the same, except that the final state is $(T; \{X', X''\})$, with a single, composite system.)

Note that the only difference between the two treatments of the system discussed above is that the order of the two operations is reversed. However, as discussed in Section A.2, both of these operations are carried out in such manners that they induce no macroscopically observable processes within the system. Thus, as mentioned in Footnote 22 on p. 35, because we regard the state of a system to be determined by only the properties intrinsic to the system itself, we conclude that the final states obtained from these two treatments must be the same. Hence, the individual systems obtained when a system is partitioned have the same temperature as the original system in the original state.

Removal of a partitioning wall as a quasi-static operation

As discussed in Section A.2, the removal of a partitioning wall is quasi-static if and only if the initial state of the system under this operation can be realized as the final state under the inverse operation. Thus, from the above result, we see that in order for the removal of a partitioning wall to be quasi-static, it is a necessary condition that the systems on either side of the wall have the same temperature.[36]

Quasi-static removal of a partitioning wall does not change the temperature

Now, let us consider the quasi-static removal of a partitioning wall separating two systems at temperature T. Because we can regard the initial state under this operation to be realized through the insertion of the wall into a system with the same temperature, and because both the insertion and removal of the wall induce no macroscopically observable processes within the system, we conclude that under the combination of these two operations, the state of the system is unchanged. Hence the temperature of the final state under the removal of the partitioning wall is also T.

---

[36] While this condition is necessary for the operation to be quasi-static, as discussed in Section 7.4, generally it is not sufficient.

3
Isothermal Operations and Helmholtz Free Energy

In this chapter, after introducing the fundamental concept of an isothermal operation, we present Kelvin's principle, which can be regarded as expressing the essential idea of thermodynamics. Next, we define the *Helmholtz free energy* through analogy to the mechanical potential energy. The Helmholtz free energy is a thermodynamic function of fundamental importance in the treatment of thermodynamic systems under isothermal conditions in situations that the collective extensive variable is a controlled quantity. For this reason, it plays a central role in this book. After discussing the properties of the Helmholtz free energy, we elucidate its relationship with the pressure, which is a directly measurable physical quantity. Finally, we introduce the concept of the *equation of state* of a thermodynamic system.

3.1 Isothermal operations

In this section, we consider the application of externally controlled operations to thermodynamic systems under isothermal conditions.

3.1.1 General isothermal operations

Let us consider a system under isothermal conditions at temperature T and suppose that at some initial time, the collective extensive variable, X, takes the value X_1, and the system is in the equilibrium state $(T; X_1)$. Then, suppose that through some operation, the value of X is changed from X_1 to a value X_2. Here, we impose no restriction on the speed with which the operation is carried out. We also allow the possibility that part or all of the system is enclosed within adiabatic walls (i.e., placed under adiabatic conditions) during some time interval or intervals after the beginning and before the end of the operation.[1] Then, after the completion of this operation, with the value of X fixed at X_2 for a sufficiently long time, in accordance with Postulate 2.3 (p. 36), the system will approach the new equilibrium state $(T; X_2)$. An operation of this type is referred to as an *isothermal operation*. Symbolically, we express the transition resulting from such an operation as

---

[1] More precisely, in order for an operation to be regarded as isothermal, it is only necessary that the system begin in an equilibrium state under isothermal conditions, end in an equilibrium state under isothermal conditions with the same temperature, and never be subjected to isothermal conditions with a different temperature at any time between.

Thermodynamics: A Modern Approach. Hal Tasaki and Glenn Paquette, Oxford University Press.
© Hal Tasaki and Glenn Paquette (2026). DOI: 10.1093/9780191878091.003.0003

$$(T; X_1) \xrightarrow{\text{i}} (T; X_2) \,. \tag{3.1}$$

In general, during the time that an isothermal operation is being carried out, the system will not be in equilibrium. For example, consider a gas inside a container with a piston at one end. When the piston is moved, this gas will generally exhibit density variations and hence material flow. As the speed of the piston's motion is increased, this flow will become increasingly complex, and eventually turbulent. Therefore, the state of the system undergoing a transition represented by (3.1) will generally be of a type that cannot be described in terms of thermodynamic theory. One aspect of such non-equilibrium behavior is that the temperature of the system as measured by a thermometer may temporarily increase or decrease in some spatially dependent manner, and even become practically unmeasurable. Thus, during an operation, the system will generally not be "isothermal" in the everyday, intuitive sense of this word.

We assume that a transition of the type (3.1) can always be realized if it is possible to carry out a mechanical manipulation through which the collective extensive variable changes from X_1 to X_2. This is clearly reasonable, because to obtain such a transition, it is only necessary to place the system in contact with an environment at temperature T, allow the system to reach equilibrium with X fixed at X_1, apply the manipulation, and then allow the system to reach equilibrium with X fixed at X_2.

3.1.2 Quasi-static isothermal operations and their reverse operations

Consider an isothermal operation carried out on a gas contained in a vessel equipped with a piston whose position determines the volume of the system. If this piston is moved sufficiently slowly, we can expect that no large currents will appear in the gas. In such a case, therefore, it is reasonable that at any time during the operation, the system will be close to an equilibrium state. An operation of this kind is referred to as a *quasi-static operation*. (More detailed discussion and a formal definition of quasi-static operations is given in Section A.2.) A quasi-static operation that is also an isothermal operation is referred to as a *quasi-static isothermal operation*.

We express the transition realized under a quasi-static isothermal operation from a state $(T; X_1)$ to a state $(T; X_2)$ as

$$(T; X_1) \xrightarrow{\text{qi}} (T; X_2) \,. \tag{3.2}$$

As in the case of (3.1), we assume that a transition of the type (3.2) is always realizable if it is possible to carry out a mechanical manipulation through which the collective extensive variable is changed from X_1 to X_2.

Let us study a quasi-static isothermal operation inducing the transition (3.2). The time-reversed counterpart to this operation obtained by simply reversing the evolution of the mechanical world (and hence the variation of each extensive variable) is another quasi-static isothermal operation, yielding the transition

$$(T; X_2) \xrightarrow{\text{qi}} (T; X_1) \,. \tag{3.3}$$

Because the system is in equilibrium at any time during which either the original operation or the reverse operation is being carried out, and thus the properties of the system at any time during either are determined entirely by the values of the extensive variables and the temperature, the behavior of the system over the entire operation in one direction is identically the time reversal of that in the other direction.[2] Also, under these conditions, the force that the system exerts on the mechanical world at any time is determined entirely by the equilibrium state. Hence, if the work performed by the system over the entire operation in one direction is W, that in the other direction is $-W$.

We use the notation

$$(T; X_1) \xleftrightarrow{\text{qi}} (T; X_2) \tag{3.4}$$

to indicate that the transitions (3.2) and (3.3) are both realizable and that they are related by a simple time reversal.

Now, recall that even in the general case of isothermal operations, changing the collective extensive variable from X_1 to X_2, we can realize the transition $(T; X_1) \xrightarrow{\text{i}} (T; X_2)$, and changing the collective extensive variable from X_2 to X_1, we can realize the transition $(T; X_2) \xrightarrow{\text{i}} (T; X_1)$. However, if these operations are not performed quasi-statically, the system will not be in equilibrium while they are being carried out, and for this reason, the properties of the system at any time will not be determined simply by the values of the extensive variables and temperature. Instead, they will depend in some complicated manner on the dynamics of the operations. Thus, even if these two operations themselves are carried out in precisely time-reversed manners, the behavior exhibited by the system under one of them will not be simply the time reversal of that exhibited by the system under the other, as in the quasi-static case. For this reason, the amounts of work performed by the system under the two operations will generally not be related in a simple manner. This can be regarded as the essential difference between quasi-static and non-quasi-static operations.

3.2 Kelvin's principle

In this section, we discuss Kelvin's principle,[3] which is one of the fundamental postulates of thermodynamics. This postulate is the formal statement within a rigorous theoretical framework of the empirical finding that heat cannot be freely converted into work. It can be said that this principle expresses the essence of thermodynamics.

Let us consider a system that is originally in an equilibrium state $(T; X)$. Then, suppose that the system is subject to an isothermal operation under which the collective extensive variable is varied in some arbitrary manner before eventually

---

[2] This requires that the states of the system that correspond to each other under the time reversal of the operation have equal values of the temperature. If the system is under isothermal conditions, then of course this is trivial. But if the system is under adiabatic conditions, it is not certain at the present point in the construction of the theory. (Recall that for an isothermal operation, there can be time intervals during which the system is under adiabatic conditions.) However, the situation is clarified in Chapter 4, because this equality follows from the result that in equilibrium, T is uniquely determined by X and U, the energy, which is implied by Result 4.6 (p. 75).

[3] Lord Kelvin, after whom this principle is named, was born William Thomson.

being returned to the value X. An isothermal operation like this, which begins and ends with the same value of the collective extensive variable, is called an *isothermal cycle*. Let us write the amount of work performed by the system on the mechanical world over this entire operation as W_cyc. In this context, Kelvin's principle takes the following form.[4]

Postulate 3.1 (Kelvin's principle) *For any isothermal cycle carried out at any temperature, the following relation holds:*

$$W_\mathrm{cyc} \leq 0 \, . \tag{3.5}$$

Kelvin's principle asserts that it is not possible for a thermodynamic system to perform a positive amount of work on the mechanical world through an isothermal cycle. In order to understand the implication of this assertion, let us suppose that Kelvin's principle did not hold. In other words, we suppose that there exists an isothermal operation beginning and ending in the same equilibrium state that performs a positive amount of work on the mechanical world. Then, consider the situation in which this cycle is repeated many times. Under such a hypothetical operation, the system and the temperature of its environment undergo no change, but the system is capable of doing an unlimited amount of work. In other words, without itself experiencing any change over many cycles, the system is able to produce mechanical energy indefinitely. A process of this kind would have the effect of converting heat (see Section 5.1) into mechanical energy, and thus it would not necessarily violate the law of energy conservation. Nonetheless, if there existed a device capable of such behavior, it would be extraordinarily useful. For example, using the atmosphere or the ocean as the environment, we would be able to generate electric power from it without the need for any fuel, and thus all of the energy problems of the world would be solved. This kind of hypothetical device is referred to as a *perpetual motion machine of the second kind*.[5]

Kelvin's principle is equivalent to the assertion that there can exist no perpetual motion machine of the second kind.[6] In this book, we formulate the theory of thermodynamics on the basis of Kelvin's principle. This principle is an empirical fact characterizing the world in which we live. For centuries, inventors have devised countless types of heat engines with the hope of creating a perpetual motion machine of the second kind. No one has succeeded. Also, although Nature is a much more ingenious craftsman than any human inventor, and the process of evolution on Earth has been proceeding for billions of years, there exists no known

---

[4] Kelvin's principle is regarded as one statement of the second law of thermodynamics. Other statements of the second law include Clausius's principle (which is not covered in this book), the principle of maximum work (Result 3.3 (p. 52)), Planck's principle (Result 6.4 (p. 116)), and the entropy principle (Result 6.5 (p. 117)). These principles can all be regarded as different ways of expressing the single underlying essence of thermodynamics. Indeed, many thermodynamics textbooks assert that (at least some of) these statements are equivalent. However, it should be noted that often, the "proofs" given of this equivalence are lacking in rigor and implicitly include various assumptions concerning the nature of thermodynamics. (The first law of thermodynamics, a law of energy conservation, is stated in Postulate 4.5 (p. 69) and discussed in Section 4.2.)

[5] A *perpetual motion machine of the first kind* is a device that violates the law of energy conservation. On the basis of our experience, a device of this type is also believed to be impossible.

[6] Let us emphasize that Kelvin's principle regards a cyclic operation. Obviously, there are many thermodynamic systems under isothermal conditions that can perform positive work on the external world through non-cyclic operations (e.g., a charged battery, a compressed gas).

organism capable of violating Kelvin's principle. While there does exist a biological process through which light energy is converted into chemical energy (i.e., photosynthesis), there exists no "thermo-synthesis" process that simply extracts thermal energy from an environment and converts it into some different form that can fuel biological activity. This fact also provides strong empirical support for the universal validity of Kelvin's principle.[7]

There have been various attempts to derive Kelvin's principle from more microscopic theories.[8] From these studies, we now know that Kelvin's principle is at least consistent with statistical mechanics based on classical mechanics or quantum mechanics in many contexts. Nevertheless, it is fair to say that a completely general and robust microscopic derivation applicable to all thermodynamic phenomena is still lacking. In this sense, Kelvin's principle remains a phenomenological law. However, the lack of a complete "proof" on the basis of some more microscopic theory does not diminish the theoretical status of Kelvin's principle.[9] In general, the fundamental postulates of science should represent universal behavior distilled through abstraction from empirical facts. In the case of Kelvin's principle, an enormous body of empirical support covering a vast range of phenomena provides very strong evidence of its universal validity. Theoretical results demonstrating its consistency with microscopic physics strengthens this support further.

Before ending this section, we discuss a simple but very important implication of Kelvin's principle.

Result 3.2 (The work performed in a quasi-static isothermal cycle) *We refer to a cycle $(T; X) \xrightarrow{\text{qi}} (T; X)$ realized under a quasi-static isothermal operation as a* quasi-static isothermal cycle. *For any quasi-static isothermal cycle realized at any temperature, the equality $W_{\text{cyc}} = 0$ holds.*

Derivation *From Kelvin's principle, we have $W_{\text{cyc}} \leq 0$. Then, because the operation considered here is quasi-static, the operation obtained by reversing the*

---

[7] It should be noted, however, that in order for a biological process to be realized in living organisms, it is not enough that this process be possible in principle. It is also necessary that the mechanism responsible for this process be sufficiently accessible (in an evolutionary sense) and provide sufficient competitive advantage in comparison with alternative mechanisms that could emerge in the same organism at any given point in its evolution. Thus, the absence of thermo-synthesis in organisms cannot be regarded as definitive evidence for the impossibility of a mechanism that could violate Kelvin's principle. It may simply be attributable to the inaccessibility of such a mechanism or the lack of comparative advantage that it could provide.

[8] It is known, for example, that mathematical relations that can be interpreted as Kelvin's principle (i.e., $W_{\text{cyc}} \leq 0$) can be derived from classical or quantum mechanics by assuming that the initial state is an ideal equilibrium state (described exactly by, e.g., the canonical distribution in equilibrium statistical mechanics) and that the operation can be modeled as a variation of parameters in the Hamiltonian. (For example, see the following: T. Sagawa, "Second Law-Like Inequalities with Quantum Relative Entropy: An Introduction," in *Lectures on Quantum Computing, Thermodynamics and Statistical Physics*, M. Nakahara and S. Tanaka eds., World Scientific, 2012 (https://arxiv.org/abs/1202. 0983); H. Tasaki, "Quantum Statistical Mechanical Derivation of the Second Law of Thermodynamics: A Hybrid Setting Approach," Physical Review Letters, vol. 116, iss. 17, 2016, 170402:1–5 (https://link. aps.org/doi/10.1103/PhysRevLett.116.170402).

[9] As an example for comparison, let us consider the inverse square law that characterizes gravity. This has been deduced through idealization from our observations of the motion of heavenly bodies and direct experimentation. However, there is no "proof" of its validity. It is true that it can be derived from general relativity, but this derivation does not constitute a proof of the validity of the inverse square law. In fact, the situation is quite the reverse. The consistency of general relativity with the inverse square law is a necessary condition for the former to be regarded as a viable theory, and hence it is this phenomenological law that provides a fundamental check of the theory of general relativity.

variation of the extensive variables is also a quasi-static isothermal cycle, and the work performed during this cycle is $-W_{\mathrm{cyc}}$. Again, from Kelvin's principle, we have $-W_{\mathrm{cyc}} \leq 0$. Combining these, we obtain $W_{\mathrm{cyc}} = 0$. ∎

In the previous section, we argued that the amounts of work performed by a system on the mechanical world in two mutually time-reversed quasi-static isothermal operations are of equal magnitude and opposite sign. Result 3.2 allows us to draw a more general conclusion. From the fact that $W_{\mathrm{cyc}} = 0$ holds for any quasi-static cycle, it follows that the work done by the system in any quasi-static isothermal operation causing a transition from one state $(T; X_1)$ to another state $(T; X_2)$ is the opposite of that done in any quasi-static isothermal operation causing a transition from $(T; X_2)$ to $(T; X_1)$. More concisely, quasi-static isothermal operations that induce inverse transitions engender opposite amounts of work. This implies a kind of path independence of the work performed in a quasi-static isothermal operation: This work depends only on the initial and final states, and is independent of how X varies between them.

3.3 Mechanical potential energy

The main purpose of this chapter is to introduce a thermodynamic function known as the *Helmholtz free energy*. This function plays a role analogous to that of the potential energy in classical mechanics, and for this reason, to motivate the construction of the former, it is useful to first briefly review the latter.

As a simple example, let us consider a system consisting of a single particle of mass m that possesses some potential energy $V(\boldsymbol{r})$, where $\boldsymbol{r}$ represents the particle's position in three-dimensional space. We assume that the particle experiences no frictional force. Suppose that at the initial time, through the application of an external force (i.e., a force distinct from that derived from $V(\boldsymbol{r})$), the particle is fixed at the position $\boldsymbol{r}_1$. Then, through the application of some additional external force, the particle is caused to move to a new position, $\boldsymbol{r}_2$, where it is again fixed. During this process, the total work performed by the particle on the agent applying the external forces is given by

$$W_{\mathrm{mech}}(\boldsymbol{r}_1 \to \boldsymbol{r}_2) = V(\boldsymbol{r}_1) - V(\boldsymbol{r}_2) . \tag{3.6}$$

Because the particle is stationary at both the initial and final times, the quantities $V(\boldsymbol{r}_1)$ and $V(\boldsymbol{r}_2)$ represent the total amounts of mechanical energy of the particle at these times, and thus (3.6) is simply an expression of the conservation of energy: The difference between the initial and final amounts of mechanical energy of the system is extracted by the mechanical world in the form of work.

The main point of this example is that the relation (3.6) holds exactly for any process through which the particle begins with 0 velocity at $\boldsymbol{r}_1$ and ends with 0 velocity at $\boldsymbol{r}_2$, no matter what positions or velocities are realized along the way. This path independence of the work for a mechanical system is analogous to that identified above in the case of a quasi-static operation for a thermodynamic system.

Now, the first important step in constructing the theoretical structure of thermodynamics is to find quantities that obey relations analogous to (3.6). For this purpose, we begin by noting that in the thermodynamic case, the variables analogous to the particle's position are those which specify its thermodynamic state, T and X. With this understanding, in Section 3.6, we derive a relation analogous to (3.6) in which W_{mech} is replaced with the *maximum work*, W_{max}, and $V(\boldsymbol{r})$ with the Helmholtz free energy, $F[T; X]$. Next, in Section 4.3, we derive a relation analogous to (3.6) in which W_{mech} is replaced with the *adiabatic work*, W_{ad}, and $V(\boldsymbol{r})$ with the energy, $U(T; X)$.

3.4 A comparison of mechanical and thermodynamic "black boxes"

Let us reconsider the expression of energy conservation for a mechanical system given in (3.6). This can also be interpreted as a relation through which the potential energy experienced by the particle, $V(\boldsymbol{r})$, can be ascertained through measurement of the work performed by the particle on the mechanical world. To understand this, let us consider Figure 3.1(a). In the system depicted there, the particle is situated inside a "black box," and it is subject to the force exerted by a spring, which is also part of the system. Attached to the particle is an externally controlled rigid rod. Using this rod, an agent can move the particle to any desired position inside the box. Measuring the work necessary to carry out such operations and using (3.6), the agent can completely determine the form of the potential energy, without ever opening the box.[10]

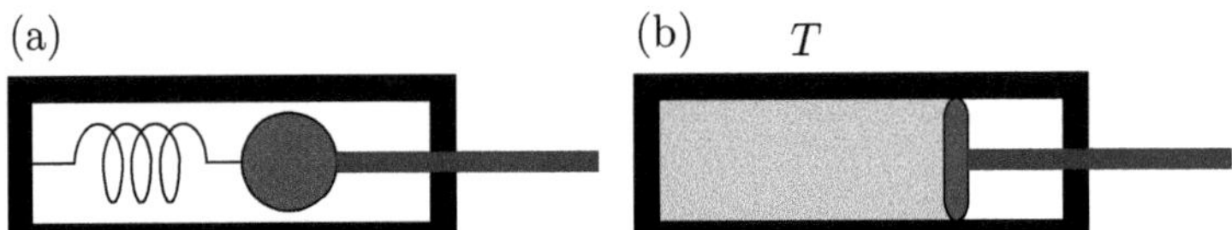

Figure 3.1 Comparing two "black boxes" whose internal systems are, respectively, mechanical and (isothermal) thermodynamic in nature, we are able to understand the essential difference between such systems. As depicted in (a), for the mechanical system, by measuring the reaction that the agent feels when moving the control rod to various positions, the potential energy of the spring can be determined directly. This relies on the fact that for this mechanical system, the amount of work performed by the agent in an operation through which the particle moves between two positions is determined by these positions alone (given that the velocity of the particle is 0 at both the initial and final times). Contrastingly, in the case of (b), where the "spring" consists of air compressed by a piston and held at constant temperature through contact with an environment, the work performed by the agent in an operation through which the piston moves between two positions is not determined by these positions alone, but also depends on the speed with which the control rod is moved during the operation.

---

[10] Of course, only differences between values of $V(\boldsymbol{r})$ at different positions can be measured in this manner, but, as we recall from introductory classical mechanics, only differences in the potential energy are physically meaningful.

Next, consider the case in which, as depicted in Figure 3.1(b), the system inside the box consists not of a particle attached to a mechanical spring but of a cylinder separated into two regions, one of which is filled with air, and the other of which is vacuum. The boundary between these regions is provided by a piston whose position is controlled by a rigid rod. Further, suppose that this entire apparatus is surrounded by an environment at fixed temperature. Noting the similarity between this "air spring" system and the ordinary spring system considered above, one might think that by measuring the work required to move the piston between various positions in arbitrary manners, the "potential energy" of the air spring could also be determined. However, this is not true. In the case of this isothermal thermodynamic system, the amount of work performed by the system on the external world during an operation through which the system undergoes a transition between two equilibrium states depends in a complicated way on the nature of the operation. This fact reflects the fundamental difference between a mechanical system and an isothermal thermodynamic system.

Let us consider a simple example. The system depicted in Figure 3.2 consists of a gas contained within a cylindrical container with a fixed wall at either end. A moveable, perfectly sealing piston separates the container into two compartments. The compartment to the left of the piston is filled with gas, while the compartment to the right is a vacuum. The entire apparatus is surrounded by an environment with fixed temperature. In the situation considered in (a), the piston is moved slowly, and the gas continuously exerts pressure on the piston. For this reason, as the piston is moved to the right and the gas expands, the gas performs work on the external world. Contrastingly, in (b), the piston is moved so rapidly that it loses contact with the gas.[11] As a result, the region just to the left of the piston temporarily becomes a vacuum, and thus the pressure exerted on the piston vanishes. Then, after the piston has reached the right end of the cylinder, the gas spreads to fill the entire cylinder, and eventually an equilibrium state is re-established. In this case, the gas does no work on the external world during the operation. However, despite this striking difference with regard to the work performed by the system, both the initial and final states of the system are identical in the two cases. It is

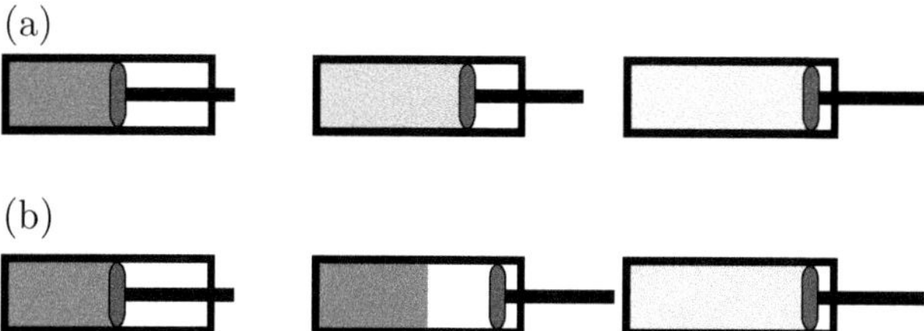

Figure 3.2 Two processes through which a gas expands. (a) In this case, the piston is moved slowly, and for this reason, the gas continuously exerts a force to the right as it expands. As a result, it performs a positive amount of work on the external world. (b) Here, the piston moves so rapidly that the gas cannot keep up, and hence it performs no work on the external world.

---

[11] Although experimentally realizing such a situation would be technically somewhat difficult, the temporary vanishing of the pressure experienced by the piston described here is merely the extreme case of a generally observed effect characterizing non-slow operations.

thus seen that the work performed in an isothermal operation that causes a transition between two given equilibrium states depends on the manner in which the operation is carried out.

3.5 Maximum work

As we saw above, in the case of a thermodynamic system under isothermal conditions, there does not exist a relation analogous to (3.6) that provides a generally valid specification of the work performed in an operation. However, if we replace the work and potential energy with the appropriate quantities, in fact there is a relation of this form that characterizes isothermal systems. These quantities are, respectively, the *maximum work* and the *free energy*.

3.5.1 Definition of the maximum work

Let us consider a system that undergoes an isothermal operation inducing some transition

$$(T; X_1) \xrightarrow{\text{i}} (T; X_2) \, . \tag{3.7}$$

As discussed above, for given T, X_1 and X_2, the amount of work performed by the system on the mechanical world in this kind of operation will depend on the manner in which the operation is carried out. Here we are interested in identifying the maximum such value. As we will see, this value plays a fundamental role in the theory of thermodynamics.

We write the maximum value of the work that can be realized in any operation yielding the transition (3.7) as $W_{\max}(T; X_1 \to X_2)$ and refer to this value as the *maximum work*. Clearly, for given initial and final states $(T; X_1)$ and $(T; X_2)$, the quantity $W_{\max}(T; X_1 \to X_2)$ is uniquely determined. The fundamental role played by $W_{\max}(T; X_1 \to X_2)$ is due to the defining property expressed by the following.

Result 3.3 (Principle of maximum work) *The maximum work $W_{\max}(T; X_1 \to X_2)$ is identical to the amount of work performed by the system on the mechanical world in an arbitrary quasi-static isothermal operation inducing the transition $(T; X_1) \xrightarrow{\text{qi}} (T; X_2)$.*

An intuitive understanding of Result 3.3 is provided by Figures 1.3 and 3.2 in the case of one simple type of system, but we wish to emphasize that it is of entirely general validity, holding for all transitions exhibited by thermodynamic systems under which the temperature is unchanged. From this principle, it is seen that the path independence of the work performed in a quasi-static isothermal operation discussed below Result 3.2, analogous to the path independence characterizing work in a mechanical system, is an attribute of the *maximal* work in a thermodynamic system.

Result 3.3 can be derived from Kelvin's principle, Postulate 3.1 (p. 47), as demonstrated below.[12]

---

[12] Also, as shown in Problem 3.2, Postulate 3.1 can be derived from Result 3.3. Hence, these two assertions are equivalent.

Derivation of Result 3.3 *Consider an arbitrary quasi-static isothermal operation under which the system makes a transition from $(T; X_1)$ to $(T; X_2)$. Let
us refer to the work performed by the system on the mechanical world during
this operation as W. Next, consider an arbitrary (not necessarily quasi-static)
isothermal operation causing a transition between the same two states. We refer
to the work performed by the system during this operation as W'. Now, suppose
that beginning in the state $(T; X_1)$, we carry out the latter operation and then
the time-reversed counterpart of the former operation. We thereby obtain the
isothermal cycle*

$$(T; X_1) \xrightarrow{\text{i}} (T; X_2) \xrightarrow{\text{qi}} (T; X_1) . \tag{3.8}$$

*Recalling that, as discussed in Section 3.2, quasi-static isothermal operations
that induce inverse transitions engender opposite amounts of work, we find that
the total work performed by the system in the isothermal cycle (3.8) is $W_{\text{cyc}} =
W' - W$. Therefore, from Kelvin's principle, we obtain the relation $W' \leq W$. It
is thus concluded that W is the maximal amount of work that can be performed
in an isothermal operation through which the system makes a transition from
$(T; X_1)$ to $(T; X_2)$.* ∎

The principle of maximum work, Result 3.3, guarantees that for any system
and any two equilibrium states of this system with the same temperature, the
maximum value of the work among all the operations that connect these states can
be determined by simply measuring the work performed under a single, arbitrary
quasi-static isothermal operation. We assume that this measurement can be made
for any such transition of interest.

3.5.2 Properties of the maximum work

Below, we elucidate the fundamental properties of the quantity $W_{\text{max}}(T; X_1 \to X_2)$.

Sign rule

For quasi-static isothermal operations that induce mutually inverse transitions,
Result 3.2 and Result 3.3 together yield the following sign rule:

$$W_{\text{max}}(T; X_1 \to X_2) = -W_{\text{max}}(T; X_2 \to X_1) . \tag{3.9}$$

Sum rule

Suppose that a system is subject to two quasi-static isothermal operations in
succession, one inducing a transition $(T; X_1) \xrightarrow{\text{qi}} (T; X_2)$, and one inducing a
transition $(T; X_2) \xrightarrow{\text{qi}} (T; X_3)$. With the understanding that consecutive operations can always be regarded as forming a single operation (see Section 2.3.1), and
using the implication of Result 3.2 that the work performed under a quasi-static
isothermal operation depends on only the initial and final states, we obtain the
following sum rule:

$$W_{\text{max}}(T; X_1 \to X_3) = W_{\text{max}}(T; X_1 \to X_2) + W_{\text{max}}(T; X_2 \to X_3) . \tag{3.10}$$

Additivity property

In Chapter 2, we introduced the concept of a composite system as a thermodynamic system that contains multiple subsystems in thermal contact. The additivity property of the maximum work describes the maximum work performed by a composite system.

Consider a composite system consisting of two subsystems with collective extensive variables X and Y that undergoes a transition

$$(T; X_1, Y_1) \xrightarrow{\text{qi}} (T; X_2, Y_2) \,. \tag{3.11}$$

In concrete terms, this transition can be realized under many actual (quasi-static, isothermal) operations. As one example, let us consider a series of operations through which first the diathermal wall separating the subsystems is replaced with an adiabatic wall (so that these subsystems become independent systems), then X and Y are varied from X_1 to X_2 and Y_1 to Y_2, yielding the independent transitions

$$(T; X_1) \xrightarrow{\text{qi}} (T; X_2), \quad (T; Y_1) \xrightarrow{\text{qi}} (T; Y_2) \,, \tag{3.12}$$

and finally, the adiabatic wall separating the systems is replaced with a diathermal wall. Then, because the work performed under a quasi-static isothermal operation depends on only the initial and final states, and because no work is performed in the operations interchanging the diathermal and adiabatic walls, we find that the maximum work satisfies the following relation:

$$W_{\max}(T; \{X_1, Y_1\} \to \{X_2, Y_2\}) = W_{\max}(T; X_1 \to X_2) + W_{\max}(T; Y_1 \to Y_2) \,. \tag{3.13}$$

Beginning instead with a composite system consisting of n subsystems with collective extensive variables $X^{(1)}, \ldots, X^{(n)}$ and proceeding analogously, we can readily extend (3.13) to the following general relation, representing the additivity property of the maximum work:

$$W_{\max}(T; \{X_1^{(1)}, \ldots, X_1^{(n)}\} \to \{X_2^{(1)}, \ldots, X_2^{(n)}\}) = \sum_{j=1}^{n} W_{\max}(T; X_1^{(j)} \to X_2^{(j)}) \,. \tag{3.14}$$

Extensivity property

We next demonstrate that the maximum work is an extensive quantity, as defined in Section 2.2.2.

Consider a thermodynamic system and a scaled copy of this system (see Section 2.2.2) with a scaling factor $\lambda \in \mathbb{R}_+$. For a quasi-static isothermal operation carried out on the former that induces a transition $(T; X_1) \xrightarrow{\text{qi}} (T; X_2)$ and a quasi-static isothermal operation carried out on the latter that induces the corresponding transition $(T; \lambda X_1) \xrightarrow{\text{qi}} (T; \lambda X_2)$, the amounts of work performed by the respective systems are related as follows:

$$W_{\max}(T; \lambda X_1 \to \lambda X_2) = \lambda W_{\max}(T; X_1 \to X_2) \,. \tag{3.15}$$

This extensivity property of the maximum work is demonstrated below.

Derivation of (3.15) *Consider an individual system with collective extensive variable Y that undergoes the transition*

$$(T; Y_1) \xrightarrow{\text{qi}} (T; Y_2) \,. \tag{3.16}$$

As one way to realize this transition, let us consider the following series of operations. First, a number of adiabatic walls are inserted into the system in such a manner that it is partitioned into n separate systems, all in the same state, $(T; Y_1/n)$. Then, with each of these systems under the original isothermal conditions, its collective extensive variable is varied quasi-statically to the value Y_2/n. In this way, each system independently undergoes the transition

$$(T; \frac{1}{n}Y_1) \xrightarrow{\text{qi}} (T; \frac{1}{n}Y_2) \,. \tag{3.17}$$

Finally, the adiabatic walls partitioning the systems are removed, and the transition (3.16) is thereby realized. Because the work performed under a quasi-static isothermal operation depends on only the initial and final states, and because no work is performed in the operations of inserting and removing the walls, we obtain the following:

$$W_{\max}(T; Y_1 \to Y_2) = nW_{\max}(T; \frac{1}{n}Y_1 \to \frac{1}{n}Y_2) \,. \tag{3.18}$$

Next, considering a second procedure just like the above, but with the original system partitioned into some different number of independent systems m, we deduce the relation

$$mW_{\max}(T; \frac{1}{m}Y_1 \to \frac{1}{m}Y_2) = nW_{\max}(T; \frac{1}{n}Y_1 \to \frac{1}{n}Y_2) \,. \tag{3.19}$$

Introducing the new collective extensive variable $X = Y/n$ and making the identifications $X_1 = Y_1/n$ and $X_2 = Y_2/n$, the above relation can be rewritten as

$$W_{\max}(T; \frac{n}{m}X_1 \to \frac{n}{m}X_2) = \frac{n}{m}W_{\max}(T; X_1 \to X_2) \,. \tag{3.20}$$

Because this holds for arbitrary integers m and n, and because with regard to the value of a collective extensive variable, experimentally it is not meaningful to distinguish between rational numbers and real numbers,[13] we conclude that (3.15) holds for arbitrary $\lambda > 0$. ∎

3.6 Helmholtz free energy

Making use of the maximum work, we can define a state function characterizing thermodynamic systems that is very similar to the mechanical potential energy. This quantity is referred to as the *Helmholtz free energy*. The Helmholtz free

---

[13] In other words, we stipulate that $W_{\max}(T; X_1 \to X_2)$ be a continuous function of X_1 and X_2.

energy is one example of a complete thermodynamic function, i.e., a function that incorporates all of the information provided by the theory of thermodynamics in regard to the properties and behavior of a given system.[14] It plays a fundamental role in the formulation of thermodynamics presented in this book.

3.6.1 Definition of the Helmholtz free energy

Consider a thermodynamic system with collective extensive variable X. For each value of T, we choose a fixed value $X_0(T)$ of X and call this function of T the *reference point* for X. We choose the values of $X_0(T)$ such that this is a continuous function. Also, considering scaled copies of this system (see Section 2.2.2), we choose their reference points such that for each T, the value of the reference point scales in proportion to the size of the system.

In the case of an individual system parameterized by $(T; V, N)$, the reference point $X_0(T)$ can be written as

$$X_0(T) = (V_0(T, N), N) , \tag{3.21}$$

where $V_0(T, N)$ is the *reference volume*. Further noting that if the size of the system changes by a factor of λ, the amount of substance changes to λN, we see that the reference volume can be written as $V_0(T, N) = v(T)N$, in terms of some function $v(T)$ depending only on T. Hence, we obtain the general form

$$X_0(T) = (v(T)N, N) . \tag{3.22}$$

At this point, the function $v(T)$ is undetermined. As discussed below, its determination is made in Section 6.1.

Now, for each value of T, we define the Helmholtz free energy as a function of X for all values of X that can be realized through an operation beginning at $X_0(T)$ as follows:[15]

$$F[T; X] := W_{\max}(T; X \to X_0(T)) . \tag{3.23}$$

For a given system, the values of T, X and $X_0(T)$ uniquely determine $W_{\max}(T; X \to X_0(T))$. Also, here we are considering the situation in which $X_0(T)$ is fixed from the outset. Hence, given a state $(T; X)$, there exists a unique value of $F[T; X]$. Because of this special property, $F[T; X]$ is referred to as a *state function* or a *thermodynamic function*.[16]

In general, in an operation under which the extensive variables are changed by a very small amount, the work performed should also be very small. From this

---

[14] In essence, this means that a complete thermodynamic function is a function from which all state functions within the thermodynamic description can be derived.

[15] In most textbooks, the Helmholtz free energy is defined by first introducing the energy, U, and the entropy, S, and then simply writing $F := U - TS$. In this book, instead, we first define F in terms of the maximum work and U in terms of the adiabatic work (see Section 4.3). Then, we introduce S through these quantities as $S := (U - F)/T$ (see Section 6.1). It is important to understand, however, that despite this difference on a formal level, the values realized by these quantities in any given case do not depend on the approach used.

[16] A comment on notation is in order here. In this book, quantities that constitute complete thermodynamic functions are indicated by the use of square brackets, as in the case of $F[T; X]$. The physical meaning of complete thermodynamic functions is discussed in detail in Section 7.1.2.

consideration, we expect $F[T; X]$ to be a continuous function of X. Further, we expect that the work performed in a given operation will change only slightly if the temperature is changed slightly. Thus, because $X_0(T)$ is a continuous function, it should be the case that $F[T; X]$ is also a continuous function of T. With this understanding based on our experience, we postulate the continuity of the Helmholtz free energy with respect to both T and X.[17]

Before moving on, let us briefly clarify the situation regarding $X_0(T)$. At this point, because $X_0(T)$ is regarded as an undetermined function, the dependence of $F[T; X]$ on T is not yet specified. However, this does not represent a deficiency of the theory constructed to this point. Because we have not yet considered operations under which the temperature is altered, there simply has been no need to consider temperature dependence. Indeed, in the context of a single value of T, the choice of $X_0(T)$ is meaningless. The functional form of $X_0(T)$ is fixed in Section 6.1, when we introduce the entropy and through this introduction connect the functions $F[T; X]$ for different values of T.

3.6.2 Properties of the Helmholtz free energy

From the extensivity of both the reference point and the maximum work (see (3.15)), we have the following:

$$F[T; \lambda X] = W_{\max}(T; \lambda X \to \lambda X_0(T)) = \lambda\, W_{\max}(T; X \to X_0(T)) = \lambda\, F[T; X]\,.$$
$$(3.24)$$

Hence, the Helmholtz free energy is also an extensive quantity.

Now, let us consider two systems described by the collective extensive variables X and Y, with the reference points $X_0(T)$ and $Y_0(T)$. Suppose that we combine these systems into a single composite system, described by the pair of extensive variables $\{X, Y\}$ with the reference point $\{X_0(T), Y_0(T)\}$. Then, from the additivity of the maximum work, expressed by (3.13), we immediately obtain

$$F[T; X, Y] = F[T; X] + F[T; Y]\,. \qquad (3.25)$$

We thus conclude that the Helmholtz free energy is additive.

Next, let us consider a system with reference point $X_0(T)$ and suppose that with the system beginning at $X = X_0(T)$, there exist operations such that $X = X_1$ and $X = X_2$ are both realizable. Then, from the definition given in (3.23) and the properties of the maximum work expressed by (3.9) and (3.10), we obtain

$$\begin{aligned}
F[T; X_1] - F[T; X_2] &= W_{\max}(T; X_1 \to X_0(T)) - W_{\max}(T; X_2 \to X_0(T)) \\
&= W_{\max}(T; X_1 \to X_0(T)) + W_{\max}(T; X_0(T) \to X_2) \\
&= W_{\max}(T; X_1 \to X_2)\,.
\end{aligned} \qquad (3.26)$$

It is thus found that the maximum amount of work that can be performed by a thermodynamic system in an isothermal operation through which it makes a

---

[17] **Advanced note**: As noted in Footnote 11 on p. 30, the description of a system in terms of states defined by $(T; V, N)$ is incomplete at the triple point. However, even at this point, $F[T; V, N]$ is a continuous function of T, V and N, and it provides a meaningful description of equilibrium states within the triple-point region (see Section 7.5.4 and Appendices D and H).

transition between two equilibrium states is equal to the difference between the values of the Helmholtz free energy in these states. This is a very important result, and for the sake of emphasis, it is worth re-expressing (3.26) as

$$W_{\max}(T; X_1 \to X_2) = F[T; X_1] - F[T; X_2] \tag{3.27}$$

and comparing this with (3.6). We thereby see explicitly that for a quasi-static isothermal operation, the Helmholtz free energy plays a role exactly analogous to that played by the potential energy in a mechanical system.

As its name suggests, the Helmholtz free energy represents the amount of energy that can be freely extracted from an isothermal thermodynamic system and used mechanically. The fact that some part of a system's energy is aptly referred to by this term suggests that the remaining energy could somehow be understood as the "non-free energy." As we see in Chapter 6, this is indeed the case, and this energy is related to a fundamental quantity know as the *entropy*.

3.7 Pressure and the equation of state

In this section, considering an isothermal operation through which the volume is changed by a small amount, we introduce the *pressure*, which is an important state function for fluid systems.

3.7.1 Pressure and the Helmholtz free energy

We begin by considering a fluid system that starts in an equilibrium state $(T; V, N)$ and then undergoes a quasi-static isothermal operation causing the transition

$$(T; V, N) \xrightarrow{\text{qi}} (T; V + \Delta V, N) , \tag{3.28}$$

through which the volume changes from V to $V + \Delta V$. We assume that ΔV is very small compared to V. As a type of mechanical device through which we could carry out such an experiment, we consider that described by Figure 3.3, in which the system is contained in a vessel equipped with a piston whose position determines the volume. For simplicity, we assume that the region to the right of the piston is a vacuum. We also assume that the system is in thermal contact with an environment at temperature T. Prior to the operation, the position of the piston is fixed through the use of some kind of latching device. Now, in an actual experiment of this kind, if the latch were simply released, the pressure exerted by the fluid on the piston would cause it to move rapidly to the right, and thus

Figure 3.3 The piston is moved through a mechanical operation, and the volume of the system changes as a result. The pressure is defined in terms of the work performed by the system on the mechanical world during this operation.

the resulting process would not be a quasi-static operation. For this reason, we consider the following scenario. At the time that the latch is released, an external force that cancels the force exerted by the fluid is applied to the piston. Then, the magnitude of this external force, $\mathcal{F}$, is decreased very slowly. In such a situation, the piston will move quasi-statically to the right. Finally, when the volume has reached $V + \Delta V$, the latch is again engaged.

From Result 3.3 (p. 52), we know that the work done by the system through the operation (3.28) is $W_{\max}(T;(V,N) \to (V+\Delta V, N))$. Then, writing the distance traveled by the piston in the operation as $\Delta\ell$ and assuming that the force exerted by the fluid on the piston changes very little during the operation,[18] we find that this work can also be expressed as $\mathcal{F}\,\Delta\ell + O((\Delta\ell)^2)$.[19] We thus obtain the relation

$$W_{\max}(T;(V,N) \to (V+\Delta V, N)) = \mathcal{F}\,\Delta\ell + O((\Delta\ell)^2)\,, \qquad (3.29)$$

equating a thermodynamic quantity and a mechanical quantity.

When a fluid exerts a force $\mathcal{F}$ on a flat surface, the magnitude of this force divided by the area of the surface, A, is regarded as the *pressure*, p, of the fluid: $p = \mathcal{F}/A$. Applying this to the situation considered above, with A interpreted as the area of the piston, we have the relation $A\,\Delta\ell = \Delta V$, and we can rewrite (3.29) as

$$W_{\max}(T;(V,N) \to (V+\Delta V, N)) = p\,\Delta V + O((\Delta V)^2)\,. \qquad (3.30)$$

Then, recalling the definition of the maximum work, we see that this equation allows us to reinterpret the pressure, introduced above as a mechanical quantity, as a thermodynamic state function, determined uniquely by the state $(T;V,N)$. Explicitly, we have

$$
\begin{aligned}
p(T;V,N) &= \lim_{\Delta V \searrow 0} \frac{W_{\max}(T;(V,N) \to (V+\Delta V, N))}{\Delta V} \\[2mm]
&= \lim_{\Delta V \searrow 0} \frac{F[T;V,N] - F[T;V+\Delta V, N]}{\Delta V} \\[2mm]
&= -\frac{\partial}{\partial V} F[T;V,N]\,,
\end{aligned}
\qquad (3.31)
$$

where (3.27) has been used. We have thus found how the pressure can be determined operationally and that it is directly connected to the Helmholtz free energy. In this book, we regard (3.31) as the definition of the pressure.

The pressure is representative of state functions that can be determined operationally, and it is of fundamental importance for understanding the nature of fluid systems. We posit that for any state $(T;V,N)$, the pressure possesses a positive value and that it is a continuous function of T, V and N.[20]

---

[18] Empirically, it is known that this condition is satisfied in "normal" situations in which the volume of the system is a continuous function of $\Delta\ell$.

[19] As depicted in Figure 3.3, the external agent pushes on the piston with a force of magnitude $\mathcal{F}$. Thus, when the position of the piston changes by $\Delta\ell$ in the direction opposite to that of this force, the work done by the mechanical world on the system is $-\mathcal{F}\Delta\ell$. The work in which we are interested, that done by the system on the mechanical world, is simply the opposite of this value.

[20] This is equivalent to the condition that $F[T;V,N]$ be differentiable with respect to V. (**Advanced note**: As discussed in Section D.3, the continuity of the pressure and the differentiability of the Helmholtz free energy with respect to V hold even at the triple point.)

If we use the relation obtained by replacing (V, N) in (3.31) with $(\lambda V, \lambda N)$, along with the extensivity property of the Helmholtz free energy, expressed by (3.24), we obtain

$$p(T; \lambda V, \lambda N) = -\frac{\partial F[T; \lambda V, \lambda N]}{\partial(\lambda V)} = -\frac{\lambda\, \partial F[T; V, N]}{\lambda\, \partial V} = p(T; V, N) \,. \qquad (3.32)$$

This confirms our intuitive understanding (as well as a fundamental principle in the theory of fluid statics) that the pressure is unchanged when a system is scaled; i.e., it is an intensive quantity.

With given, arbitrary T and N, if we integrate (3.31) from the reference point (3.22) to some arbitrary V, we obtain the Helmholtz free energy at $(T; V, N)$ (up to the arbitrariness of $v(T)$):

$$F[T; V, N] = -\int_{v(T)N}^{V} dV'\, p(T; V', N) \,. \qquad (3.33)$$

It is thus seen how the pressure, which is a directly measurable quantity, provides a concrete method for obtaining the Helmholtz free energy.

3.7.2 Equation of state

For a given system parameterized by $(T; V, N)$, the equation that expresses the functional dependence of p on T, V and N in equilibrium is known as the *equation of state*.[21] Equations of state are specific to the actual thermodynamic systems under consideration. In other words, they are not determined in some universal manner by a theory of general applicability. This point warrants particular stress: The equation of state for an arbitrary system cannot be determined from the general theoretical framework of thermodynamics. Deducing equations of state is not a task addressed by this framework. Thermodynamics is concerned with the universal principles and laws that transcend the individual properties of particular systems. Therefore, rather than providing the specific equations of state for individual systems, it provides the universal conditions that such equations must satisfy. In the application of thermodynamics, equations of state are used together with the general theoretical framework to produce many kinds of powerful results. The best method for determining the equation of state for a particular type of system is to carry out many experiments, employing many different values of the temperature and concentration, and to thereby construct a phenomenological functional dependence of the pressure. On the other hand, in some cases, after constructing an appropriate model of a system, methods of statistical mechanics can be used to theoretically derive an equation of state.

It has been found through countless experimental investigations that for gas systems at ordinary temperatures with sufficiently large volumes, the pressure is

---

[21] This expression is sometimes used in reference to other kinds of equations as well. For example, an equation that expresses the energy of a system in terms of T, V and N is often referred to by the same name. In this book, however, we reserve the term *equation of state* for an equation expressing the pressure in terms of the state variables.

given to great precision by the relation

$$p(T; V, N) \simeq \frac{NRT}{V}. \tag{3.34}$$

Here, R is the *gas constant*. In general, we would expect R to depend on the type of gas, but if we use moles as the unit for the amount of substance, R is the same for all gases. In units of joules (J) for energy, kelvins (K) for temperature, and moles (mol) for amount of substance, we have $R \simeq 8.31\,\mathrm{J\,K^{-1}mol^{-1}}$. (The exact value of R is given in Footnote 2 on p. 2.)

The particular form of the equation of state of course depends explicitly on the choice of the temperature scale. The simple form (3.34) reflects the fact that we are using the Kelvin scale.

Because it provides a simple model that is quite accurate for "ordinary" gases, it is useful to introduce the concept of a theoretical gas for which the equality

$$p(T; V, N) = \frac{NRT}{V} \tag{3.35}$$

holds exactly for all values of T, V and N. This is referred to as an *ideal gas*. (We studied an ideal gas in Section 1.1.) In fact, there is no actual gas that behaves exactly in accordance with (3.35).[22] Furthermore, all known gases undergo phase transitions to liquid or solid states at low temperatures, and thus the behavior exhibited by actual gases is even qualitatively very different from that described by (3.35). An ideal gas should be regarded as a simplistic model that is useful for investigating the qualitative properties of gases under high-temperature, low-concentration conditions. Given this status, there has been some objection to using the concept of an ideal gas in the construction of the theoretical framework of thermodynamics.[23] However, because there is a definite physical regime in which the equation of state (3.35) provides a realistic model,[24] as long as its limitations are understood, there is no reason to refrain from its investigation. In this book, ideal gas systems are often used to provide explicit illustrative examples of universal behavior derived in general form. Within the structure of the theory, the ideal gas is used only to establish the Kelvin temperature scale, as described in detail in Section A.3. As discussed there, when conceived of properly, its use in this role is understood to be of entirely general validity.

Let us derive the Helmholtz free energy of an ideal gas. Substituting the equation of state (3.35) into (3.33) and integrating, we obtain

$$F[T; V, N] = -\int_{v(T)N}^{V} dV' \, \frac{NRT}{V'} = -NRT \log \frac{V}{v(T)N}. \tag{3.36}$$

---

[22] One model of gases that is somewhat more realistic than (3.35) is the van der Waals equation of state (see Problem 3.3).

[23] For example, it has been noted that although Clausius made explicit use of the concept of an ideal gas as a demonstrative tool in constructing the framework of thermodynamics, Kelvin was strongly opposed to its use [15].

[24] Indeed, because the behavior of any gas at any fixed temperature approaches that described by (3.35) in the limit of vanishing pressure, in this limit, the description provided by this equation of state becomes quantitatively exact.

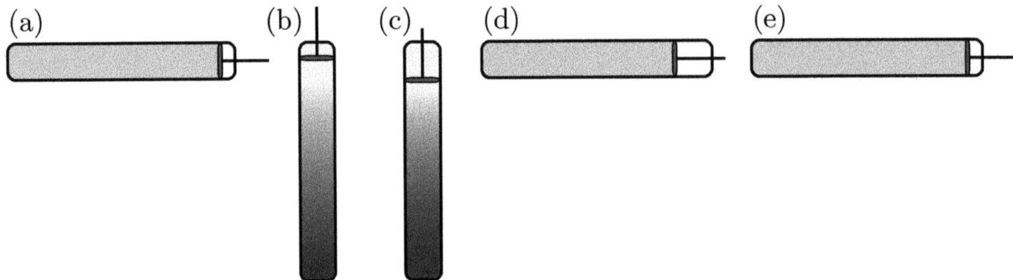

Figure 3.4 A perpetual motion machine of the second kind that utilizes a density variation created by gravity.

We determine the form of $v(T)$ for an ideal gas after we introduce the entropy, in Chapter 6 (see (6.30)). The final form of the Helmholtz free energy for an ideal gas is given in (7.9).

Recall that the Helmholtz free energy can be regarded as the energy of a system that is freely usable in mechanical form. Now, note that the Helmholtz free energy given in (3.36) is a monotonically decreasing function of V. We thus see that, given a fixed temperature, if the volume of a gas increases, the amount of "freely useable energy" possessed by the gas decreases.[25] This characterization is clearly reasonable, as we all know from experience that a compressed gas possesses the potential to perform mechanical work if allowed to expand.

Problems 3

3.1 (Section 3.2) Below we present an idea for a perpetual motion machine of the second kind that utilizes gravity. Discuss why this idea would not work.

Consider the system depicted in Figure 3.4, consisting of gas contained in a very long, narrow cylindrical container. Suppose that this system undergoes the following cycle under isothermal conditions. (a) Initially, the container is oriented horizontally. (b) With the right end fixed, the left end is allowed to move downward until the container is oriented vertically. The container is assumed to be very long, and therefore the influence of gravity causes the gas to be more concentrated near the bottom, with the upper region becoming a nearly perfect vacuum. (c) An external agent presses down on the piston. Because there is almost no gas near the top, this requires effectively no work, and the distribution of gas particles is essentially unchanged as a result. (d) The lower end is raised so that the container once again becomes oriented horizontally. (e) Finally, the external agent pulls the piston out in order to return the system to its original state. During this operation, thermal energy is converted into mechanical energy.

Note that the potential energies of the apparatus as a whole in (a) and at the completion of (d) are identical. Therefore, the work done by gravity in step (b) and the work that must be performed by the external agent in step (d) appear to be the same. Thus, if we were to use the work gained

---

[25] This example provides a good way of understanding the difference between a system's energy and its free energy: While the free energy for an ideal gas system with constant temperature is a decreasing function of V, the energy is independent of V. (See Section 4.4 for further discussion.)

in (b) in order to carry out (d),[26] it would seem that these reorientation operations could be carried out with no external energy source. Also, note that whereas (essentially) no work is needed to compress the piston in step (c), the gas does positive work on the mechanical world in step (d), while absorbing heat from the environment. Hence, considering the cycle as a whole, it would appear that there is a finite amount of net work done on the mechanical world.

3.2 (Section 3.5.1) Show that Kelvin's principle, Postulate 3.1 (p. 47), follows from the principle of maximum work, Result 3.3 (p. 52).

3.3 (Section 3.7.2) The following is known as the van der Waals equation:[27]

$$\tilde{p}(T; V, N) = \frac{NRT}{V - bN} - \frac{aN^2}{V^2} \quad (\text{with } V > bN) . \tag{3.37}$$

This pressure formula provides a more realistic description of a gas than the ideal gas equation of state, (3.35). In this equation, the parameter $a > 0$ represents the effect of the interactions between gas molecules, while the parameter $b > 0$ represents the effect created by their finite volumes. In the following, we take N to be fixed.

Let us regard T to be given, and treat $\tilde{p}(T; V, N)$ as a function of V alone. Show that there exists a value $T_c > 0$ such that for $T > T_c$, the pressure $\tilde{p}(T; V, N)$ is a decreasing function of V, while for $T < T_c$ it is not. Clearly, in the low-temperature regime, $\tilde{p}(T; V, N)$ does not represent a physically meaningful pressure.[28]

Next, sketch graphs of $\tilde{p}(T; V, N)$ as a function of V for various values of T. This can be done by computer, but in that case, you should first think about what behavior you wish to see, and choose the parameter values and the bounds of the plot accordingly.

Finally, use (3.33) to obtain the Helmholtz free energy $\widetilde{F}[T; V, N]$ corresponding to (3.37). Just as $\tilde{p}(T; V, N)$ represents an unphysical pressure below T_c, the free energy obtained in this manner is a kind of unphysical "pseudo free energy." (For related discussion, see Problem 7.8 and Section 10.3.)

---

[26] For example, this could be accomplished by using two identical systems of this type that are coupled together in such a way that (b) in one and (d) in the other are carried out simultaneously, with the energy gained from the former being used to realize the latter.

[27] The use of the tilde here indicates that the pressure $\tilde{p}(T; V, N)$ exhibits non-physical behavior, as described in the main text.

[28] Problem 7.8 considers how a physically meaningful pressure can be obtained in this case.

4

Adiabatic Operations and Energy

In this chapter, carrying out a detailed investigation of operations applied to systems under adiabatic conditions, we introduce the concept of the *energy* of a thermodynamic system. As in other fields of physics, energy is a fundamental quantity in thermodynamics. After elucidating the basic properties of the energy, we introduce an important quantity called the *heat capacity at constant volume*. We then discuss the meaning of the law of energy conservation from a general point of view. Finally, we consider adiabatic operations applied to ideal gas systems.

4.1 Adiabatic operations

In the previous chapter, we introduced isothermal operations, which are carried out on systems in contact with fixed-temperature environments. In the present chapter, we introduce adiabatic operations, which are carried out on systems insulated from their environments by adiabatic walls. All of the operations that we consider in this book are either isothermal or adiabatic. This follows the approach developed by Carnot, in which the actual, complicated operations observed in the real world are idealized and represented by these two simple types, corresponding to opposing limits with regard to the influence of the environment.[1] This idealization is one of the simplifications that allows for the universality inherent in thermodynamic behavior to be clearly manifested.

4.1.1 General adiabatic operations and quasi-static adiabatic operations

Let us consider a thermodynamic system in an equilibrium state $(T_1; X_1)$. Suppose that this system is isolated from the environment by adiabatic walls. Then, let us consider the situation in which the system is subject to an arbitrary operation consisting of mechanical manipulations of one or more of the types described in Section 2.3 through which the collective extensive variable changes from X_1 to X_2. This operation can be carried out in any manner and at any speed. The only stipulation is that the system remain under adiabatic conditions throughout the operation. After the completion of this operation, if the system is left undisturbed for a long time with the adiabatic walls left in place, then in accordance with Postulate 2.3 (p. 36), it will approach a new equilibrium state, $(T_2; X_2)$. In this situation, although we can choose the final value of the collective

---

[1] Although formulations of thermodynamics that employ both adiabatic and isothermal operations have historically been the norm, there are formulations (e.g., Ref. [9]) in which isothermal operations are never considered explicitly.

Thermodynamics: A Modern Approach. Hal Tasaki and Glenn Paquette, Oxford University Press.
© Hal Tasaki and Glenn Paquette (2026). DOI: 10.1093/9780191878091.003.0004

extensive variable, X_2, we cannot freely choose the final value of the temperature, T_2. Rather, the system chooses this value, in response to the operation. This type of operation is referred to as an *adiabatic operation*. We represent a transition realized under an adiabatic operation as follows:[2]

$$(T_1; X_1) \xrightarrow{\text{a}} (T_2; X_2) \,. \tag{4.1}$$

The concept of a quasi-static operation is essentially the same in the adiabatic case as in the isothermal case, considered in Section 3.1.2, with the only difference being the type of external conditions. (See Section A.2 for detailed discussion of quasi-static operations.) A quasi-static operation carried out under adiabatic conditions is termed a *quasi-static adiabatic operation*. As in the isothermal case, the distinctive feature of a quasi-static adiabatic operation is that the system is in equilibrium at all times during its performance. A transition from a state $(T_1; X_1)$ to a state $(T_2; X_2)$ realized through a quasi-static adiabatic operation is represented by the expression

$$(T_1; X_1) \xrightarrow{\text{qa}} (T_2; X_2) \,. \tag{4.2}$$

We can anticipate that if we slowly change an extensive variable by a very small amount, the temperature of the system will also change only very slightly. For this reason, it is reasonable to hypothesize that the value of T_2 appearing in (4.2) is a continuous function of X_2.

We assume the realizability of adiabatic transitions in direct analogy to the case of isothermal transitions, as discussed in the previous chapter. More precisely, we assume the following. First, suppose that for the system under consideration, it is possible to carry out a mechanical manipulation through which the collective extensive variable is varied from X_1 to X_2. Then, for any temperature T_1, with the system initially in the state $(T_1; X_1)$, if we carry out such a manipulation while the system is surrounded by adiabatic walls, we realize an adiabatic transition, represented by (4.1), and in the case that this operation is carried out quasi-statically, this is a quasi-static adiabatic transition, represented by (4.2).

As in the case of a quasi-static isothermal operation, for any transition realizable under a quasi-static adiabatic operation, the inverse transition is realized under the reverse operation.[3] In other words, such an operation is *reversible*. Further, the quantities of work done by the system on the mechanical world in two mutually time-reversed quasi-static adiabatic operations are of equal magnitude and opposite sign. We use the notation

$$(T_1; X_1) \xleftrightarrow{\text{qa}} (T_2; X_2) \tag{4.3}$$

to express the meaning that both the transition represented by (4.2) and its inverse are realizable.

---

[2] In some textbooks, an adiabatic operation is understood to necessarily be a slow operation. We do not use such a definition. Rather, in this book, *adiabatic* describes only the condition of isolation from the environment through the use of adiabatic walls.

[3] See the discussion in Section 3.1.2, including Footnote 2 on p. 46.

4.1.2 What kinds of adiabatic operations are possible?

The amount by which the temperature of a system varies under a given adiabatic operation depends not only on the nature of the operation, but also on the material composition and structure of the system in question, and hence, there are innumerable possibilities. However, from our experience, we know that no matter what type of system we consider, for operations through which the mechanical world does work on the system through frictional or stirring processes, it is possible to increase the temperature by any desired amount. In this book, we take this empirical fact as a fundamental postulate.

Postulate 4.1 (Existence of adiabatic operations that merely increase T)
Consider a system in an arbitrary equilibrium state, $(T_1; X)$. Then, for any temperature T_2 satisfying $T_2 > T_1$, there exists an adiabatic operation under which the temperature of the system increases to T_2 while the collective extensive variable is unchanged. In other words, the transition

$$(T_1; X) \xrightarrow{\text{a}} (T_2; X) \tag{4.4}$$

is realizable for any X and any $T_2 > T_1$. Further, in any operation inducing this transition, the mechanical world performs a positive amount of work on the system.

One example of such a temperature-increasing adiabatic operation is treated in Problem 1.1. In that operation, the volume of the vessel containing a gas is first very rapidly increased and then very slowly decreased to its initial value (see Figure 1.5 in Section 1.1). The temperature increase resulting from an operation of this type is easily calculated. More generally, we can consider the "pumping" of the fluid in a vessel resulting from repeatedly increasing and decreasing its volume in an arbitrary manner. In such a situation, macroscopic flow will develop in the fluid, and this flow will, through viscosity and eventually friction on a microscopic scale, be dissipated into "frictional heat." The result is an increase in the temperature of the fluid.

From Postulate 4.1 and the fundamental nature of quasi-static adiabatic operations, we are led to the following important result concerning the realizability of transitions under adiabatic conditions.

Result 4.2 (Possibility to connect any two states under an adiabatic operation) *Consider an arbitrary thermodynamic system, and assume that for two values of the collective extensive variable X_1 and X_2, there exist operations through which each can be realized from the other. Then, for arbitrary values of the temperature T_1 and T_2, it is necessarily the case that there exists an adiabatic operation under which at least one of the following two transitions can be realized:*

$$(T_1; X_1) \xrightarrow{\text{a}} (T_2; X_2) \,, \quad (T_2; X_2) \xrightarrow{\text{a}} (T_1; X_1) \,. \tag{4.5}$$

Derivation Beginning in the state $(T_1; X_1)$, carry out a quasi-static adiabatic operation through which the collective extensive variable changes to X_2. Let T_3 be *the value of the temperature realized at the end of this operation. Thus, we have the quasi-static adiabatic transition $(T_1; X_1) \xrightarrow{\text{qa}} (T_3; X_2)$. (Note that, in general, we do not know the value of T_3.) First, let us assume the relation $T_2 \geq T_3$. In this case, combining this first operation with an operation inducing a transition of the type (4.4), whose existence is guaranteed by Postulate 4.1, we obtain*

$$(T_1; X_1) \xrightarrow{\text{qa}} (T_3; X_2) \xrightarrow{\text{a}} (T_2; X_2) . \tag{4.6}$$

Hence, the transition $(T_1; X_1) \xrightarrow{\text{a}} (T_2; X_2)$ is realizable. Next, let us assume the relation $T_2 \leq T_3$. In this case, employing the time-reversed counterpart of the original quasi-static adiabatic operation, we can construct the following:

$$(T_2; X_2) \xrightarrow{\text{a}} (T_3; X_2) \xrightarrow{\text{qa}} (T_1; X_1) . \tag{4.7}$$

Hence, the transition $(T_2; X_2) \xrightarrow{\text{a}} (T_1; X_1)$ is realizable. ∎

Postulate 4.1 asserts that it is possible to realize any desired temperature increase in a system without altering its extensive variables. The following result asserts the opposite with regard to a temperature decrease, i.e., that it is impossible to realize a temperature decrease through an adiabatic operation under which the extensive variables are unaltered. This restriction on the realizability of adiabatic transitions forms the basis of the important discussion given in Section 6.2 concerning reversibility and irreversibility in thermodynamics.

Result 4.3 (Non-existence of adiabatic operations that merely decrease T) *Consider a system in an arbitrary equilibrium state, $(T_1; X)$. Then, for any temperature T_2 satisfying $T_2 < T_1$, there exists no adiabatic operation that realizes the transition*

$$(T_1; X) \xrightarrow{\text{a}} (T_2; X) . \tag{4.8}$$

The above result is proved in Section 4.2.2 on the basis of Kelvin's principle, Postulate 3.1 (p. 47), and the law of energy conservation, Postulate 4.5 (p. 69). It implies, for example, that if we act on a gas in an insulated container by changing the positions of its walls, as long as we eventually return them to their original positions, the final temperature of the gas cannot be lower than its initial temperature.

The situation regarding adiabatic operations as described by Postulate 4.1 and Result 4.3 is clearly consistent with our everyday experience, as we know that through mechanical operations performing work against friction, an object can be heated, but it cannot be cooled.

As an application of Result 4.3, we next demonstrate the following important property of quasi-static adiabatic operations.

Result 4.4 (Quasi-static adiabatic operations yield minimum final temperatures) *For an arbitrary thermodynamic system in an initial state $(T_1; X_1)$, consider an arbitrary quasi-static adiabatic operation and its resulting transition,*

$$(T_1; X_1) \xrightarrow{\text{qa}} (T_2; X_2) \,. \tag{4.9}$$

Then, for any adiabatic operation inducing a transition

$$(T_1; X_1) \xrightarrow{\text{a}} (T_3; X_2) \,, \tag{4.10}$$

the relation $T_2 \leq T_3$ necessarily holds.

We thus find that among all possible adiabatic operations applied to a given initial state with a given final value of the collective extensive variable, the minimum value of the final temperature is realized in the case of a quasi-static operation.[4] This also implies that for a quasi-static adiabatic operation, the final temperature of the system is uniquely determined by the initial state and the final value of the collective extensive variable.

Derivation of Result 4.4 *Because the transition (4.9) is realized under a quasi-static operation, applying its time-reversed counterpart to the state $(T_2; X_2)$, we realize the transition $(T_2; X_2) \xrightarrow{\text{qa}} (T_1; X_1)$. Combining this transition with (4.10), we obtain the transition*

$$(T_2; X_2) \xrightarrow{\text{a}} (T_3; X_2) \,. \tag{4.11}$$

The relation $T_2 \leq T_3$ then follows from Result 4.3. ∎

4.2 The law of energy conservation in thermodynamics and adiabatic work

4.2.1 The law of energy conservation

In Section 3.4, we discussed the difference between mechanical and thermodynamic systems with regard to the performance of work. There, we noted that while the work done by a mechanical system on the external agent in any operation depends only on the initial and final states of the system, the work done by a thermodynamic system on the external agent in an isothermal operation in general depends on the manner in which the operation is carried out, not only on the initial and final states. Interestingly, the situation is different in the case of an adiabatic operation. Without exception, the work performed by a perfectly insulated thermodynamic system in an operation connecting two given equilibrium states is independent of the path taken between these states. In this book, we employ this important empirical fact as a fundamental principle in the construction of our thermodynamic formalism. Its use as such implies an understanding of the concept of adiabaticity that is in some sense independent of the concept of heat.

---

[4] Combining Result 4.4, Postulate 4.1 and Postulate 4.5 (p. 69), we also find that among all possible adiabatic operations applied to a given initial state with a given final value of the collective extensive variable, the work performed by the system is maximized in the case of a quasi-static operation. Thus, the situation for adiabatic operations is analogous to that for isothermal operations.

Postulate 4.5 (The law of energy conservation in thermodynamics) *The work performed on the mechanical world by a thermodynamic system in an arbitrary adiabatic operation is determined entirely by the initial and final equilibrium states of the system. Specifically, this work is independent of the manner in which the operation is carried out and the states realized by the system between these two endpoints.*

This postulate is often referred to as the *first law of thermodynamics*. We use this term interchangeably with *law of energy conservation*.

We wish to emphasize that Postulate 4.5 holds even in the case that the operation is not performed quasi-statically. The reason that this is termed the *law of energy conservation* is elucidated in the next section. This law was established empirically by James Prescott Joule in the mid-19th century through the performance of many cleverly designed experiments.[5] At the time that Joule was performing these experiments, Mayer, Helmholtz and others were obtaining similar results, largely from a theoretical point of view. In addition, Carnot apparently had arrived at a form of the idea of energy conservation at an earlier time and considered including this in his book (see Section 1.1.3), but omitted it.[6]

Joule's law of energy conservation, Postulate 4.5, and Kelvin's principle, Postulate 3.1 (p. 47), represent the two great pillars of thermodynamics, the first and second laws of thermodynamics, respectively.[7] It can be said that Joule's law expresses the essential point of commonality between thermodynamics and mechanics, while Kelvin's principle expresses the essential point of distinction.

As one application of Postulate 4.5, let us consider the special class of adiabatic operations through which the system undergoes a cycle, with identical initial and final states. Of course, the simplest such operation is that in which the system is merely left undisturbed. In this case, obviously, no work is performed. With this observation, Postulate 4.5 implies that for any adiabatic operation under which the system begins and ends in the same state, the work vanishes. We express this as follows (see Section 4.2.2):

$$W_{\mathrm{ad}}((T;X)\longrightarrow(T;X)) = 0 \ . \tag{4.12}$$

Recall that Result 3.2 (p. 48) asserts the analogous relation for W_{max}. We thus see that in the cases of both adiabatic operations and quasi-static isothermal

---

[5] Even before Joule's time, countless people invested great efforts into devising perpetual motion machines of the first kind (i.e., machines that are able to produce energy without input of any kind, including heat), only to face bitter disappointment. However, these efforts and their universal failure can be regarded as the experiences that laid the foundation for the first law of thermodynamics. Indeed, strictly speaking, the reason that modern physicists categorically dismiss perpetual motion machines devised by pseudo-scientists is *not* because the existence of such a machine would contradict the first law of thermodynamics. Rather, our confidence in the universal validity of this law and the impossibility of perpetual motion machines derives from the fact that no experimentally reproducible counterexample has ever been realized.

[6] For a brief discussion of this and Carnot's numerous contributions to thermodynamics, see J. Young, "Heat, Work and Subtle Fluids: A Commentary on Joule (1850) 'On the Mechanical Equivalent of Heat'," Philosophical Transactions of the Royal Society A, vol. 373, iss. 2039, 2015 (https://doi.org/10.1098/rsta.2014.0348).

[7] It should be noted, however, that there are more than just two or three fundamental postulates of thermodynamics. For this reason, it is perhaps best to avoid such "nth law" terminology. (The so-called third law, also known as the Nernst-Planck postulate, is discussed in Section 6.1.4.)

operations, if the initial and final states realized under an operation are the same, then the work performed by the system vanishes.

With (4.12), we are now able to demonstrate Result 4.3.

Derivation of Result 4.3 *Suppose that there exists an operation under which a transition $(T_1; X) \xrightarrow{a} (T_2; X)$ with $T_2 < T_1$ is realizable. Then, by Postulate 4.1, we can realize the sequence of transitions $(T_1; X) \xrightarrow{a} (T_2; X) \xrightarrow{a} (T_1; X)$. From (4.12), we know that the total work performed in the operations resulting in these transitions is 0. Also, from Postulate 4.1, we know that the work performed by the system in the operation yielding $(T_2; X) \xrightarrow{a} (T_1; X)$ is negative. Thus, the work performed by the system in the operation yielding $(T_1; X) \xrightarrow{a} (T_2; X)$ is positive. Now, we can define an isothermal cycle $(T_1; X) \xrightarrow{a} (T_2; X) \xrightarrow{i'} (T_1; X)$ formed by starting with the system under isothermal conditions at temperature T_1, placing adiabatic walls around the system, carrying out the operation varying the temperature from T_1 to T_2, and then removing the adiabatic walls to re-establish the original isothermal conditions. (This last operation is a generalized isothermal operation.[8]) Because no work is performed in the operations through which the adiabatic walls are added and removed, we obtain an isothermal cycle in which the total work performed by the system is positive. This violates Kelvin's principle, Postulate 3.1. We therefore conclude that no transition $(T_1; X) \xrightarrow{a} (T_2; X)$ with $T_2 < T_1$ is realizable.* ∎

4.2.2 Adiabatic work and its properties

According to the law of energy conservation, if an adiabatic transition $(T_1; X_1) \xrightarrow{a} (T_2; X_2)$ can be realized, then the work performed by the system on the mechanical world during an operation yielding this transition is determined by $(T_1; X_1)$ and $(T_2; X_2)$ alone. We refer to the work performed in an adiabatic operation as *adiabatic work* and express it as $W_{\mathrm{ad}}((T_1; X_1) \to (T_2; X_2))$. Below, we establish several properties of adiabatic work. It is instructive to compare these properties with the corresponding properties of the maximum work, given in Section 3.5.2. While these two sets of properties have many similarities, there are also some important differences, most significantly with regard to the existence of transitions.

Sum rule

For a given system, let us consider values of the temperature T_1, T_2 and T_3 and values of the collective extensive variable X_1, X_2 and X_3 such that the transitions

$$(T_1; X_1) \xrightarrow{a} (T_2; X_2) , \quad (T_2; X_2) \xrightarrow{a} (T_3; X_3) \tag{4.13}$$

are both realizable. Carrying out operations that induce these transitions in succession, we realize the transition $(T_1; X_1) \xrightarrow{a} (T_3; X_3)$. Interpreting these operations alternatively as two separate operations and as a single operation (see Section 2.3.1)

---

[8] This class of operations is treated systematically in Problem 5.2.

and using Postulate 4.5, we obtain the following sum rule for the adiabatic work:

$$W_{\mathrm{ad}}((T_1; X_1) \to (T_3; X_3)) = W_{\mathrm{ad}}((T_1; X_1) \to (T_2; X_2)) + W_{\mathrm{ad}}((T_2; X_2) \to (T_3; X_3)) \,.$$
$$(4.14)$$

Sign rule

For a given system, let us suppose that for some values T and T' of the temperature and X and X' of the collective extensive variable, the mutually inverse transitions $(T; X) \xrightarrow{\mathrm{a}} (T'; X')$ and $(T'; X') \xrightarrow{\mathrm{a}} (T; X)$ are both realizable. Then, applying (4.14) in the particular case with $(T_1; X_1) = (T_3; X_3) = (T; X)$ and $(T_2; X_2) = (T'; X')$, and using (4.12), we obtain the following sign rule for the adiabatic work:

$$W_{\mathrm{ad}}((T; X) \to (T'; X')) = -W_{\mathrm{ad}}((T'; X') \to (T; X)) \,. \qquad (4.15)$$

However, it must be kept in mind that even in the case that some adiabatic transition $(T; X) \xrightarrow{\mathrm{a}} (T'; X')$ is realizable, it is not necessarily the case that its inverse, $(T'; X') \xrightarrow{\mathrm{a}} (T; X)$, is also realizable. Of course, we can always simply reverse the variation of the collective extensive variable itself, retracing its values backward from X' to X. However, representing the resulting adiabatic transition as $(T'; X') \xrightarrow{\mathrm{a}} (T''; X)$, T'' will generally differ from T. Indeed, the relation $T'' \geq T$ follows from Result 4.4, with equality holding in the quasi-static case.

Additivity property

Consider a composite system consisting of two subsystems separated by a diathermal wall in a state $(T_1; X_1, Y_1)$. Suppose that we apply to this system the following series of operations. First, the diathermal wall is replaced with an adiabatic wall, and we thereby obtain two independent systems in the states $(T_1; X_1)$ and $(T_1; Y_1)$. Next, we carry out arbitrary adiabatic operations on these two independent systems with the stipulation that the temperatures of their final states be equal. We thus obtain transitions $(T_1; X_1) \xrightarrow{\mathrm{a}} (T_2; X_2)$ and $(T_1; Y_1) \xrightarrow{\mathrm{a}} (T_2; Y_2)$. Finally, we replace the adiabatic wall with a diathermal wall. In this way, we have realized the transition $(T_1; X_1, Y_1) \xrightarrow{\mathrm{a}} (T_2; X_2, Y_2)$. Applying Postulate 4.5 and using the fact that no work is performed in the operations exchanging the isothermal and adiabatic walls, we obtain the relation

$$W_{\mathrm{ad}}((T_1; X_1, Y_1) \to (T_2; X_2, Y_2))$$
$$= W_{\mathrm{ad}}((T_1; X_1) \to (T_2; X_2)) + W_{\mathrm{ad}}((T_1; Y_1) \to (T_2; Y_2)) \,. \qquad (4.16)$$

This relation applies to all states $(T_1; X_1, Y_1)$ and $(T_2; X_2, Y_2)$, subject to the condition that the transitions $(T_1; X_1) \xrightarrow{\mathrm{a}} (T_2; X_2)$ and $(T_1; Y_1) \xrightarrow{\mathrm{a}} (T_2; Y_2)$ be realizable.

Beginning instead with a composite system consisting of n subsystems and proceeding analogously, we can readily extend the above to the following general relation, representing the additivity property of the adiabatic work:

$$W_{\mathrm{ad}}((T_1; X_1^{(1)}, \ldots, X_1^{(n)}) \to (T_2; X_2^{(1)}, \ldots, X_2^{(n)}))$$
$$= \sum_{j=1}^{n} W_{\mathrm{ad}}((T_1; X_1^{(j)}) \to (T_2; X_2^{(j)})) \,. \qquad (4.17)$$

As above, this relation is subject to the condition that the transitions $(T_1; X_1^{(j)}) \to (T_2; X_2^{(j)})$ be realizable for all j.

Extensivity property

The extensivity property of the adiabatic work is as follows. Consider an arbitrary thermodynamic system and a scaled copy of this system with a scaling factor $\lambda \in \mathbb{R}_+$. Given an adiabatic operation on the former that induces a transition $(T_1; X_1) \xrightarrow{\text{a}} (T_2; X_2)$, there exists an adiabatic operation on the latter that induces the transition $(T_1; \lambda X_1) \xrightarrow{\text{a}} (T_2; \lambda X_2)$, and the amounts of work performed in these operations satisfy the relation

$$W_{\text{ad}}((T_1; \lambda X_1) \to (T_2; \lambda X_2)) = \lambda \, W_{\text{ad}}((T_1; X_1) \to (T_2; X_2)) \,. \tag{4.18}$$

We demonstrate these assertions below.

Derivation of (4.18) *The proof here is very similar to that for the extensivity of W_{max}, but in the present case, because we are considering adiabatic operations rather than isothermal operations, the realizability of the transitions of interest is non-trivial. We first treat this point.*

Consider a system in a state $(T_1; X_1)$ and the collection of n systems in the state $(T_1; X_1/n)$ obtained from it through an appropriate introduction of adiabatic walls. We are interested in pairs of transitions $(T_1; X_1) \xrightarrow{\text{a}} (T_2; X_2)$ and $(T_1; X_1/n) \xrightarrow{\text{a}} (T_2; X_2/n)$ in the original system and constituent systems, which can be understood as scaled copies of each other. Below we show that the realizability of one member of such a pair implies that of the other.

First, suppose that the transition $(T_1; X_1/n) \xrightarrow{\text{a}} (T_2; X_2/n)$ is realizable in the constituent systems. Then, we can obtain the transition $(T_1; X_1) \xrightarrow{\text{a}} (T_2; X_2)$ in the original system by first partitioning it as described above, carrying out an operation inducing the transition $(T_1; X_1/n) \xrightarrow{\text{a}} (T_2; X_2/n)$ in each constituent system, and then removing the partitioning walls.

Next, suppose that the transition $(T_1; X_1) \xrightarrow{\text{a}} (T_2; X_2)$ is realizable in the original system. Now, consider the following sequence of operations. We first partition the original system as described above. Then, on each constituent system, we carry out a quasi-static operation under which the collective extensive variable is varied from X_1/n to X_2/n, with each system thus undergoing a transition $(T_1; X_1/n) \xrightarrow{\text{qa}} (T_3; X_2/n)$. If we were to next remove the partitioning walls, we would realize the transition $(T_1; X_1) \xrightarrow{\text{qa}} (T_3; X_2)$ in the original system. Result 4.4 thus implies the relation $T_3 \leq T_2$. Therefore, from Postulate 4.1, we know that in each constituent system, there exists an operation yielding the transition $(T_3; X_2/n) \xrightarrow{\text{a}} (T_2; X_2/n)$. In this way, we can obtain the transition $(T_1; X_1/n) \xrightarrow{\text{a}} (T_2; X_2/n)$.

The above results apply to systems related by an integer-valued scaling factor. We can easily generalize this treatment to the case of a rational-valued scaling factor. Doing so, we obtain the result that if a transition $(T_1; X_1) \xrightarrow{\text{a}} (T_2; X_2)$ is realizable in a given system, then the transition $(T_1; mX_1/n) \xrightarrow{\text{a}} (T_2; mX_2/n)$ is realizable in a scaled copy with the scaling factor m/n, where m and n are arbitrary positive integers.

With the necessary realizability of transitions thus established, we can now follow the proof of the extensivity of the maximum work, given in Section 3.5.2. For *this purpose, it is necessary only to replace the quasi-static isothermal operations and corresponding amounts of work W_{max} there with arbitrary adiabatic operations and the corresponding amounts of work W_{ad}, replace T there with T_1 and T_2 for the initial and final states of the resulting transitions, and replace the path independence of W_{max} provided by Result 3.2 (p. 48) with the path independence of W_{ad} provided by Postulate 4.5. We have thus established the extensivity property of the adiabatic work.* ∎

4.3 Energy

Just as we were able to use the maximum work to define the Helmholtz free energy (see (3.23)), we are able to use the adiabatic work to define the *energy* as a state function. Energy plays a fundamental role in every area of modern physics, including thermodynamics.

4.3.1 Definition and fundamental properties of energy

For a given system, let us choose some (arbitrary) reference values of the temperature and collective extensive variable, which we write T^* and X^*. We stipulate that the reference value of the temperature be intensive, while the reference value of the collective extensive variable be extensive, so that if we consider a scaled version of the original system, obtained by merely changing its size by a factor of λ, then the reference values of the temperature and collective extensive variable for this scaled system are T^* and λX^*. It follows that in the case of an individual system parameterized by $(T; V, N)$, X^* necessarily takes the form

$$X^* = (v^* N, N) \,, \tag{4.19}$$

where v^* is an arbitrary constant.

Now, suppose that through some operation, it is possible to change the value of the collective extensive variable from the reference value X^* to a value X. Then, from Result 4.2 (p. 66), we know that for an arbitrary temperature T, at least one of the transitions $(T; X) \xrightarrow{\mathrm{a}} (T^*; X^*)$ and $(T^*; X^*) \xrightarrow{\mathrm{a}} (T; X)$ is realizable. In the case that the first transition is realizable, we define the *energy* (or *internal energy*) of the state $(T; X)$ as

$$U(T; X) := W_{\mathrm{ad}}((T; X) \to (T^*; X^*)) \,, \tag{4.20}$$

while if the second transition is realizable, we define this energy as

$$U(T; X) := -W_{\mathrm{ad}}((T^*; X^*) \to (T; X)) \,. \tag{4.21}$$

In the case that both transitions are realizable, from (4.15) it is seen that the above two definitions are equivalent. We have thus defined the state function $U(T; X)$.[9]

---

[9] The energy is a complete thermodynamic function only in the case that it is regarded as a function of the entropy, S, and the collective extensive variable, X. To make this point explicit, only in this case do we display the arguments of the energy in square brackets, writing $U[S, X]$. (See Appendix F for related discussion.)

In consideration of this definition, we expect that $U(T; X)$ will change only slightly if T or X is changed by a small amount. With this intuitive understanding, we postulate that $U(T; X)$ is a continuous function of both T and X.[10]

Note that the definition above does not determine $U(T; X)$ in an absolute sense. There remains the indeterminacy corresponding to the choice of the reference state and the value of the energy assigned to that state. In other words, when considering the energies of a system in various states, we are free to uniformly add or subtract any constant value. This type of indeterminacy is necessarily possessed by all potential-type quantities.[11]

From the extensivity of adiabatic work, expressed by (4.18), the extensivity of X^*, and the intensivity of T^*, it is seen that energy possesses the extensivity property

$$U(T; \lambda X) = \lambda U(T; X) . \tag{4.22}$$

Also, from the additivity of adiabatic work, expressed by (4.16), it is seen that energy possesses the additivity property

$$U(T; X, Y) = U(T; X) + U(T; Y) . \tag{4.23}$$

(Of course, we have implicitly assumed consistent choices of X^* and Y^* here.)

Differences in the energy are related to the adiabatic work in the same way that differences in the Helmholtz free energy are related to the maximum work, as expressed in (3.27). Explicitly, when an adiabatic transition $(T_1; X_1) \stackrel{a}{\longrightarrow} (T_2; X_2)$ is realizable, the following relation holds:

$$W_{\mathrm{ad}}((T_1; X_1) \to (T_2; X_2)) = U(T_1; X_1) - U(T_2; X_2) . \tag{4.24}$$

The meaning of this relation is that when a thermodynamic system makes a transition from one state to another through an adiabatic operation, the amount of work performed by the system on the mechanical world is equal to the energy in the initial state minus the energy in the final state. Note that (4.24) has the same form as (3.6), which relates the work and the potential energy for a mechanical system. It is thus seen that thermodynamic systems are characterized by the same type of energy conservation law as mechanical systems. This is the reason that Postulate 4.5 (p. 69) is called the *law of energy conservation.* We also sometimes use this term in reference to (4.24).

Derivation of (4.24) *It is necessary to consider three cases, representing the possible ways for the reference state, $(T^*; X^*)$, and the two states under consideration, $(T_1; X_1)$ and $(T_2; X_2)$, to be connected by adiabatic operations:*

$$(T^*; X^*) \stackrel{a}{\longrightarrow} (T_1; X_1) \stackrel{a}{\longrightarrow} (T_2; X_2) , \tag{4.25}$$

$$(T_1; X_1) \stackrel{a}{\longrightarrow} (T^*; X^*) \stackrel{a}{\longrightarrow} (T_2; X_2) , \tag{4.26}$$

$$(T_1; X_1) \stackrel{a}{\longrightarrow} (T_2; X_2) \stackrel{a}{\longrightarrow} (T^*; X^*) . \tag{4.27}$$

---

[10] As discussed in Appendix D, this does not hold at the triple point.

[11] For example, recall the cases of the potential energy in classical mechanics and the electric potential in electromagnetism.

From Result 4.2 (p. 66), regarding the existence of adiabatic operations, we know that in any given situation, at least one of these transitions is realizable. Let us first consider the case of (4.25). Then, using (4.21) and the sum rule for the adiabatic work, given in (4.14), we find

$$U(T_1; X_1) - U(T_2; X_2) = -W_{\mathrm{ad}}((T^*; X^*) \to (T_1; X_1)) + W_{\mathrm{ad}}((T^*; X^*) \to (T_2; X_2))$$

$$= -W_{\mathrm{ad}}((T^*; X^*) \to (T_1; X_1))$$

$$+ \left\{ W_{\mathrm{ad}}((T^*; X^*) \to (T_1; X_1)) + W_{\mathrm{ad}}((T_1; X_1) \to (T_2; X_2)) \right\}$$

$$= W_{\mathrm{ad}}((T_1; X_1) \to (T_2; X_2)) \,. \tag{4.28}$$

The same result can be derived similarly in the remaining two cases with the appropriate application of (4.20) and (4.21). ∎

The following is a fundamental result.

Result 4.6 (Energy as a function of temperature) *Consider an arbitrary thermodynamic system. For any fixed value of X, the energy of this system, $U(T; X)$, is an increasing function of T.*

Derivation *According to Postulate 4.1 (p. 66), there exist adiabatic operations whose effect is to merely raise the temperature of the system, and for any such operation, the work performed by the system on the mechanical world is negative. Thus, from (4.24), we immediately obtain the following for arbitrary X and any temperatures T_1 and T_2 satisfying $T_1 < T_2$:*

$$U(T_1; X) - U(T_2; X) = W_{\mathrm{ad}}((T_1; X) \to (T_2; X)) < 0 \,. \; \blacksquare \tag{4.29}$$

4.3.2 Heat capacity at constant volume

The *heat capacity at constant volume* of a thermodynamic system, written $C_{\mathrm{v}}(T; X)$, is the rate of change of the energy as a function of the temperature:[12]

$$C_{\mathrm{v}}(T; X) = \frac{\partial}{\partial T} U(T; X) \,. \tag{4.30}$$

Below, we consider in simple terms how we might measure this quantity.

Suppose that we carry out an operation on an adiabatic system for which the initial and final values of the collective extensive variable are identical and that in this operation an infinitesimal amount of work ΔW is performed on the system by the mechanical world. Then, with ΔT representing the increase in temperature experienced by the system as a result of this operation, the heat capacity is given

---

[12] The name *heat capacity* is a historical artifact that reflects the outdated understanding that matter "stores" heat—a relic of the caloric theory. With this understanding, the heat capacity is viewed as a proportionality constant representing the amount of heat needed to raise the temperature by a given amount. However, as we see in Chapter 5, heat is simply one form of transferred energy, and thus the term *heat capacity* is something of a misnomer. Instead, it would perhaps be most appropriate to call this quantity the *internal energy capacity*.

by $C_{\mathrm{v}}(T; X) = \Delta W/\Delta T$. There are countless possibilities for the method that could be used to carry out an operation of this type. However, the equilibrium state that the system would eventually realize after the operation is independent of this method, depending only on the amount of work done on the system. This is guaranteed by the law of energy conservation, Postulate 4.5 (and the fact that temperature is an increasing function of energy). As an actual operation of the type considered here, one could use the simple procedure described in the discussion of Postulate 4.1, in which the volume of a gas system is repeatedly increased and decreased through alternating motion of a piston, or one could, like Joule himself, stir a fluid system by rotating an impeller located inside it. As another type of procedure, one could place an electric heating element inside the system and force a small amount of current through it, as depicted in Figure 4.1. In this case, ΔW would be the electric work performed by the external electric power source on the heating element. (Because this work is performed on the heating element and not the system, this kind of procedure is not an operation as defined in this book.) The fact that the law of energy conservation asserted in Postulate 4.5 applies also to the case of electric work was confirmed by Joule in his pioneering series of experiments. Fittingly, a process through which heat is generated by a heating element (or, more correctly, a process through which energy is transferred to a thermodynamic system from a heating element as a result of electric work applied to the heating element) is termed *Joule heating*.

Because energy is an extensive quantity and temperature is an intensive quantity, $C_{\mathrm{v}}(T; X)$, given by the derivative of the former with respect to the latter, is an extensive quantity. It is thus proportional to the amount of substance of the system

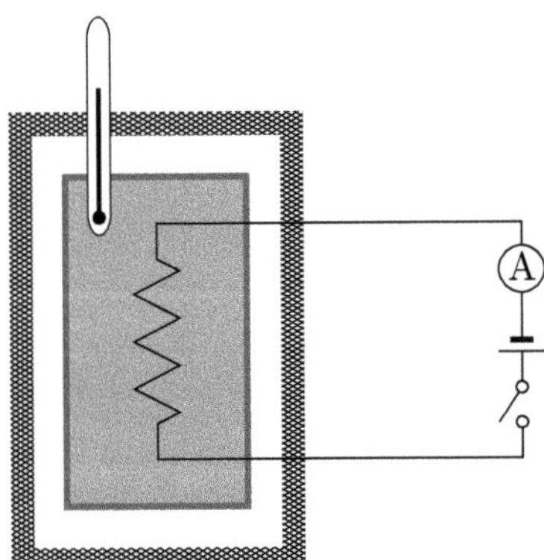

Figure 4.1 One method for measuring heat capacity. A heating element is placed within a system surrounded by adiabatic walls. Then, with ΔT denoting the increase in temperature experienced by the system when an amount of work ΔW is performed on the heating element by an external power source, the heat capacity at constant volume of the system is given by $C_{\mathrm{v}}(T; X) \simeq \Delta W/\Delta T$. Using the same apparatus, if instead of keeping the volume fixed, we allow the volume to vary in such a manner that the pressure remains fixed, the quantity $\Delta W/\Delta T$ would represent the heat capacity at constant pressure, $C_{\mathrm{p}}(T, p; N)$. In many actual experimental situations, it is this quantity rather than the heat capacity at constant volume that can be directly measured. (See Section 8.3 for discussion of this point.) Today, there are many technologically advanced methods for measuring the heat capacity that facilitate extremely precise measurements for systems over very broad temperature ranges.

in question. Hence, in order to obtain an intensive quantity that reflects the intrinsic properties of the material composing the system, we must divide $C_v(T; X)$ by another extensive quantity. It is standard to use the amount of substance in units of moles for this purpose. The quantity thus obtained, $c_v(T; X) = C_v(T; X)/N$, is termed the *molar specific heat at constant volume.*

4.3.3 The meaning of the law of energy conservation

Before ending this section, we briefly discuss the meaning of the law of energy conservation.

Those readers familiar with kinetic molecular theory may interpret $U(T; X)$ as the total mechanical energy (i.e., the sum of the potential and kinetic energies) of all of the particles composing a thermodynamic system. Of course, this is correct. When interpreted in this manner, the relation between work and energy given in (4.24) is simply a statement of the law of energy conservation within mechanics. This too is, in some sense, correct. Given these observations, the reader may wonder why Postulate 4.5 (p. 69) has been presented here as a postulate based on empirical findings. In other words, if this law can be "proven" as a simple result of the mechanical law of energy conservation, why is it necessary to assume its validity here?

Certainly, it is the conventional understanding of modern physics that the macroscopic systems treated by thermodynamics are composed of particles that obey the laws of mechanics, and hence any such system as a whole must obey the mechanical law of energy conservation. Indeed, we do not necessarily doubt that this is true. However, we must ask how this conventional reasoning has been established. Does it follow from the development of classical mechanics (or quantum mechanics) as a mathematical model that energy is necessarily conserved in macroscopic physical systems? Of course it does not.

Although it is often regarded as a fact that the motion of each of the great multitude of particles composing a macroscopic system is described by Newton's second law (or the Schrödinger equation), this has never been tested directly. It is far beyond present human capability to directly test the validity of such a postulate, and this will certainly remain the case far into the future. For many reasons (both technical and fundamental), to experimentally observe each of the particles contained in a macroscopic physical system is an impossible task. In addition, even if we were somehow able to obtain such data, because it would be entirely unfeasible to carry out a computation using Newton's second law to derive a prediction for the trajectories of all of the particles in a macroscopic system, we would not be able to compare these experimental data with the theoretical model. Why then, in spite of the lack of any kind of direct proof, do we believe that Newton's second law (or the Schrödinger equation) applies to the multitudinous microscopic particles composing a macroscopic system? The reason is that the empirical facts regarding thermodynamic behavior (most notably, Joule's law of energy conservation), which have been established through exhaustive experimentation with macroscopic systems, are entirely consistent with this belief. From the empirical point of view, the situation here can be summarized as follows: Combining the thermodynamic law of energy conservation established as an empirical fact with the picture furnished by kinetic molecular theory, we obtain very strong evidence that mechanical theories provide valid descriptions of macroscopic systems.

4.4 Adiabatic operations on ideal gas systems

In this section, we investigate energy and quasi-static adiabatic operations in the context of ideal gases.

4.4.1 Energy of an ideal gas

We begin our treatment by considering idealizations of the experiments on ideal gases carried out by Gay-Lussac and (later) Joule, in which they studied the expansion of gases under adiabatic conditions.

As depicted in Figure 4.2, we consider a vessel of volume V separated into two compartments of volumes V' and $V - V'$. We introduce an amount N of a single species of gas into the compartment of volume V', and then allow it to reach an equilibrium state while in contact with an environment at temperature T. The compartment of volume $V - V'$ is left as a vacuum. After equilibrium has been established, the entire vessel is enclosed within adiabatic walls. Finally, the wall separating the two compartments is removed. Through this operation, the following adiabatic transition is realized:

$$(T; V', N) \xrightarrow{\text{a}} (T'; V, N) \,. \tag{4.31}$$

Because no work is performed by the system on the mechanical world during this operation, from (4.24), expressing conservation of energy, we obtain

$$U(T'; V, N) = U(T; V', N) \,. \tag{4.32}$$

In actual adiabatic free expansion experiments carried out with gases at moderate temperatures and pressures, the temperatures before and after the expansion are nearly identical.[13] Idealizing this empirical result and adopting the postulate

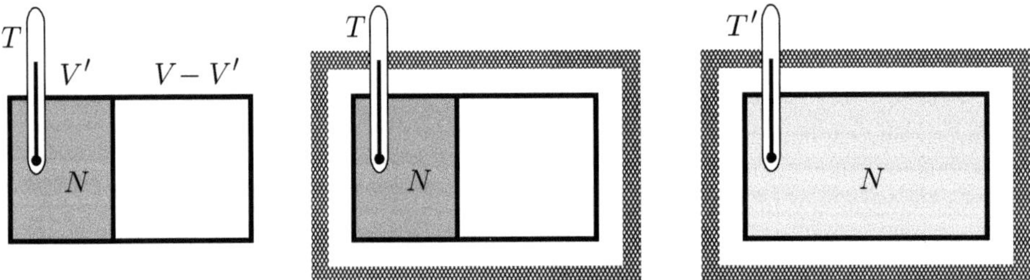

Figure 4.2 The adiabatic expansion experiment carried out by Gay-Lussac. Gas of temperature T contained in a vessel surrounded by adiabatic walls is allowed to freely expand into a vacuum. After the expansion, the system is allowed to re-attain equilibrium, and the temperature is then remeasured. Gay-Lussac found that this final temperature is almost identical to the initial temperature. Idealizing this empirical result, we posit that the energy of an ideal gas does not depend on its volume.

---

[13] Gay-Lussac reported that in these experiments, after the wall was removed, the temperature reading on the thermometer initially decreased and then gradually increased, eventually reaching the original value, where it remained. The transient behavior that he observed can be understood as follows. Immediately after the removal of the wall, macroscopic-scale currents are formed in the gas.

that the relation $T = T'$ holds exactly in such experiments, from (4.32) we obtain

$$U(T; V, N) = U(T; V', N) \, . \tag{4.33}$$

We regard this relation, asserting that the energy of an ideal gas does not depend on its volume, to be a fundamental property of ideal gases.[14]

Measurements of the heat capacity at constant volume for ideal gases reveal that over very broad ranges of the temperature and pressure, this quantity is independent of the temperature, and it can be expressed as

$$C_{\mathrm{v}}(T; V, N) \simeq cNR \, . \tag{4.34}$$

Here, R is the gas constant (see Section 3.7), and c is a constant whose value is determined by the particular properties of the species of gas under consideration.[15] The following values of c can be determined from models based on the kinetic theory of gases and statistical mechanics:

$$c = \begin{cases} 3/2 & \text{for a monatomic (ideal) gas} \, , \\ 5/2 & \text{for a diatomic (ideal) gas} \, . \end{cases} \tag{4.35}$$

For a certain range of temperatures, these theoretical values provide good approximations of the actually measured values. On the basis of these considerations, through idealization of (4.34), we postulate that the relation

$$C_{\mathrm{v}}(T; V, N) = \frac{\partial}{\partial T} U(T; V, N) = cNR \, , \tag{4.36}$$

with constant c, holds exactly for an ideal gas.

If we fix the amount of substance, then from (4.33) it is seen that $U(T; V, N)$ depends on T alone. Thus, the functional form of $U(T; V, N)$ can be determined from (4.36). Integrating this relation, we obtain

$$U(T; V, N) = cNRT + Nu \, . \tag{4.37}$$

This equation is called the *energy formula for an ideal gas*. The quantity u appearing here is a constant whose value is determined by the choice of the reference point of the energy. (This constant plays no substantial role in this book until we consider chemical reactions.)

The relations (3.35) and (4.37), which respectively regard the pressure and energy, provide a complete description of the thermodynamic properties of an ideal gas.

Because these currents possess kinetic energy, the internal energy of the gas (i.e., that reflected by the thermometer reading) temporarily decreases [15]. Obviously, the extremely non-equilibrium behavior exhibited by the system immediately after the removal of the wall cannot be treated with equilibrium thermodynamic theory. However, despite this fact, if we consider only the equilibrium states of the system existing before and after the operation, this theory can indeed be applied, and it provides a precise quantitative description of these equilibrium states.

[14] In Section 7.2.3, we find that this property can in fact be obtained as a necessary implication of the equation of state (3.35).

[15] It is conventional to express this relation in terms of the constant $\gamma = 1 + c^{-1}$ instead of c. The reason for this is clarified in Footnote 17 on p. 81.

4.4.2 Quasi-static adiabatic operations on ideal gas systems

Using the functional form of the energy of an ideal gas obtained above, let us now investigate quasi-static adiabatic operations carried out on ideal gas systems.

In a quasi-static adiabatic operation causing the transition

$$(T; V, N) \xrightarrow{\text{qa}} (T'; V', N) \,, \tag{4.38}$$

the temperature changes together with the volume. In order to ascertain the relation between these quantities, let us consider the case of a very small change in the volume, ΔV. Then T should also change by a small amount, which we write ΔT. Thus, we consider (4.38) with

$$V' = V + \Delta V, \quad T' = T + \Delta T \,. \tag{4.39}$$

First, recalling the relation between the pressure and force exerted by a gas, we find that the amount of work performed by the system on the mechanical world through the operation yielding (4.38) is given by[16]

$$\Delta W = p(T; V, N)\Delta V + O((\Delta V)^2) = \frac{NRT\Delta V}{V} + O((\Delta V)^2) \,. \tag{4.40}$$

The second equality here is provided by the equation of state (3.35). Then, from the energy conservation law (4.24) and the energy formula (4.37), we find that ΔW can also be expressed as

$$\Delta W = U(T; V, N) - U(T + \Delta T; V + \Delta V, N) = -cNR\Delta T \,. \tag{4.41}$$

Setting (4.40) and (4.41) equal, we obtain

$$\frac{\Delta T}{\Delta V} = -\frac{T}{cV} + O(\Delta V) \,. \tag{4.42}$$

In the limit $\Delta V \to 0$, this becomes the differential equation

$$\frac{dT}{dV} = -\frac{T}{cV} \,. \tag{4.43}$$

Fixing the initial values of V and T, this equation uniquely determines T as a function of V. Thus, we find that in the case of a quasi-static adiabatic operation, T can be expressed as a function of V.

---

[16] **Advanced note**: Actually, it is not entirely obvious that the pressure measured when the volume of a system is changed by a small amount through an adiabatic operation is identical to the pressure $p(T; V, N)$ defined in the context of isothermal operations. This point (which was brought to the authors' attention by Shin-ichi Sasa) is considered in Problem 4.2.

Applying the standard method of separation of variables to (4.43), we obtain the following integral equation:

$$c \int \frac{dT}{T} = - \int \frac{dV}{V} \,. \tag{4.44}$$

Integrating both sides, we find the relation $c \log T = - \log V + (\text{constant})$, which can be re-expressed in the simpler form[17]

$$T^c V = (\text{constant}) \,. \tag{4.45}$$

This is referred to as a *Poisson relation.*

Summarizing the result obtained here, in the case of a quasi-static adiabatic operation carried out on an ideal gas, V and T vary in such a manner that they satisfy (4.45) throughout the operation. The constant appearing in this formula is determined by the initial values of V and T alone.

Problems 4

4.1 (Section 4.1) Show that for any quasi-static adiabatic operation under which the collective extensive variable is unchanged, the temperature is also unchanged.

4.2 (Section 4.4.2) Here we investigate the point raised in Footnote 16 on p. 80. Consider a quasi-static isothermal operation and a quasi-static adiabatic operation carried out on identical, arbitrary fluid systems. Suppose that these operations result in the transitions $(T; V, N) \xrightarrow{\text{qi}} (T; V + \Delta V, N)$ and $(T; V, N) \xrightarrow{\text{qa}} (T + \Delta T; V + \Delta V, N)$, respectively, where ΔV may be positive or negative. We assume that the amounts of work performed by the systems on the mechanical world during these operations are $\Delta W_\text{i} = p_\text{i}(T; V, N)\Delta V + O((\Delta V)^2)$ and $\Delta W_\text{a} = p_\text{a}(T; V, N)\Delta V + O((\Delta V)^2)$. (We can regard $p_\text{i}(T; V, N)$ and $p_\text{a}(T; V, N)$ here as the "isothermal pressure" and "adiabatic pressure.") Demonstrate the identity $p_\text{i}(T; V, N) = p_\text{a}(T; V, N)$ using Kelvin's principle. (This problem is taken from Ref. [11].)

4.3 (Section 4.4.2) Consider an individual system consisting of an ideal gas under adiabatic conditions. Compute the work done by the system in a quasi-static operation under which it makes a transition from a state $(T_1; V_1, N)$ to a state $(T_2; V_2, N)$. First, express this work in terms of V_1, V_2 and T_1. Next, express it in terms of V_1, V_2 and the initial and final pressures, p_1 and p_2.

4.4 (Section 4.4.2) Suppose that the system considered in Problem 4.3 is in thermal contact with a second ideal gas system consisting of the same amount, N, of the same substance, and this combined system is surrounded by adiabatic walls. Let us refer to the second system as the "heat bath." Consider the situation in which the original system is prepared with the same initial state as in the previous problem, $(T_1; V_1, N)$, and then, while remaining in

---

[17] Using the ideal gas equation of state, (4.45) can also be written in the form $pV^\gamma = (\text{constant})$, where $\gamma = 1 + c^{-1}$. This expression in terms of p and V is often more useful.

thermal contact with the heat bath, it again undergoes a quasi-static operation under which its volume changes from V_1 to V_2 (while the heat bath undergoes no operation). Derive an expression in terms of V_1, V_2 and T_1 for the work performed by the system in this operation. Do the same in the case that the heat bath consists of n ideal gas systems, each containing the same amount of the same substance as the system. Finally, show that in the $n \to \infty$ limit, the form of the expression for this work reduces to that for an ideal gas system under isothermal conditions.

This problem illustrates explicitly how isothermal conditions are realized for a system when its heat bath becomes very large (and thus approaches a fixed-temperature environment) in the simple case of an ideal gas.

5

Heat and Carnot's Theorem

In this chapter, we introduce the concept of the *heat* absorbed by a system from its environment in an isothermal operation. Then we present Carnot's theorem, which plays an essential role in thermodynamics from both historical and theoretical points of view. We also treat the efficiency of heat engines, which was the problem that originally motivated Carnot himself. The most important application of Carnot's theorem is in the formulation of the concept of entropy, which is carried out in the next chapter.

5.1 Exchange of heat between a system and its environment

Although we have occasionally considered the concept of heat in the preceding chapters, it has not played any role in the formal construction of the theory presented to this point. In this section, heat makes its first explicit appearance within the theory itself. First, heat is formally defined as the form of energy transferred between a system and its environment during an isothermal operation. Next, an important quantity termed the *maximum heat* is introduced.

Before proceeding to the main discussion of this section, we present a brief comment concerning the role played by the concept of heat in this book. Let us first point out that we do not attempt to give a definition of heat that is universally applicable to all systems and behavior that can be treated within the theory of thermodynamics.[1] With the approach taken in this book, such a definition is unnecessary, because within our formulation of thermodynamics, the concept of heat plays no fundamental role. In the treatment presented here, consideration of heat is kept to a minimum, and in the end, we obtain a theoretical framework that contains no explicit dependence on this concept whatsoever.

5.1.1 Heat absorbed by a system from its environment

Let us consider an arbitrary thermodynamic system that experiences an arbitrary isothermal operation resulting in the transition

$$(T; X_1) \overset{\text{i}}{\longrightarrow} (T; X_2) . \tag{5.1}$$

Let W denote the work done by the system on the mechanical world through this operation. As a result of this operation, the system loses an amount of energy $U(T; X_1) - U(T; X_2)$. If the system under consideration were a purely mechanical

---

[1] Indeed, we do not know whether formulating a universally applicable definition is even possible. (The definition of heat is investigated further in Problems 5.1 and 5.2.)

Thermodynamics: A Modern Approach. Hal Tasaki and Glenn Paquette, Oxford University Press.
© Hal Tasaki and Glenn Paquette (2026). DOI: 10.1093/9780191878091.003.0005

system, or if this were a thermodynamic system subject to an adiabatic operation, then, in accordance with (4.24), we would have the relation

$$W = U(T; X_1) - U(T; X_2) \, . \tag{5.2}$$

However, in the case of an isothermal operation, (5.2) generally does not hold. In fact, if we consider the expansion of an ideal gas, W will necessarily be positive, but as seen from (4.37), in this case we have $U(T; X_1) - U(T; X_2) = 0$. More generally, for an isothermal operation, while $U(T; X_1) - U(T; X_2)$ will always be determined by only the initial and final states of the system, W will depend on the manner in which the operation is carried out (see Section 3.4).

In the case that (5.2) does not hold, obviously there is a discrepancy between the mechanical energy transferred from the system to the external world, W, and the amount of energy that the system loses, $U(T; X_1) - U(T; X_2)$. One might conclude from this that the law of energy conservation is violated in this situation. However, rather than drawing such a radical conclusion, let us assume that energy is conserved even in the case that (5.2) does not hold. Thus, in this case we must conclude that, through some mechanism, the thermodynamic system directly exchanges energy with its environment. Indeed, there can be no other source of energy that allows the law of energy conservation to hold. As emphasized in Section 2.3, we can determine the amount of mechanical energy exchanged through an operation by simply observing the behavior of the mechanical world. Contrastingly, the direct exchange of energy that we are considering now apparently takes place in some "hidden" manner, through a process that we cannot directly observe. We refer to this form of energy exchange as *heat*, and express it as Q. We introduce the heat by simply adding a term to (5.2):[2]

$$W = U(T; X_1) - U(T; X_2) + Q \, . \tag{5.3}$$

Formally, we define the heat as follows.

Definition 5.1 (Definition of the heat absorbed in an isothermal operation) *Consider an arbitrary isothermal operation causing a transition* $(T; X_1) \overset{\mathrm{i}}{\longrightarrow} (T; X_2)$, *and suppose that during this operation, the system performs an amount of work W on the mechanical world. Then the amount of energy in the form of heat imported by the system from the environment during this transition is given by*[3]

$$Q := W + U(T; X_2) - U(T; X_1) \, . \tag{5.4}$$

In order to obtain a clearer understanding of the quantity Q, let us rewrite (5.4) as

$$U(T; X_2) - U(T; X_1) = -W + Q \, . \tag{5.5}$$

The left-hand side of this equation is the change in the energy of the system as a result of the transition (5.1), while the right-hand side represents the total energy

---

[2] Equation (5.3) is often referred to as the *first law of thermodynamics.*
[3] Simple generalizations of this definition are considered in Problems 5.1 and 5.2.

received by the system from the external world. This equality makes explicit the idea that we split the energy imported by the system into two pieces, the work done by the mechanical world, $-W$, and the remainder, Q, which is regarded as the heat absorbed from the environment.[4] Because the work is the total amount of energy exchanged between the system and the external world that can be directly observed, the heat is that amount of this energy that cannot be directly observed.[5] Only if we know both the change in energy of the system and the work that it performs can we determine the amount of heat exchanged by the system and its environment.

Through the preceding discussion, we have established the concept of heat within our theoretical framework of thermodynamics. However, note that (5.4) describes nothing more than a form of energy transfer between a thermodynamic system and its environment during an isothermal operation. In particular, it is important to understand that this equation does not define heat as a state function; i.e., it does not provide a function $Q(T; X)$.[6] The realization of this fact was essential for the establishment of the modern theory of thermodynamics.

5.1.2 Heat and molecular theory

From the point of view of molecular theory, even energy transfer in the form of heat is fundamentally regarded as the exchange of mechanical energy between the system and the external world. However, energy transfer of this kind takes place on a microscopic level, as kinetic energy is exchanged among molecules through their collisions. Because we are unable to observe such processes, within thermodynamics this energy transfer is interpreted as imperceptible "heat." Of course,

---

[4] There is no definitive method for distinguishing work and heat, and indeed, there are situations in which there is room for debate regarding their proper distinction. We believe that the method used in this book is best suited to our formulation of thermodynamics: With the idealization that the mechanical world is a purely mechanical system that interacts with the system strictly through macroscopic forces, the work performed by the system during some process is regarded as the total increase in energy experienced by the mechanical world. However, there are types of processes accompanied by energy transfer that we interpret as work which, from a different point of view, could reasonably be interpreted as heat. For example, consider an operation in which an impeller is used to stir a fluid. Because the impeller can be operated in a purely mechanical manner, and the energy that it transfers to the system takes place through the macroscopic forces exerted by the impeller blades on the fluid, we interpret this as work. However, this type of work is of a very different nature than that performed by a piston (or through any continuous extensive operation), because the work performed by an impeller stirring a fluid is always positive, and hence irreversible. Thus, this transfer of energy is essentially the same as the transfer of energy in the form of heat that takes place when two systems of unequal temperatures are placed in thermal contact.

[5] Let us note, however, that there do exist experimental methods that allow for the indirect observation of the amount of transferred heat. These include methods employing a temperature sensor array to measure the heat gradient in the region of contact between the system and environment and methods employing a calorimeter to measure very small changes in the temperature of the environment.

[6] To illustrate this point, let us consider a transition $(T_1; X) \to (T_2; X)$ through which the temperature increases, while the collective extensive variable remains fixed. Suppose that this transition is realized through a simple generalized isothermal process (see Problem 5.2) in which the system is initially in the state $(T_1; X)$ and is then placed in contact with an environment at temperature T_2. In this case, heat is the only form of energy transferred to the system during the transition. It is thus tempting to conclude that this transition results from the "accumulation" of heat in the system, and thus that thermodynamic states are characterized by corresponding amounts of heat "contained" by the system. However, it is easily seen that this understanding is erroneous, because, as asserted by Postulate 4.1 (p. 66), the transition $(T_1; X) \to (T_2; X)$ is also realizable under an adiabatic operation, which by definition involves no heat whatsoever.

if we were able to directly observe phenomena on the molecular scale and hence could recognize mechanical work as the mechanism responsible for this exchange of energy, then the concept of heat would be unnecessary, and all energy transfer would be interpreted as taking place by means of work. For this reason, there would also be no need to introduce the idea of an environment, and the world beyond the boundary of the system would consist only of the mechanical world.

The above considerations may lead one to believe that our approach to the treatment of thermodynamic systems—in which we split energy into two types that are treated in very different manners—is simply an artifact of the finiteness of human knowledge and our limited ability to observe and measure physical phenomena. However, this is only partly true. The success of the thermodynamic approach can be attributed to two facts: that on a macroscopic scale, there are two clearly distinguishable types of energy transfer, and that there exist universal thermodynamic laws that govern these phenomena. Therefore, even if humans were somehow able to perceive microscopic processes, it is quite likely that we would still distinguish the types of energy transport that take place on microscopic and macroscopic scales and that we would have developed a theory of thermodynamics that describes the universal structure of macroscopic phenomena.

Prior to the development of the modern theory of thermodynamics, most scientists regarded heat as consisting of a hypothetical fluid known as *caloric* contained in matter. Within the so-called caloric theory, the total amount of this fluid is a conserved quantity. Interpreted favorably, this seems to suggest that, consistently with the modern theory, scientists at that time did recognize the internal energy, $U(T; X)$. However, they regarded this quantity as being the heat possessed by the system. Also, scientists at that time apparently did not understand that variation of $U(T; X)$ through interaction with the external world is mediated by two types of mechanisms, one of which can be directly observed, and one of which cannot.

5.1.3 Maximum heat

The definition (5.4) expresses the heat imported by a system during an isothermal operation in terms of the work done by the system on the mechanical world. Now, in general, there will exist innumerably many possible individual operations yielding transitions that possess the same initial and final states, $(T; X_1)$ and $(T; X_2)$, and the amount of heat absorbed by the system will vary among these operations. For given $(T; X_1)$ and $(T; X_2)$, let us consider the maximal value of Q that can be realized in such an operation. Because the change in total energy, $U(T; X_2) - U(T; X_1)$, is determined by the initial and final states alone, the maximal value of Q is realized by maximizing W. Recall that the maximum value of W, i.e., $W_{\max}(T; X_1 \to X_2)$, is realized in any quasi-static isothermal operation inducing the transition from $(T; X_1)$ to $(T; X_2)$ (see Result 3.3 (p. 52)). Using similar notation for the quantity that we term the *maximum heat* (or, more precisely, *maximum absorbed heat*), we thus have

$$Q_{\max}(T; X_1 \to X_2) = W_{\max}(T; X_1 \to X_2) + U(T; X_2) - U(T; X_1) . \qquad (5.6)$$

Then, using the relation between the maximum work and the Helmholtz free energy given in (3.27), this can be written as follows:

$$Q_{\max}(T; X_1 \to X_2) = F[T; X_1] - F[T; X_2] + U(T; X_2) - U(T; X_1) \,. \qquad (5.7)$$

From this expression, we see that in order to determine the maximum heat, it is not necessary to measure the heat itself. Instead, this can be done by measuring the work performed in a single quasi-static isothermal operation and a single adiabatic operation each connecting $(T; X_1)$ and $(T; X_2)$ (going in either direction).

We now state some basic properties of the maximum heat.

First, we immediately see from (5.7) that the maximum heat satisfies the sign rule

$$Q_{\max}(T; X_1 \to X_2) = -Q_{\max}(T; X_2 \to X_1) \,. \qquad (5.8)$$

Now, consider some values X_1, X_2 and X_3 of the collective extensive variable that can be reached from each other operationally. From the sum rule (3.10) for the maximum work, we obtain the following sum rule for the maximum heat:

$$Q_{\max}(T; X_1 \to X_3) = Q_{\max}(T; X_1 \to X_2) + Q_{\max}(T; X_2 \to X_3) \,. \qquad (5.9)$$

Next, consider a composite system characterized by the collective extensive variable $\{X, Y\}$ and an operation inducing the transition

$$(T; X_1, Y_1) \xrightarrow{\;i\;} (T; X_2, Y_2) \,. \qquad (5.10)$$

Then, from the additivity laws for the free energy and the energy, given in (3.25) and (4.23), we obtain the following additivity law for the maximum heat:

$$Q_{\max}(T; (X_1, Y_1) \to (X_2, Y_2)) = Q_{\max}(T; X_1 \to X_2) + Q_{\max}(T; Y_1 \to Y_2) \,. \quad (5.11)$$

Finally, the extensivity relation

$$Q_{\max}(T; \lambda X_1 \to \lambda X_2) = \lambda Q_{\max}(T; X_1 \to X_2) \qquad (5.12)$$

follows from the corresponding relations for the free energy and the energy, (3.24) and (4.22).

5.1.4 Isothermal transitions with vanishing maximum heat

Before ending this section, we present an important result regarding isothermal operations that yield transitions characterized by vanishing maximum heat. Below, we show that for any such operation, inducing a transition between arbitrary equilibrium states, there exists a corresponding quasi-static adiabatic operation that induces a transition between the same two states. This result is interesting in its own right, but it also plays an essential role in the proof of Carnot's theorem, given in Section 5.2.3.[7]

Result 5.2 (Vanishing maximum heat and adiabatic transitions) *Consider an arbitrary thermodynamic system subject to a quasi-static isothermal operation inducing the transition*

---

[7] This result is due to Takuma Tanaka.

$$(T; X_1) \xrightarrow{\text{qi}} (T; X_2) \,, \tag{5.13}$$

and suppose that for this operation we have $Q = 0$, which implies $Q_{\max}(T; X_1 \to X_2) = 0$. In this situation, there also exists a quasi-static adiabatic operation that induces the transition

$$(T; X_1) \xrightarrow{\text{qa}} (T; X_2) \,. \tag{5.14}$$

Furthermore, the amounts of work done by the system on the mechanical world in all operations that yield (5.13) and (5.14) are identical.

Because there is no net heat transferred to the system in an operation that causes the transition (5.13), this isothermal transition is, in some sense, similar to an adiabatic transition. Indeed, in many thermodynamics textbooks, this naive analogy is used as the basis for asserting the existence of an adiabatic operation yielding (5.14) (see Footnote 12 on p. 95). Actually, however, the existence of such an adiabatic operation is far from obvious. This can be understood by noting that even for an isothermal operation under which the total heat vanishes, in general, there will be some finite exchange of heat between the system and the environment during the course of the operation. In other words, generally, the vanishing of the total heat results not from a complete lack of exchanged heat but from the canceling of the total amounts of heat imported and exported during the course of the operation.[8]

To derive Result 5.2, we invoke Kelvin's principle.

Derivation of Result 5.2   *From Result 4.2 (p. 66), we know that at least one of the transitions $(T; X_1) \xrightarrow{\text{a}} (T; X_2)$ and $(T; X_2) \xrightarrow{\text{a}} (T; X_1)$ can be realized in an adiabatic operation. Without loss of generality, let us assume that the former is realizable. Next, let us write the transition realized under a quasi-static adiabatic operation through which the collective extensive variable is changed from X_1 to X_2 as*

$$(T; X_1) \xrightarrow{\text{qa}} (T'; X_2) \,, \tag{5.15}$$

where the final temperature, T', is unknown. Below, we show that the relation $T' = T$ necessarily holds. This implies that (5.15) is precisely the desired transition, (5.14).

We first assume $T' > T$. Then, combining the inverse of the transition (5.15) with the assumed adiabatic transition, $(T; X_1) \xrightarrow{\text{a}} (T; X_2)$, we find that the transition

$$(T'; X_2) \xrightarrow{\text{qa}} (T; X_1) \xrightarrow{\text{a}} (T; X_2) \tag{5.16}$$

is realizable. However, the operation resulting in this combined transition is a temperature-lowering adiabatic operation under which the collective extensive

---

[8] Even in the unusual case of an isothermal operation in which the incoming and outgoing heat fluxes cancel at every instant, it is still not obvious that there exists an adiabatic operation that yields the corresponding adiabatic transition, because we cannot *a priori* rule out the possibility that, despite this continuous balance, there is a non-negligible influence of the fixed-temperature environment on the behavior of the system and the transitions it can undergo. Thus, the problem here is much deeper than generally recognized.

variable is unchanged. By Result 4.3 (p. 67), we know that no such operation exists. We thus conclude the inequality $T' \leq T$.

We next assume $T' < T$. Then, combining the inverse of the original isothermal transition, (5.13), and the quasi-static adiabatic transition (5.15), we obtain the following isothermal cycle:

$$(T; X_2) \xrightarrow{\text{qi}} (T; X_1) \xrightarrow{\text{qa}} (T'; X_2) \xrightarrow{\text{i}'} (T; X_2) \,. \tag{5.17}$$

The final transition here is realized through a generalized isothermal operation in which the adiabatic walls surrounding the system are removed. The total work performed by the system on the mechanical world in this cycle can be evaluated by using (5.6), (4.24), and the fact that no work is performed during a generalized isothermal operation. We obtain

$$\begin{aligned} W_{\text{cyc}} &= U(T; X_2) - U(T; X_1) + Q_{\max}(T; X_2 \to X_1) + U(T; X_1) - U(T'; X_2) \\ &= U(T; X_2) - U(T'; X_2) \,, \end{aligned} \tag{5.18}$$

where we have used $Q_{\max}(T; X_1 \to X_2) = 0$ to obtain the final relation. Now, recall that Result 4.6 (p. 75) implies that the right-hand side of (5.18) is strictly positive. Hence, we obtain the relation $W_{\text{cyc}} > 0$. However, this contradicts Kelvin's principle, Postulate 3.1 (p. 47). We thus conclude the inequality $T' \geq T$.

The identity of the amounts of work performed in the two operations under consideration follows simply from the fact that the isothermal transition (5.13) involves no net heat transfer. Proceeding more formally, first note that the work performed in the quasi-static isothermal operation inducing (5.13) is $W_{\max}(T; X_1 \to X_2)$. Then, using (5.6) and the fact that $Q_{\max}(T; X_1 \to X_2)$ vanishes, we see that this is simply $U(T; X_1) - U(T; X_2)$, which is identically the work performed in the adiabatic operation inducing (5.14). ■

5.2 Carnot's theorem

In this section, we present and prove Carnot's theorem, which is of fundamental importance in the theory of thermodynamics. An interesting process called a *Carnot cycle*, which is an idealized model of a heat engine, plays an essential role in the proof of the theorem.

5.2.1 Carnot's theorem: universality of the ratio
of values of the maximum heat

Here we investigate an arbitrary system prepared separately under isothermal conditions at two temperatures and compare the values of the maximum heat for transitions occurring at these temperatures. We consider transitions $(T; X_0) \xrightarrow{\text{qi}} (T; X_1)$ and $(T'; X_0') \xrightarrow{\text{qi}} (T'; X_1')$, where T, T', X_0, X_1, X_0' and X_1' are arbitrary

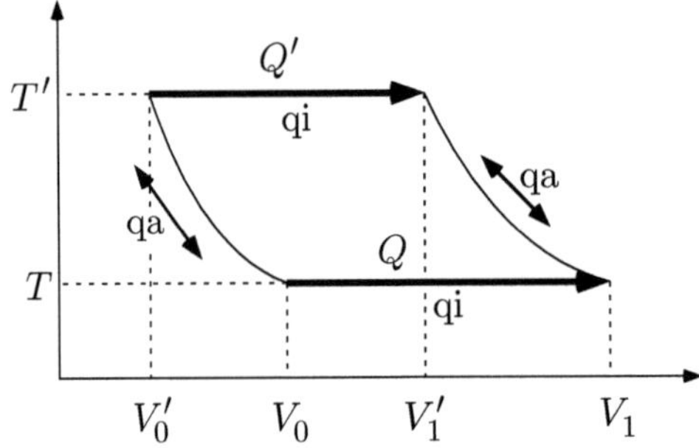

Figure 5.1 We consider quasi-static isothermal operations at temperatures T and T' with initial and final values of X denoted by X_0 and X_1 for temperature T and X_0' and X_1' for temperature T'. We write the values of the maximum heat for these operations as $Q = Q_{\max}(T; X_0 \to X_1)$ and $Q' = Q_{\max}(T'; X_0' \to X_1')$. The ratio of these values of the maximum heat is a universal quantity determined by only the two temperatures involved. Explicitly, we have $Q'/Q = T'/T$. This is the assertion of Carnot's theorem. The figure describes the case of an individual fluid system with the collective extensive variable $X = (V, N)$. The amount of substance is understood to be fixed, and thus we can write $X_0 = (V_0, N)$, $X_1 = (V_1, N)$, $X_0' = (V_0', N)$ and $X_1' = (V_1', N)$. The evolution of the system as it undergoes operations through which it makes transitions among these states is depicted in the T-V plane.

values subject only to the conditions that $Q_{\max}(T; X_0 \to X_1)$ be positive[9] and that there exist quasi-static adiabatic operations yielding the transitions

$$(T; X_0) \xleftrightarrow{\text{qa}} (T'; X_0'), \quad (T; X_1) \xleftrightarrow{\text{qa}} (T'; X_1') . \tag{5.19}$$

We then compare the values of the maximum heat for these transitions, $Q_{\max}(T; X_0 \to X_1)$ and $Q_{\max}(T'; X_0' \to X_1')$ (see Figure 5.1).

The following theorem is one of the most fundamental results in thermodynamics.

Result 5.3 (Carnot's theorem) *In the situation described above, the ratio*

$$f(T', T) = \frac{Q_{\max}(T'; X_0' \to X_1')}{Q_{\max}(T; X_0 \to X_1)} \tag{5.20}$$

is independent of the thermodynamic system and the choice of the reference values X_0, X_1, X_0' and X_1' (subject to the realizability of the transitions in (5.19)), and is determined by only T and T'. For the temperature scale that we use in this book (the Kelvin scale), this universal function takes the specific form

---

[9] For Carnot's theorem to hold, the condition $Q_{\max}(T; X_0 \to X_1) > 0$ is not necessary. We only need $Q_{\max}(T; X_0 \to X_1) \neq 0$. However, because the discussion in this section is somewhat simpler with the condition $Q_{\max}(T; X_0 \to X_1) > 0$, we use this. With several very straightforward modifications, this discussion can be applied to the case $Q_{\max}(T; X_0 \to X_1) < 0$ as well. Let us also note here that for any $T > 0$ and any X_0, there necessarily exists X_1 for which the condition $Q_{\max}(T; X_0 \to X_1) > 0$ holds. (See Problem 5.3 for investigation of this point.)

$$f(T', T) = \frac{T'}{T} \, . \tag{5.21}$$

The ratio of values of the maximum heat given in (5.20) is referred to as the *Carnot function*. The import of Carnot's theorem is that this ratio is a universal quantity that depends on neither the nature of the thermodynamic system in question nor the particular initial and final states of the transitions considered. The fact that the actual value of this ratio happens to be precisely T'/T simply reflects the fact that we have chosen the most natural temperature scale with regard to universal thermodynamic phenomena.

Carnot worked within the framework of the caloric theory and conducted research through which he derived the maximal efficiency of a heat engine. Although Result 5.3 cannot be attributed entirely to Carnot, he is responsible for the ingenious idea at its core.

A full proof of Carnot's theorem is presented in Section 5.2.3. Below we demonstrate the validity of the equality $f(T', T) = T'/T$ by assuming the main part of the theorem, i.e., that the Carnot function is independent of the nature of the system.

Derivation of (5.21) *Because the Carnot function $f(T', T)$ is (by assumption) independent of the thermodynamic system under consideration, the general result can be obtained by studying the simple case of an ideal gas.*

Using (5.7), which expresses the maximum heat in terms of the Helmholtz free energy and the energy, along with the explicit forms (3.36) and (4.37) for the free energy and the energy of an ideal gas, we readily obtain

$$Q_{\max}(T; (V_0, N) \to (V_1, N)) = F[T; V_0, N] - F[T; V_1, N] = NRT \log \frac{V_1}{V_0} \, . \tag{5.22}$$

Next, for an arbitrary temperature T', we choose V_0' and V_1' such that there exist quasi-static adiabatic operations that realize the transitions

$$(T; V_0, N) \xleftrightarrow{\text{qa}} (T'; V_0', N), \quad (T; V_1, N) \xleftrightarrow{\text{qa}} (T'; V_1', N) \, . \tag{5.23}$$

As above, we have

$$Q_{\max}(T'; (V_0', N) \to (V_1', N)) = NRT' \log \frac{V_1'}{V_0'} \, . \tag{5.24}$$

Then, using the equalities $T^c V_0 = T'^c V_0'$ and $T^c V_1 = T'^c V_1'$ provided by the Poisson relation (4.45), we find $V_1/V_0 = V_1'/V_0'$. Combining this relation with (5.22) and (5.24), we obtain

$$f(T', T) = \frac{Q_{\max}(T'; (V_0', N) \to (V_1', N))}{Q_{\max}(T; (V_0, N) \to (V_1, N))} = \frac{T'}{T} \, . \quad \blacksquare \tag{5.25}$$

5.2.2 Carnot cycle

As preparation for proving Carnot's theorem, here we study a sequence of quasi-static operations causing transitions that form a *Carnot cycle*, which is composed of transitions connecting four states of the kind discussed in the previous section:

$$(T'; X_0') \xrightarrow[\text{(a)}]{\text{qi}} (T'; X_1') \xrightarrow[\text{(b)}]{\text{qa}} (T; X_1) \xrightarrow[\text{(c)}]{\text{qi}} (T; X_0) \xrightarrow[\text{(d)}]{\text{qa}} (T'; X_0') . \qquad (5.26)$$

The letters (a), (b), (c) and (d) below the arrows, which serve as the names of both the respective transitions and the operations inducing them, correspond to the letters in Figures 5.2 and 5.3. They are included only for the sake of clarity. Also, note that in the ensuing discussion, we assume the relation $Q_{\max}(T; X_0 \to X_1) > 0$.

We now describe each leg of the cycle. Beginning in the state $(T'; X_0')$, the system undergoes the quasi-static isothermal operation (a) at temperature T', which yields the state $(T'; X_1')$. Next, under the quasi-static adiabatic operation (b), the system makes a transition to the state $(T; X_1)$. Then, the system undergoes the isothermal operation (c) at this temperature, through which it realizes the state $(T; X_0)$. Finally, the system returns to its initial state under the quasi-static adiabatic operation (d). Thus this sequence of transitions forms a cycle. However, because in this cycle the system interacts with two separate environments at two distinct temperatures, it is not an isothermal cycle (see Figure 5.2). Carnot introduced this cycle for the purpose of investigating the efficiency of a heat engine. Discussion of that investigation and its results are presented in Section 5.3.

As described above, a Carnot cycle includes two quasi-static isothermal operations carried out at different temperatures. In the Carnot cycle considered here, during the quasi-static isothermal operation (a), causing the transition $(T'; X_0') \xrightarrow{\text{qi}} (T'; X_1')$, the system absorbs the quantity of heat $Q_{\max}(T'; X_0' \to X_1') > 0$ from the environment at temperature T'.[10] During the quasi-static isothermal operation (c), causing the transition $(T; X_1) \xrightarrow{\text{qi}} (T; X_0)$, the system absorbs the quantity of heat $Q_{\max}(T; X_1 \to X_0) = -Q_{\max}(T; X_0 \to X_1) < 0$ from the

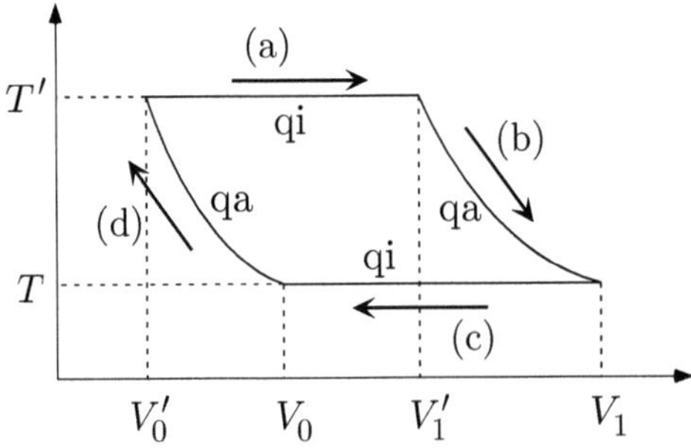

Figure 5.2 Example of a Carnot cycle undergone by an individual fluid system. The quantities appearing here have the same meanings as the corresponding ones in Figure 5.1.

---

[10] Because T and T' are both positive, and because we have assumed that $Q_{\max}(T; X_0 \to X_1)$ is positive, it follows from Carnot's theorem (proved below) that $Q_{\max}(T'; X_0' \to X_1')$ is also positive. (Of course, we do not assume that $Q_{\max}(T'; X_0' \to X_1')$ is positive in the proof of Carnot's theorem itself.)

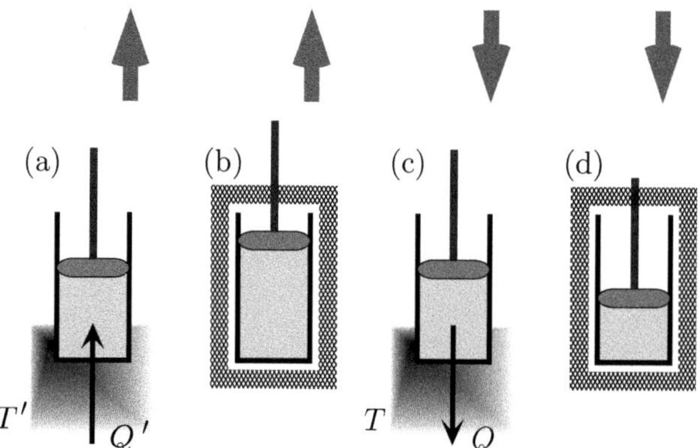

Figure 5.3 Illustration of the Carnot cycle depicted graphically in Figure 5.2. The cycle is realized under the following individual operations: (a) isothermal expansion at T'; (b) adiabatic expansion; (c) isothermal compression at temperature T; (d) adiabatic compression. At the conclusion of the four operations, the system returns to its original state. During (a), the system absorbs an amount of heat $Q' = Q_{\max}(T'; X_0' \to X_1') > 0$ from the environment at temperature T', while during (c), it expels an amount of heat $Q = Q_{\max}(T; X_0 \to X_1) > 0$ into the environment at temperature T. The difference between these quantities, $Q' - Q = W_{\mathrm{cyc}}$, is the amount of work performed by the system on the mechanical world in one Carnot cycle. This illustration makes intuitive the interpretation of a Carnot cycle as a model of a heat engine.

environment at temperature T. Equivalently, we can say that the system expels the quantity of heat $Q_{\max}(T; X_0 \to X_1) > 0$ into the environment (see Figure 5.3).

Let W_{cyc} denote the amount of work performed by the system on the mechanical world in one complete Carnot cycle. From the discussion given to this point, we know that this must be equal to the total amount of heat absorbed by the system during the same time, and thus we have

$$W_{\mathrm{cyc}} = Q_{\max}(T'; X_0' \to X_1') - Q_{\max}(T; X_0 \to X_1) \,. \tag{5.27}$$

Although the above result can be obtained immediately from the definition of heat given in (5.3), it is instructive to derive it in steps by considering the amount of work performed in each leg of the cycle. Recalling the definitions of the maximum work and the adiabatic work, we have

$$\begin{aligned}
W_{\mathrm{cyc}} = {}& W_{\max}(T'; X_0' \to X_1') + W_{\mathrm{ad}}((T'; X_1') \to (T; X_1)) \\
& + W_{\max}(T; X_1 \to X_0) + W_{\mathrm{ad}}((T; X_0) \to (T'; X_0')) \,.
\end{aligned} \tag{5.28}$$

Applying the energy conservation law (4.24) to this expression, we obtain

$$\begin{aligned}
W_{\mathrm{cyc}} = {}& W_{\max}(T'; X_0' \to X_1') + U(T'; X_1') - U(T; X_1) \\
& + W_{\max}(T; X_1 \to X_0) + U(T; X_0) - U(T'; X_0') \\
= {}& Q_{\max}(T'; X_0' \to X_1') + Q_{\max}(T; X_1 \to X_0) \,,
\end{aligned} \tag{5.29}$$

where we have used the expression for $Q_{\max}$ given in (5.6).

The above treatment describes the basic functioning of any heat engine: It imports heat from one environment, exports heat to another environment, and performs an amount of work on the mechanical world that is equal to the difference between these amounts of heat.

5.2.3 Proof of Carnot's theorem

The proof given here is based on Kelvin's principle, Postulate 3.1 (p. 47). However, its application is not entirely straightforward, as we now explain.

Kelvin's principle applies to isothermal cycles, which necessarily involve only a single environment at a single temperature, while a Carnot cycle contains two isothermal operations with two environments at different temperatures. However, if we consider two systems undergoing Carnot cycles that employ the same two environments and combine them in the appropriate manner (as described below), we can regard the composite system as effectively interacting with only a single environment. This composite system then undergoes what is in essence an isothermal cycle while the two constituent systems undergo their respective Carnot cycles. For this reason, Kelvin's principle can be applied to the composite system. This very powerful idea is due to Carnot.

To implement this approach, we now introduce a second arbitrary thermodynamic system, with collective extensive variable Y (the "Y system"), to go along with the system considered to this point (the "X system"). Then, as in the case of the X system, we consider four equilibrium states of the Y system, two at temperature T, $(T; Y_0)$ and $(T; Y_1)$, and two at temperature T', $(T'; Y_0')$ and $(T'; Y_1')$. Here too, these states are chosen such that there exist quasi-static isothermal operations under which the transitions

$$(T; Y_0) \xleftrightarrow{\text{qi}} (T; Y_1) \quad \text{and} \quad (T'; Y_0') \xleftrightarrow{\text{qi}} (T'; Y_1') \tag{5.30}$$

are realized and quasi-static adiabatic operations under which the transitions

$$(T; Y_0) \xleftrightarrow{\text{qa}} (T'; Y_0') \quad \text{and} \quad (T; Y_1) \xleftrightarrow{\text{qa}} (T'; Y_1') \tag{5.31}$$

are realized. Also, as in the case of the original system, we assume $Q_{\max}(T; Y_0 \to Y_1) > 0$.

For the Y system, we also consider the *reverse Carnot cycle* composed of the following sequence of transitions:

$$(T'; Y_1') \xrightarrow[\text{(ā)}]{\text{qi}} (T'; Y_0') \xrightarrow[\text{(d̄)}]{\text{qa}} (T; Y_0) \xrightarrow[\text{(c̄)}]{\text{qi}} (T; Y_1) \xrightarrow[\text{(b̄)}]{\text{qa}} (T'; Y_1') . \tag{5.32}$$

The labels (ā), (b̄), (c̄) and (d̄) appearing below the arrows are regarded as the names of the respective transitions and the operations causing them. These names correspond to those appearing in Figure 5.2, with the line above the letters in the present case expressing the meaning that the operations are carried out in reverse order and direction with respect to those depicted in the figure. The sequence of transitions in (5.32) is exactly what we would obtain if we took the original Carnot cycle, replaced X with Y, and carried out the sequence of operations in reverse.

In this cycle, the system absorbs an amount of heat $Q_{\max}(T; Y_0 \to Y_1)$ from the environment at temperature T and expels an amount of heat $Q_{\max}(T'; Y_0' \to Y_1')$ into the environment at temperature T' during its two isothermal operations, (c̄) and (ā).[11]

Next, we introduce a third system (the "αY system"), obtained from the Y system by scaling its extensive variables by the quantity α defined as follows:

$$\alpha = \frac{Q_{\max}(T; X_0 \to X_1)}{Q_{\max}(T; Y_0 \to Y_1)} \,.$$ (5.33)

For this system, the transition corresponding to (c̄) is $(T; \alpha Y_0) \xrightarrow{\text{qi}} (T; \alpha Y_1)$, and from the extensivity of the maximum heat, expressed by (5.12), we know that the amount of heat absorbed in this transition is $Q_{\max}(T; \alpha Y_0 \to \alpha Y_1) = \alpha Q_{\max}(T; Y_0 \to Y_1)$. Note that the value of α has been chosen such that this quantity of heat absorbed by the αY system is equal to the quantity of heat expelled by the X system in operation (c), namely $Q_{\max}(T; X_0 \to X_1)$. This implies that if the isothermal operation (c) on the X system and the isothermal operation (c̄) on the αY system are carried out simultaneously, then the total amount of heat transferred between the two systems and the environment vanishes.

With the above observation, we introduce a composite system whose collective extensive variable is $(X, \alpha Y)$. By simultaneously applying the quasi-static isothermal operations (c) and (c̄) to the X and αY subsystems, respectively, we obtain the transition

$$(T; X_1, \alpha Y_0) \xrightarrow{\text{qi}} (T; X_0, \alpha Y_1) \,.$$ (5.34)

Formally applying the additivity of the maximum heat, expressed by (5.11), we find that indeed the total heat absorbed by the composite system in this transition vanishes:[12]

$$Q_{\max}(T; (X_1, \alpha Y_0) \to (X_0, \alpha Y_1)) = Q_{\max}(T; X_1 \to X_0) + Q_{\max}(T; \alpha Y_0 \to \alpha Y_1) = 0 \,.$$ (5.35)

Result 5.2 then allows us to conclude that there exists a quasi-static adiabatic operation yielding the transition

$$(T; X_1, \alpha Y_0) \xrightarrow{\text{qa}} (T; X_0, \alpha Y_1)$$ (5.36)

and that the amount of work done by the composite system in this operation is the same as that done in the operation yielding (5.34).

The remainder of the proof is quite straightforward. Combining operations (a) (yielding $(T'; X_0') \xrightarrow{\text{qi}} (T'; X_1')$) and (ā) (yielding $(T'; Y_1') \xrightarrow{\text{qi}} (T'; Y_0')$), we

---

[11] A reverse Carnot cycle acts as a refrigerator (see Problem 5.5).

[12] While (5.34) is similar to an adiabatic transition in the sense that it is accompanied by no net exchange of heat with the environment, as discussed below Result 5.2 (p. 87), it differs from an adiabatic transition in an essential way. For this reason, Result 5.2 is necessary to rigorously complete the argument here.

obtain a quasi-static isothermal operation inducing the transition

$$(T'; X'_0, \alpha Y'_1) \xrightarrow{\text{qi}} (T'; X'_1, \alpha Y'_0) \,, \tag{5.37}$$

combining (b) (yielding $(T'; X'_1) \xrightarrow{\text{qa}} (T; X_1)$) and ($\bar{\text{d}}$) (yielding $(T'; Y'_0) \xrightarrow{\text{qa}} (T; Y_0)$), we obtain a quasi-static adiabatic operation inducing the transition

$$(T'; X'_1, \alpha Y'_0) \xrightarrow{\text{qa}} (T; X_1, \alpha Y_0) \,, \tag{5.38}$$

and combining (d) (yielding $(T; X_0) \xrightarrow{\text{qa}} (T'; X'_0)$) and ($\bar{\text{b}}$) (yielding $(T; Y_1) \xrightarrow{\text{qa}} (T'; Y'_1)$), we obtain a quasi-static adiabatic operation inducing the transition

$$(T; X_0, \alpha Y_1) \xrightarrow{\text{qa}} (T'; X'_0, \alpha Y'_1) \,. \tag{5.39}$$

Then, connecting the four transitions (5.37), (5.38), (5.36) and (5.39) in a sequence, we construct the cycle

$$(T'; X'_0, \alpha Y'_1) \xrightarrow[(\text{a},\bar{\text{a}})]{\text{qi}} (T'; X'_1, \alpha Y'_0) \xrightarrow[(\text{b},\bar{\text{d}})]{\text{qa}} (T; X_1, \alpha Y_0) \xrightarrow[(\text{c},\bar{\text{c}})]{\text{qa}} (T; X_0, \alpha Y_1) \xrightarrow[(\text{d},\bar{\text{b}})]{\text{qa}} (T'; X'_0, \alpha Y'_1) \,.$$

$$\tag{5.40}$$

This cycle can also be regarded as consisting of the Carnot cycle (5.26) and the scaled version of the reverse Carnot cycle (5.32) carried out individually but coincidently, with the $(\text{c},\bar{\text{c}})$ part being replaced by an adiabatic transition. Of the four transitions forming (5.40), all but the isothermal transition at temperature T' (i.e., $(\text{a},\bar{\text{a}})$) are adiabatic. In addition, all of these transitions are quasi-static. It follows that (5.40) as a whole constitutes a quasi-static isothermal cycle at temperature T'. Hence, Result 3.2 (p. 48), which is a direct consequence of Kelvin's principle, applies to it. We therefore conclude that the work performed by the composite system on the external world during this cycle, W_{cyc}, is exactly 0.

Let us next evaluate the work W_{cyc} by considering the cycles undergone by the X and αY systems individually. From Result 5.2, we know that the amounts of work done by the composite system in the operations inducing the quasi-static isothermal transition (5.34) and the quasi-static adiabatic transition (5.36) are identical. We thus find that W_{cyc} is given by the sum of the work done in the original Carnot cycle (5.26) and the work done in the reverse Carnot cycle (5.32). Then, using the expression for the work performed in a single Carnot cycle appearing in (5.27), we obtain

$$\begin{aligned}
W_{\text{cyc}} &= Q_{\max}(T'; X'_0 \to X'_1) - Q_{\max}(T; X_0 \to X_1) \\
&\quad - \alpha Q_{\max}(T'; Y'_0 \to Y'_1) + \alpha Q_{\max}(T; Y_0 \to Y_1) \\
&= Q_{\max}(T'; X'_0 \to X'_1) - \frac{Q_{\max}(T; X_0 \to X_1)}{Q_{\max}(T; Y_0 \to Y_1)} Q_{\max}(T'; Y'_0 \to Y'_1) \,, \tag{5.41}
\end{aligned}$$

where we have used the definition of α, (5.33). Substituting this into the equality $W_{\text{cyc}} = 0$, we arrive at the final result:

$$\frac{Q_{\max}(T'; X'_0 \to X'_1)}{Q_{\max}(T; X_0 \to X_1)} = \frac{Q_{\max}(T'; Y'_0 \to Y'_1)}{Q_{\max}(T; Y_0 \to Y_1)} \,. \tag{5.42}$$

Because the above derivation is entirely general, we conclude that the Carnot function, $f(T', T)$, depends on neither the type of system studied nor on the states involved in the transitions under consideration (as long as they satisfy (5.19)). We have thus demonstrated Carnot's theorem. ∎

5.3 Heat engines and the upper limit of their efficiencies

In this section, we consider the problem that motivated Carnot's research, understanding heat engines and their efficiencies.[13] Interestingly, the maximum efficiency of a heat engine is determined by only the temperatures of the environments with which it interacts. As we see below, the laws of thermodynamics (which, as far as we know, are universally valid) dictate that no matter what kind of elaborate design one might use, it is not possible to engineer a heat engine whose efficiency exceeds this limit. This result was reported in Carnot's seminal book (see Section 1.1.3).

5.3.1 Heat engines and their efficiencies

In simple terms, a heat engine is a device that converts heat into mechanical energy.[14] Heat engines played a central role in the development of modern civilization, as they provided a convenient means of obtaining a usable form of energy from an unusable form. Today, heat engines are most notably used in power plants (even nuclear power plants) to generate electricity, as indoor heating and cooling devices, and as refrigerators. (Air conditioners and refrigerators are heat engines operating in reverse.[15])

Let us consider a thermodynamic system acting as a heat engine. We write the temperatures of the environments with which the system alternatingly interacts as T_H and T_L, where $T_H > T_L$ (see Figure 5.4). The system is initially in an equilibrium state with temperature T_H. Then, while in contact with the high-temperature environment, it undergoes an isothermal operation (or, a generalized isothermal operation). During this operation, the system absorbs a positive amount of heat Q_H. Next, the system is subject to an adiabatic operation through which its temperature decreases to T_L. Then, while in contact with the low-temperature environment, it undergoes a second isothermal operation (or generalized isothermal operation). Through this operation, the system expels a positive amount of heat Q_L into the environment. Finally, the system again experiences an adiabatic operation, at the end of which it has returned to its initial state. This is the cycle undergone by a heat engine.[16]

---

[13] This section can be skipped without creating a logical gap.

[14] Internal combustion engines, which power most automobiles, are perhaps the type of engines most familiar to us. However, they are not heat engines in the strict sense used in this book. Rather, they are *chemical engines*, which convert chemical energy into mechanical energy. However, in the literature, engines of this type are sometimes referred to as "heat engines," because they utilize the heat of combustion to generate mechanical energy.

[15] Heat pumps and refrigerators are considered in Problem 5.5.

[16] It is also possible to consider a system that interacts with its two environments through processes more complicated than that considered here. See Appendix C for the most general treatment of heat engines.

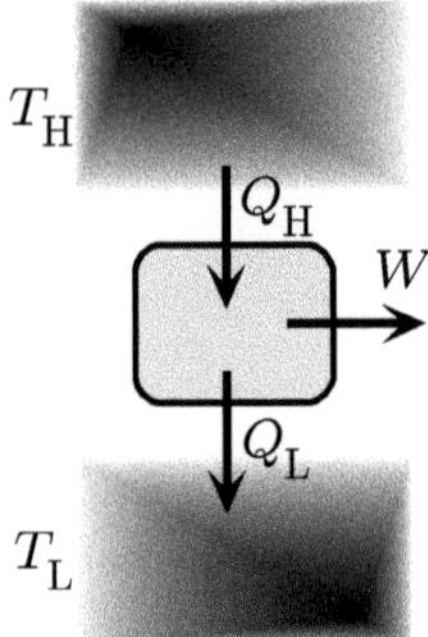

Figure 5.4 Depiction of a heat engine. In one cycle, the engine absorbs an amount of heat Q_H from the high-temperature environment (heat source) and expels a smaller amount of heat Q_L into the low-temperature environment (heat sink). Hence, it performs an amount of work $W = Q_\mathrm{H} - Q_\mathrm{L}$ on the mechanical world. The *efficiency* of this engine is defined as the ratio of the work performed to the heat absorbed from the heat source: $\epsilon = W/Q_\mathrm{H}$.

From the law of energy conservation, we immediately find that the amount of work performed by a heat engine in one cycle is $W = Q_\mathrm{H} - Q_\mathrm{L}$. We define the *efficiency* of a heat engine to be the ratio of this amount of work to the amount of heat imported by the system in the high-temperature isothermal operation, Q_H. Thus, the efficiency is given by

$$\epsilon = \frac{W}{Q_\mathrm{H}} = 1 - \frac{Q_\mathrm{L}}{Q_\mathrm{H}} . \tag{5.43}$$

5.3.2 Carnot cycle as a heat engine

Although the Carnot cycle is a theoretical construct, it is the archetype of a heat engine. In this case, for the quantities of imported and exported heat, Q_H and Q_L, Carnot's theorem (see (5.20) and (5.21)) implies the relation $Q_\mathrm{H}/Q_\mathrm{L} = T_\mathrm{H}/T_\mathrm{L}$. Thus, for a Carnot cycle, the efficiency (5.43) is given by

$$\epsilon_0 = 1 - \frac{T_\mathrm{L}}{T_\mathrm{H}} . \tag{5.44}$$

It is thus seen that the efficiency of a Carnot cycle is necessarily less than 1. For example, in the case $T_\mathrm{H} = 2T_\mathrm{L}$, ϵ_0 is only $1/2$. Therefore, with such temperatures, only half of the heat absorbed by the system can be converted to work, with the other half being wasted. Only in the limit $T_\mathrm{L}/T_\mathrm{H} \to 0$ does the efficiency approach 1.

As a particular example, here let us consider the case of a Carnot cycle undergone by an individual gas system. In this case, the Carnot cycle consists of the following legs (see Figures 5.2 and 5.3):

$$(T_\mathrm{H}; V_0', N) \xrightarrow[\text{(a)}]{\text{qi}} (T_\mathrm{H}; V_1', N) \xrightarrow[\text{(b)}]{\text{qa}} (T_\mathrm{L}; V_1, N) \xrightarrow[\text{(c)}]{\text{qi}} (T_\mathrm{L}; V_0, N) \xrightarrow[\text{(d)}]{\text{qa}} (T_\mathrm{H}; V_0', N) . \tag{5.45}$$

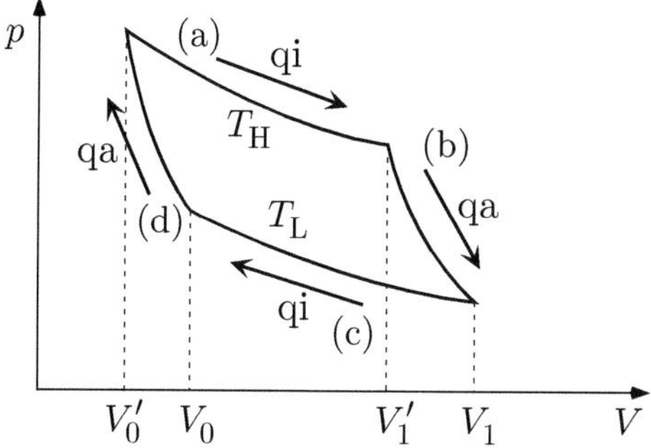

Figure 5.5 Schematic depiction of a Carnot cycle in the p-V plane.

In the investigation of cycles involving fluids, it is customary to plot the path followed by the state of the system in a p-V graph, with p and V represented by the vertical and horizontal axes, respectively, as in Figure 5.5.[17] For a graph of this type, the area enclosed within the curve is equal to the work done by the system on the mechanical world in one quasi-static cycle. In the case of an ideal gas, the curves (a) and (c) in the figure, which represent quasi-static isothermal operations, are determined by the relations $pV = NRT_H$ and $pV = NRT_L$. Also in this case, (b) and (d), which represent quasi-static adiabatic operations, are determined by the relation $p^c V^{c+1} = $ (constant), which can be derived from the Poisson relation (4.45) by writing T in terms of p and V using the ideal gas equation of state (see Section 4.4.2).

5.3.3 Otto cycle

Historically, the investigation of heat engines has been a major area of research in thermodynamics, and many types of engines have been developed and examined. Here we consider one particularly important example among these, represented by the Otto cycle, which is an idealization of a four-stroke internal combustion engine. It should be noted, however, that there is one large difference between internal combustion engines and the Otto cycle. An internal combustion engine uses a single environment (generally, the atmosphere) at a single temperature and generates work through the combustion of fuel. Contrastingly, an Otto cycle is a heat engine that employs two fixed environments at different temperatures and generates work by exploiting this difference.

Consider a system consisting of an ideal gas of amount N. We study the situation in which this gas undergoes the following cycle (in which V and V' are fixed volumes satisfying $V' > V$):

$$(T'_H; V, N) \xrightarrow[\text{(a)}]{i'} (T_H; V, N) \xrightarrow[\text{(b)}]{qa} (T'_L; V', N) \xrightarrow[\text{(c)}]{i'} (T_L; V', N) \xrightarrow[\text{(d)}]{qa} (T'_H; V, N) \, .$$

$$(5.46)$$

---

[17] However, it is important to understand that the curve in this kind of p-V graph does not uniquely specify an operation. In general, there are many actual, realizable operations corresponding to any given curve in the p-V plane (or any more general phase space). For related discussion, see Section 6.3.3, and in particular, Figure 6.5.

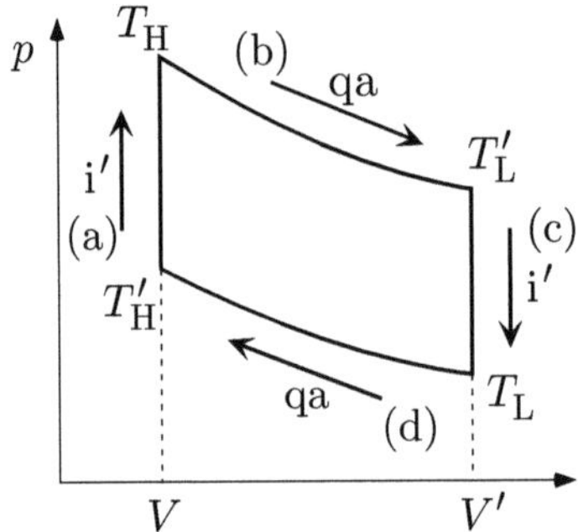

Figure 5.6 Depiction of the Otto cycle, which is an idealization of an internal combustion engine, in a p-V graph. This cycle consists of the following four operations undergone by the system, composed of a gas: (a) a temperature-increasing operation; (b) an adiabatic expansion; (c) a temperature-decreasing operation; (d) an adiabatic compression. The processes corresponding to these operations that take place in an actual internal combustion engine are as follows. During (a), there is a very rapid increase in the pressure and temperature of the gas resulting from the combustion of fuel. (Hence, the temperature-increasing process in the Otto cycle resulting from contact with a hot environment is replaced in an internal combustion engine by the combustion of fuel.) After the combustion, during (b), the gas expands rapidly as a result of the pressure increase experienced in (a). During (c), the gas is cooled. During (d), a new gas mixture consisting of air and fuel vapor is compressed. In an actual engine, between (c) and (d) there are two additional processes, namely, the expulsion of the exhaust gas produced in the combustion, and the intake of the new air-fuel mixture.

The p-V graph corresponding to this cycle appears in Figure 5.6. Initially, the gas is in the equilibrium state $(T_H'; V, N)$. In the operation corresponding to (a), the system is placed in contact with an environment at temperature T_H. During this operation, the volume of the gas remains constant, while its temperature increases from T_H' to T_H. This is an example of a generalized isothermal operation (see Problem 5.2). This operation models the combustion process that occurs in an engine as an effectively instantaneous process in which the pressure and temperature of the gas in the cylinder increase to their maximum values while the piston has yet to move. In the quasi-static adiabatic operation corresponding to (b), the volume of the gas increases from V to V', and through this expansion, the gas performs work on the mechanical world. This operation models the violent expansion of the high-pressure, high-temperature gas (the "power stroke") occurring in an engine after the combustion of the fuel. Next, in the generalized isothermal operation corresponding to (c), with the volume fixed, the temperature of the system decreases to T_L following contact with an environment at this temperature. (The two low temperatures here are related as $T_L < T_L'$.) Finally, in the quasi-static adiabatic operation corresponding to (d), the volume decreases from V' to V, and as a result, the system returns to its initial state. This operation models the process in an engine in which a gas mixture consisting of fuel vapor and air is compressed (the "compression stroke"). In an actual four-stroke internal combustion engine, between (c) and (d) there are two processes (two "strokes"). In the first of these, the exhaust produced by the combustion is expelled (the "exhaust stroke"), and in the second, a new mixture of fuel and air is introduced into the cylinder (the "intake stroke").

Let us next investigate the efficiency of an Otto cycle. In the transition (a), the amount of energy gained by the system in the form of heat absorbed from the environment at temperature T_H is $Q_H = U(T_H; V, N) - U(T'_H; V, N) = cNR(T_H - T'_H)$, while in the transition (c), the amount of energy lost by the system in the form of heat expelled into the environment at temperature T_L is $Q_L = U(T'_L; V', N) - U(T_L; V', N) = cNR(T'_L - T_L)$.[18] Hence, using (5.43), we find that the efficiency of this Otto cycle is given by

$$\epsilon = 1 - \frac{Q_L}{Q_H} = 1 - \frac{T'_L - T_L}{T_H - T'_H} \, . \tag{5.47}$$

This expression can be simplified using the following equations derived from the Poisson relation (4.45):

$$T'_L = \left(\frac{V}{V'} \right)^{1/c} T_H \, , \quad T_L = \left(\frac{V}{V'} \right)^{1/c} T'_H \, . \tag{5.48}$$

Substituting these into (5.47), we obtain

$$\epsilon = 1 - \left(\frac{V}{V'} \right)^{1/c} = 1 - \frac{T_L}{T'_H} \, . \tag{5.49}$$

Finally, with the relation $T'_H < T_H$, this yields

$$\epsilon < 1 - \frac{T_L}{T_H} = \epsilon_0 \, . \tag{5.50}$$

We conclude that the efficiency of an Otto cycle is strictly less than that of a Carnot cycle.

5.3.4 Universal upper limit on the efficiency of a heat engine

From the results (5.44) and (5.49), we have found that both the Carnot cycle and the Otto cycle are characterized by efficiencies that are necessarily less than 1. One may wonder if this property is due to the particular natures of these cycles and whether, with some intricately designed operations, it might be possible to devise a heat engine with a higher efficiency. Carnot investigated this problem himself, and he derived the following decisive conclusion.

Result 5.4 (Upper limit on the efficiency of a heat engine) *The efficiency ϵ of any heat engine satisfies the inequality*

$$\epsilon \leq 1 - \frac{T_L}{T_H} = \epsilon_0 \, , \tag{5.51}$$

where T_L and T_H (with $T_L < T_H$) are the temperatures of the environments used.

---

[18] The exchange of heat that takes place in a generalized isothermal operation is considered in Problem 5.2.

This result asserts that no heat engine can operate with an efficiency greater than that of a Carnot cycle employing the same environments. From this result, it is seen that the inferior efficiency of the Otto cycle in comparison with the Carnot cycle, expressed by (5.50), is simply one example of a universal relationship. The inequality (5.51) is a mathematically rigorous, universally valid result that reveals one aspect of thermodynamic behavior in extraordinary clarity. This result demonstrates the unique character of thermodynamics, in which quantitatively exact results with universal applicability can be derived by combining a small number of basic hypotheses.

Here, rather than carrying out a rigorous derivation of (5.51), we present an intuitive argument that follows Carnot's manner of thinking.[19] Appendix C contains a rigorous proof of (5.51) that applies to a situation more general than that considered here.

To begin, let us assume that there exists a heat engine whose efficiency does not satisfy the inequality (5.51). As above, we write the heat absorbed in one cycle from the environment at temperature T_{H} as Q_{H} and the heat expelled in one cycle into the environment at temperature T_{L} as Q_{L}. Then, again, the work performed by the system on the mechanical world in one cycle is given by $W = Q_{\mathrm{H}} - Q_{\mathrm{L}}$.[20] Thus, here we are supposing that the efficiency of the heat engine under consideration, $\epsilon = W/Q_{\mathrm{H}}$, is greater than that of a Carnot cycle employing the same environments, ϵ_0.

Next, we consider a system undergoing a reverse Carnot cycle that in one cycle expels an amount of heat $\tilde{Q}_{\mathrm{H}}$ into the environment at temperature T_{H} and absorbs an amount of heat $\tilde{Q}_{\mathrm{L}}$ from the environment at temperature T_{L}. In the case of this reverse Carnot cycle, the mechanical world performs the amount of work $W_0 = \epsilon_0 \tilde{Q}_{\mathrm{H}}$ on the system. We choose this system to have such properties that the quantity of expelled heat, $\tilde{Q}_{\mathrm{H}}$, is equal to the quantity of absorbed heat for the original system, Q_{H}.

Let us now investigate the system obtained when the two systems introduced above are combined. This composite system is understood to consist of the hypothetical heat engine and the reverse Carnot cycle operating simultaneously, as depicted in Figure 5.7. Then, because the amount of heat absorbed by the hypothetical heat engine at T_{H} and the amount of heat expelled by the reverse Carnot cycle at T_{H} are equal, this composite system effectively exchanges no heat with the environment at this temperature. Therefore, in one cycle, it merely absorbs the quantity of heat $\tilde{Q}_{\mathrm{L}} - Q_{\mathrm{L}}$ from the environment at temperature T_{L} and performs the amount of work

$$W - W_0 = \epsilon Q_{\mathrm{H}} - \epsilon_0 \tilde{Q}_{\mathrm{H}} = (\epsilon - \epsilon_0)Q_{\mathrm{H}} > 0 \tag{5.52}$$

on the mechanical world. This composite cycle thus functions as a perpetual motion machine of the second kind, which contradicts Kelvin's principle, Postulate 3.1 (p. 47). The validity of the inequality (5.51) has thus been demonstrated.[21]

---

[19] Comparing this argument with the proof of Carnot's theorem presented in Section 5.2.3, it is seen how the latter formalizes Carnot's manner of thinking.

[20] Throughout the entire treatment here, the quantities Q_{H}, Q_{L}, $\tilde{Q}_{\mathrm{H}}$, $\tilde{Q}_{\mathrm{L}}$, W and W_0 are all regarded to be positive.

[21] However, as mentioned above, this argument is not rigorous, because there is nothing corresponding to Result 5.2 (p. 87), which guarantees the validity of replacing the isothermal operation at temperature T_{H} with an adiabatic operation.

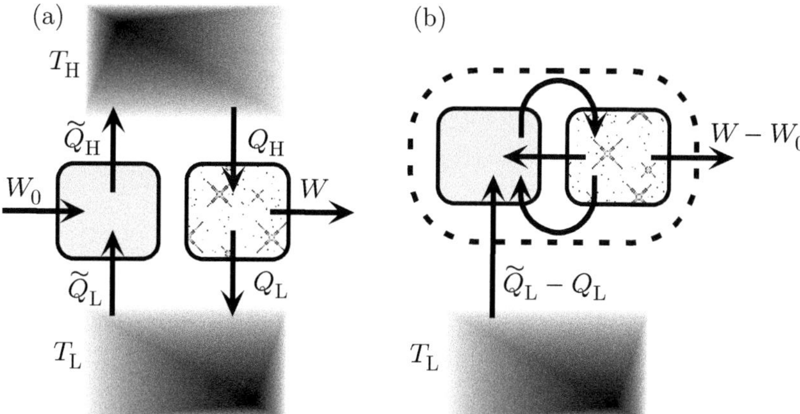

Figure 5.7 Carnot's idea for deriving the upper limit on the efficiency of a heat engine. All quantities appearing in the figure, Q_{H}, Q_{L}, $\tilde{Q}_{\mathrm{H}}$, $\tilde{Q}_{\mathrm{L}}$, W and W_0, are regarded as positive. (a) A reverse Carnot cycle (left) and the hypothetical heat engine of efficiency $\epsilon > \epsilon_0$ (right). (b) The sizes of the systems are chosen so as to realize the equality $\tilde{Q}_{\mathrm{H}} = Q_{\mathrm{H}}$. We regard these two systems as forming a single composite system. In one cycle, this composite system absorbs the total amounts of heat $\tilde{Q}_{\mathrm{L}} - Q_{\mathrm{L}}$ from the environment at temperature T_{L} and $Q_{\mathrm{H}} - \tilde{Q}_{\mathrm{H}} = 0$ from the environment at temperature T_{H}. Hence it effectively undergoes an isothermal cycle. Thus, because this system performs the amount of work $W - W_0 > 0$ on the mechanical world, it constitutes a perpetual motion machine of the second kind. This contradicts Kelvin's principle.

5.3.5 Power and efficiency of a heat engine

Finally, we briefly discuss the *performance* of heat engines. In an actual application, it is clearly desirable for a heat engine to efficiently convert absorbed heat into (usable) work. In other words, we wish to have a high efficiency, $\epsilon = W/Q_{\mathrm{H}}$. But from a practical point of view, the power is also important. (The power of a heat engine is defined as $P = W/\tau$, where W is the work done by the engine in a single cycle, and τ is the period of a cycle.) For practical applications, it is clearly best to characterize the performance of a heat engine by both its efficiency and its power.

With this characterization in mind, let us first recall that a true Carnot cycle is composed of quasi-static transitions, which are realized in the limit of infinitesimally slow operations. Therefore an engine that realizes a Carnot cycle possesses an infinitesimally small power and hence is not useful for practical purposes. It is natural to ask if it is possible to construct a heat engine that realizes the maximum efficiency, ϵ_0, but also realizes non-vanishing power.

It seems clear that any heat engine employing a mechanism similar to that of the Carnot cycle would not be capable of realizing this kind of performance, but what about a heat engine that employs an entirely different mechanism? Interestingly, this question cannot be answered within the standard framework of thermodynamics, because in thermodynamics, change is understood solely in terms of operations and transitions, which lack a quantitative concept of time. But recently, a definitive answer was obtained within the framework of non-equilibrium statistical physics.

It was shown that, under suitable assumptions regarding the microscopic structure of the heat engine, any heat engine with maximum efficiency, ϵ_0, necessarily also has vanishing power.[22] Determining a general relation between the power and efficiency of a heat engine is still an unsolved problem.

Problems 5

5.1 (Section 5.1.1) Consider two systems under adiabatic conditions in states $(T; X)$ and $(T'; Y)$, with $T \neq T'$. Suppose that these systems undergo an operation through which the adiabatic wall separating them is replaced with a diathermal wall. They thereby come to form a composite system under adiabatic conditions. According to the "zeroth law of thermodynamics," Result 2.6 (p. 42), the temperatures of the two systems will eventually become equal. During this process of equilibration, heat should "flow" from the high temperature system to the low temperature system. Formalize this process and determine the quantity of heat that is transferred between the two systems, assuming that the values of X and Y are held fixed throughout.

5.2 We have encountered the concept of a *generalized isothermal operation* several times (see, e.g., (6.17), (5.17) and (5.46)). Here, we formalize this concept and investigate the properties of such operations.

Consider a system that is initially in an equilibrium state $(T_1; X_1)$. It is then enclosed within adiabatic walls and placed in an environment at temperature T_2. Suppose that the system next undergoes an operation through which the value of the collective extensive variable is changed in some arbitrary manner from X_1 to X_2 and the adiabatic walls are removed in some arbitrary manner. These walls may be partially or completely removed and then partially or completely replaced any number of times. It is only necessary that eventually all walls are removed and left off for the remainder of the operation. Then, without further perturbation, a sufficiently long time after the final removal of the walls, the system will reach the equilibrium state $(T_2; X_2)$. We write the transition resulting from this combined operation as

$$(T_1; X_1) \xrightarrow{i'} (T_2; X_2) . \tag{5.53}$$

We term this a *generalized isothermal transition* and the operation causing it a *generalized isothermal operation*. As a special case of this type of operation, in several places in this and the preceding chapter, we have considered operations consisting merely of the removal of the adiabatic walls, with X fixed.

Let W be the work done by the system on the mechanical world in a generalized isothermal operation under which the transition (5.53) is realized. Show that for an operation of this kind, it is appropriate to define the heat

---

[22] More precisely, it was demonstrated that the power, P, and the efficiency, ϵ, of an arbitrary heat engine satisfy the inequality $P \leq$ (const.) $\epsilon (\epsilon_0 - \epsilon)$, where the constant is proportional to the kinetic energy of the engine. (For details, see N. Shiraishi, K. Saito and H. Tasaki, "Universal Trade-Off Relation between Power and Efficiency for Heat Engines," Physical Review Letters, vol. 117, iss. 19, 2016, 190601:1–6 (https://link.aps.org/doi/10.1103/PhysRevLett.117.190601).)

Q absorbed by the system from the environment through the relation

$$Q = W + U(T_2; X_2) - U(T_1; X_1) \,, \tag{5.54}$$

analogously to the situation described by (5.4).

Next, choose some X_3 such that there exist quasi-static adiabatic operations under which the transitions $(T_1; X_1) \overset{\text{qa}}{\longleftrightarrow} (T_2; X_3)$ are realized. Show that the maximum amount of work done by the system on the mechanical world and the maximum amount of heat that the system absorbs from the environment in the generalized isothermal transition (5.53) are respectively equal to the work W' and heat Q' accompanying the quasi-static transition $(T_1; X_1) \overset{\text{qa}}{\longrightarrow} (T_2; X_3) \overset{\text{qi}}{\longrightarrow} (T_2; X_2)$.

5.3 (Section 5.2.1) Show that for arbitrary T and X_0, there exists an X_1 such that the relation $Q_{\max}(T; X_0 \to X_1) > 0$ holds.[23]

Hint: First, for some $T' < T$, choose X_1 such that there exists an operation under which $(T; X_0) \overset{\text{qa}}{\longrightarrow} (T'; X_1)$ is realized. Then, expressing the work W_{cyc} performed by the system during the isothermal transition[24] $(T; X_1) \overset{\text{qi}}{\longrightarrow} (T; X_0) \overset{\text{qa}}{\longrightarrow} (T'; X_1) \overset{\text{i}'}{\longrightarrow} (T; X_1)$ in terms of the Helmholtz free energy and the energy, use Kelvin's principle and the inequality $U(T'; X_1) < U(T; X_1)$.

5.4 (Section 5.3.1) Derive the efficiency of the heat engine depicted in Figure 1.6. (See Problem 1.2 for related considerations.)

Hint: Be sure to include the work done by the system on the gas external to the engine, which exerts a nonzero pressure.

5.5 (Section 5.3.1) A refrigerator is a heat engine operating in reverse. Domestic refrigeration appliances and air conditioners are examples of actual devices of this kind.[25]

As depicted in Figure 5.8, in one cycle of a refrigerator, a positive amount of work W is performed on the system by the mechanical world, while the system absorbs a positive amount of heat Q_{L} from the low-temperature environment, at T_{L}, and expels a positive amount of heat Q_{H} into the high-temperature environment, at T_{H}.[26] From energy conservation, we have $W + Q_{\text{L}} = Q_{\text{H}}$. The performance of a refrigerator is measured by the *coefficient of performance*, defined as $\omega = Q_{\text{L}}/W$.[27] Derive the value of ω for a refrigerator operating as a reverse Carnot cycle. Devise a refrigerator based on something other than a reverse Carnot cycle. Is there a universal upper limit on ω

---

[23] This problem is from Ref. [11].

[24] The final transition here results from a generalized isothermal operation in which the adiabatic walls enclosing the system are removed while the collective extensive variable is held fixed.

[25] Refrigerators are sometimes referred to as *heat pumps*. Originally, these terms were strictly synonymous, but in modern usage *heat pump* possesses a broader meaning. Today, this term is used generally in reference to devices operating as reverse heat engines, whether they are used for heating or cooling.

[26] In the case of a refrigerator, T_{L} is the temperature inside the refrigerator, and T_{H} is the room temperature (or, more precisely, the temperature of the air outside of the refrigerator's exhaust vent, which is generally located on the back of the appliance). In recent years, dual heating and cooling heat pump appliances have become quite common. When these devices are used for cooling, T_{L} is the room temperature and T_{H} is the outside temperature, while when used for heating, this is reversed. In the latter case, an amount of energy $Q_{\text{H}} = W + Q_{\text{L}}$ is released into the room as heat. By comparison, for an electric heater, the heat released into the room is simply equal to the work performed on the heating element. (This is Joule heat, discussed in Section 4.3.) For this reason, in principle, a heat pump can heat a room using less electric power (which supplies the work in either case) than an electric heater.

[27] In the case of a heat pump used for heating, the coefficient of performance is $\omega = Q_{\text{H}}/W$.

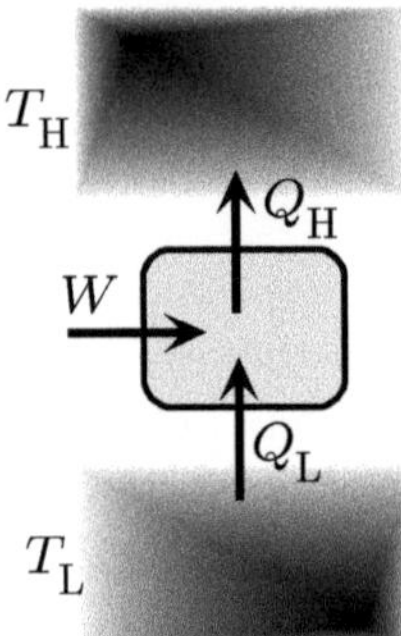

Figure 5.8 Schematic depiction of a refrigerator. (This figure should be compared with Figure 5.4, which describes a heat engine.) In one cycle, the mechanical world performs an amount of work W on the system, while the system absorbs an amount of heat Q_{L} from the cold environment and expels an amount of heat Q_{H} into the hot environment.

analogous to that on the efficiency of a heat engine?[28] With this question in mind, extend Carnot's intuitive argument outlined in Section 5.3 and the more rigorous argument presented in Appendix C.

---

[28] Even today, technological advances continue to yield increases in the performance of refrigerators and heat-pump air conditioners, and as a result, the electric power needed for their operation continues to decrease. However, this is due chiefly to improvements in insulation technology, rather than heat pump technology.

6
Entropy

In this chapter, employing only the operational concepts developed to this point, we introduce *entropy*, a quantity of fundamental importance in thermodynamics. After demonstrating how entropy is incorporated into the theoretical framework, we elucidate its basic properties. We next discuss how the essence of entropy becomes clearest when it is considered in connection to the reversibility and irreversibility of adiabatic operations. Then, investigating the relationship between entropy and heat, we elucidate its connection to the reversibility and irreversibility of isothermal operations. Next, reflecting on these results, we discuss the law of increasing entropy and its implications from a general perspective. Finally, we briefly study the entropy of compound states. With the treatment of entropy given in this chapter, the theoretical framework of thermodynamics is completed.

6.1 Introduction of entropy into the thermodynamic theory

The introduction of entropy is generally regarded as the most essential step in the construction of the theoretical framework of thermodynamics. With the development of the theory carried out in the preceding chapters, we are now ready to take this step.

6.1.1 Definition of entropy

Let us consider an arbitrary thermodynamic system characterized by the collective extensive variable X. For two unequal temperatures T and T', we choose the values X_1, X_2, X_1' and X_2' of X such that there exist quasi-static adiabatic operations under which the following transitions are realizable:

$$(T; X_1) \xleftrightarrow{\text{qa}} (T'; X_1'), \quad (T; X_2) \xleftrightarrow{\text{qa}} (T'; X_2') . \tag{6.1}$$

Then, from Carnot's theorem (specifically, the relations (5.20) and (5.21)) we have the equality

$$\frac{Q_{\max}(T; X_1 \to X_2)}{T} = \frac{Q_{\max}(T'; X_1' \to X_2')}{T'} . \tag{6.2}$$

Using (5.7), which expresses the maximum heat in terms of the Helmholtz free energy and the energy, this equation can be rewritten as

$$\frac{F[T; X_1] - F[T; X_2] + U(T; X_2) - U(T; X_1)}{T}$$
$$= \frac{F[T'; X_1'] - F[T'; X_2'] + U(T'; X_2') - U(T'; X_1')}{T'} . \tag{6.3}$$

Thermodynamics: A Modern Approach. Hal Tasaki and Glenn Paquette, Oxford University Press.
© Hal Tasaki and Glenn Paquette (2026). DOI: 10.1093/9780191878091.003.0006

Grouping terms by state, we obtain the following:

$$\frac{U(T'; X_1') - F[T'; X_1']}{T'} - \frac{U(T; X_1) - F[T; X_1]}{T}$$
$$= \frac{U(T'; X_2') - F[T'; X_2']}{T'} - \frac{U(T; X_2) - F[T; X_2]}{T} \ . \tag{6.4}$$

Judging from its suggestive form and universal validity, it would seem that (6.4) has a very deep meaning. With this observation, we define the quantity

$$S(T; X) := \frac{U(T; X) - F[T; X]}{T} \tag{6.5}$$

and rewrite (6.4) as

$$S(T'; X_1') - S(T; X_1) = S(T'; X_2') - S(T; X_2) \ . \tag{6.6}$$

The state function $S(T; X)$ introduced here is the *entropy*.[1,2] From (5.7) and (6.5), we obtain the following expression, relating the maximum heat and the difference between the entropies of two (arbitrary) states with the same temperature:

$$Q_{\max}(T; X_1 \to X_2) = T\{S(T; X_2) - S(T; X_1)\} \ . \tag{6.7}$$

Recall that the temperature dependence of the Helmholtz free energy is not yet determined, as it contains the unfixed temperature-dependent reference point $X_0(T)$, introduced in the definition (3.23). The entropy, as defined in (6.5), inherits this indeterminacy. We next impose a condition that entirely removes the indeterminacy in the temperature dependence of the entropy, determining the entropy up to an overall additive constant.

For simplicity, let us first consider the case of an individual system parameterized by $(T; V, N)$, for which we have $X = (V, N)$ and $X_0(T) = (v(T)N, N)$. Then, from the equality (6.6), we find that for any two states $(T; V, N)$ and $(T'; V', N)$ that can be reached from one another under a quasi-static adiabatic operation, the quantity $S(T'; V', N) - S(T; V, N)$ is (for given N) independent of V and V'. We can thus write

$$S(T'; V', N) - S(T; V, N) = g(T'; N) - g(T; N) \ , \tag{6.8}$$

for some undetermined function $g(T, N)$.[3] As seen below, we can fix the free energy in such a manner that the quantity $g(T'; N) - g(T; N)$ vanishes for all T and T'.

---

[1] When expressed as a function of the energy, U, and the collective extensive variable, X, the entropy is a complete thermodynamic function. Only in this case do we write it as $S[U, X]$, using square brackets (see Appendix F).

[2] The definition of the entropy given above is quite different from that due to Clausius, which is the definition used in most textbooks. Despite this formal difference, however, the quantities defined in these two manners are identical. A brief discussion of the relation between the above definition and that given by Clausius is presented in Section 6.3.

[3] The fact that $S(T'; V', N) - S(T; V, N)$ is independent of V and V' implies that there is a function $\tilde{g}(T', T; N)$ satisfying $S(T'; V', N) - S(T; V, N) = \tilde{g}(T', T; N)$. Then, considering a third state, $(T''; X'')$, that can be reached from $(X; T)$ and (X', T') under quasi-static adiabatic operations, and noting that the equalities $\tilde{g}(T', T; N) = \{S(T'; V', N) - S(T''; V'', N)\} - \{S(T; V, N) - S(T''; V'', N)\} = \tilde{g}(T', T''; N) - \tilde{g}(T, T''; N)$ hold for arbitrary T'', we see that $\tilde{g}(T', T; N)$ must take the form $g(T'; N) - g(T; N)$.

Recall that in Section 4.3.1, when defining the energy, we chose reference values of the temperature and collective extensive variable, T^* and $X^* = (v^*N, N)$. Let us consider the set of all states that can be reached from the reference state $(T^*; X^*)$ under quasi-static adiabatic operations. For the present purposes, we refer to this set of states as the "reference trajectory." For any state $(T; X)$ contained in this trajectory, we are free to assign any (finite) value to its entropy, $S(T; X)$, through the choice of $X_0(T)$. We make the assignment $S(T; X) = S^*(N)$ for every such state, where $S^*(N) = s^*N$ is chosen as the reference value of the entropy.[4] In other words, we stipulate that the entropy be constant along the reference trajectory and equal to the reference value. With this stipulation, we see from (6.8) that $g(T; N) - g(T'; N) = 0$ for all T and T'. This implies that the entropy is constant along *every* trajectory consisting of states that can be reached from one another under quasi-static adiabatic operations. Then, the unique value of the entropy for each such trajectory is determined by that for the reference trajectory through (6.7).

Because $U(T; V, N)$, $F[T; V, N]$, $X_0(T) = (v(T)N, N)$ and $S^*(N) = s^*N$ are all extensive quantities, it is clear that with the appropriate $v(T)$, the above construction is valid for any N, and the resulting entropy, $S(T; V, N)$, is an extensive thermodynamic function.

In order to extend the above procedure to the general case, we decompose the collective extensive variable as $X = (Y, Z)$, where Y denotes the collection of variables that (like V) can be controlled operationally, and Z denotes the collection of variables that (like N) are fixed during any operation. Then we can derive the same result as above for the general case by following nearly the same argument,[5] simply replacing the explicit forms of X, $X_0(T)$, X^* and S^*, namely, $X = (V, N)$, $X_0(T) = (v(T)N, N)$, $X^* = (v^*N, N)$ and $S^* = s^*N$, by their generalizations, $X = (Y, Z)$, $X_0(T) = (Y_0(T; Z), Z)$, $X^* = (Y^*(Z), Z)$ and $S^* = S^*(Z)$, and replacing the function $g(T; N)$ by a function $g(T; Z)$. In this case, however, in order for the resulting entropy to be extensive, we must make the additional stipulation that $Y_0(T; Z)$ and $S^*(Z)$ be extensive.

We have thus removed the indeterminacy in the temperature dependence of the entropy and thereby determined it up to the choice of the single constant S^*. This has been accomplished by realizing the invariance under quasi-static adiabatic operations that we originally intended for the entropy. With this fixing of the entropy, the indeterminacy in the Helmholtz free energy has been reduced to two undetermined overall additive terms, one being $-TS^*$ (as can be understood

---

[4] Perhaps it is most elegant to determine the value of S^* through the natural choice $X_0(T^*) = X^*$ (thereby unifying the choices of the reference values introduced with the energy and free energy). However, the resulting theory is not changed if we assign some other value to S^* and then choose all $X_0(T)$ in accordance with this value.

[5] There is one point, however, that requires somewhat more care in the general treatment. In the simple case considered above, on a given trajectory, for each value of T there corresponds a single value of V. This is the reason that for any state $(T; V, N)$ on the reference trajectory, we were able to assign any desired value to $S(T; V, N)$ through the choice of $X_0(T) = (v(T)N, N)$. But in the general case, for a given value of T, there may be multiple states $(T; Y, Z)$ (with different values of Y) on the reference trajectory, and hence we cannot assign the values of the entropy for these states independently. Choosing a value for $X_0(T) = (Y_0(T; Z), Z)$ simultaneously fixes $S(T; Y, Z)$ for each one of these states. However, we know from the generalized form of (6.8),

$$S(T'; Y', Z) - S(T; Y, Z) = g(T'; Z) - g(T; Z) \, ,$$

that for all such states $(T; Y, Z)$, the values of $S(T; Y, Z)$ realized through any choice of $X_0(T)$ are all equal.

from (6.5)), and the other being a constant corresponding to the freedom to assign any value of the energy to the reference state. With the function $X_0(T)$ determined through the above procedure, $S(T;X)$ and $F[T;X]$ become continuous functions of T and X, except in very special situations.[6]

The name *entropy* is due to Clausius, the person who introduced this concept. He believed that entropy is a fundamental physical quantity on the same level as energy, and it is thought that he deliberately gave it a similar name to emphasize this point.[7] Clausius's belief is certainly consistent with the modern view of entropy, as there is a common understanding among scientists today that entropy and energy are similarly important, similarly fundamental physical quantities.[8] With this in mind, it is worth reflecting on the fact that the concepts of entropy and energy illuminate physical phenomena from entirely different points of view.

The definition of entropy given in (6.5) is rich in suggestion. As we have discussed in preceding chapters, it is an essential characteristic of thermodynamics that there are two types of energy characterizing thermodynamic systems: the Helmholtz free energy and the internal energy. Now, from (6.5) we see that from the difference between these two types of energy, we obtain another fundamental state function.

Note that our definition of entropy employs only concepts that we have derived from the operational treatment presented in this book, and particularly, that of operationally defined work. This is one demonstration of the point made numerous times in the preceding chapters, that it is work, not heat, that plays the main role in our theory.

In the remainder of this chapter, we present a thorough treatment elucidating the meaning and significance of entropy.

6.1.2 Properties of entropy

Because both the Helmholtz free energy and the energy are extensive, additive quantities (see (3.24), (3.25), (4.22) and (4.23)), it is clear from the definition (6.5) that the entropy also possesses extensivity,

$$S(T; \lambda X) = \lambda S(T; X) \,, \tag{6.9}$$

and additivity,

$$S(T; X, Y) = S(T; X) + S(T; Y) \,. \tag{6.10}$$

In Section 6.1.1, we removed the indeterminacy in the entropy (up to an overall additive constant) by stipulating that it be invariant under quasi-static adiabatic operations. Because this property is of fundamental importance, we present it here as a formal result.

---

[6] **Advanced note**: Like the energy $U(T; V, N)$, the entropy $S(T; V, N)$ changes discontinuously as a function of T as the triple point is crossed (see Appendix D).

[7] In German, these terms are *Entropie* and *Energie*. Clausius took *Entropie* from the Greek-derived word *entropia*, roughly meaning "transformation within."

[8] However, while the concept of energy plays a major role throughout the entirety of modern physics, entropy appears only within certain contexts.

Result 6.1 (Quasi-static adiabatic operations and entropy) *For any two states $(T_1; X_1)$ and $(T_2; X_2)$ connected by realizable quasi-static adiabatic transitions $(T_1; X_1) \xleftrightarrow{\text{qa}} (T_2; X_2)$, we have the equality*

$$S(T_1; X_1) = S(T_2; X_2) \,. \tag{6.11}$$

Hence, the entropy is invariant under quasi-static adiabatic operations.

This result represents a special case of a more general and fundamental result known as the *entropy principle* (Result 6.5 (p. 117)), which is investigated in detail in the next section.

The following is the converse of Result 6.1.

Result 6.2 (Entropy and quasi-static adiabatic operations) *Consider two arbitrary values of the collective extensive variable, X_1 and X_2, that can be reached from each other operationally, and two arbitrary temperatures, T_1 and T_2. If the relation*

$$S(T_1; X_1) = S(T_2; X_2) \tag{6.12}$$

holds, then there exist quasi-static adiabatic operations that realize the transitions

$$(T_1; X_1) \xleftrightarrow{\text{qa}} (T_2; X_2) \,. \tag{6.13}$$

Before deriving the above result, we first demonstrate the following property.

Result 6.3 (Entropy as a function of temperature) *Entropy is an increasing function of temperature. More precisely, for arbitrary T and T' satisfying $T < T'$ and arbitrary X, the relation*

$$S(T; X) < S(T'; X) \tag{6.14}$$

necessarily holds. Furthermore, if the entropy $S(T; X)$ and energy $U(T; X)$ are differentiable with respect to T at given values of T and X, then they satisfy the relation

$$\frac{\partial U(T; X)}{\partial T} = T \frac{\partial S(T; X)}{\partial T} \,. \tag{6.15}$$

Derivation *Choose two arbitrary temperatures T and T' satisfying $T' > T$. Then, consider three values of the collective extensive variable, X_1, X_2 and X_3, that can all be reached from one another operationally. Given these conditions, we assume that there exist quasi-static adiabatic operations under which the transitions*

$$(T; X_2) \xleftrightarrow{\text{qa}} (T'; X_1), \quad (T; X_1) \xleftrightarrow{\text{qa}} (T'; X_3) \tag{6.16}$$

are realized.

Now, let us consider the following sequence of transitions (see Figure 6.1):

$$(T; X_1) \xrightarrow{\text{qi}} (T; X_2) \xrightarrow{\text{qa}} (T'; X_1) \xrightarrow{\text{i}'} (T; X_1) \,. \tag{6.17}$$

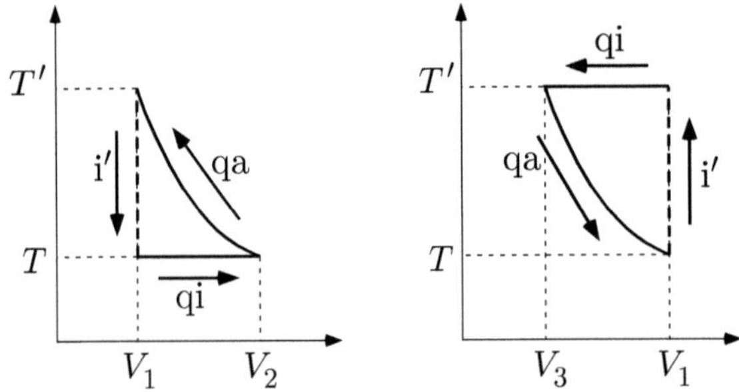

Figure 6.1 Transitions used in the investigation of the temperature dependence of the entropy. We consider an individual fluid system and three values of the collective extensive variable, $X_1 = (V_1, N)$, $X_2 = (V_2, N)$ and $X_3 = (V_3, N)$. The horizontal and vertical axes correspond to the volume and temperature, respectively.

The final transition here results from a generalized isothermal operation consisting only of the removal of the adiabatic walls that surrounded the system during the second operation (see Problem 5.2). In this operation, the system performs no work. Because the system contacts only one environment (of fixed temperature) during the entire sequence of operations, and because the initial and final states are identical, this sequence constitutes an isothermal cycle. We can express the work performed during the course of one cycle in terms of the entropy as follows:

$$
\begin{aligned}
W_{\mathrm{cyc}} &= W_{\max}(T; X_1 \to X_2) + W_{\mathrm{ad}}((T; X_2) \to (T'; X_1)) + 0 \\
&= F[T; X_1] - F[T; X_2] + U(T; X_2) - U(T'; X_1) \\
&= -T\{S(T; X_1) - S(T; X_2)\} + U(T; X_1) - U(T'; X_1) \,.
\end{aligned}
\tag{6.18}
$$

In the last step, we have rewritten $F[T; X_1]$ and $F[T; X_2]$ using the definition of the entropy, (6.5). Applying (6.11) and Kelvin's principle to this relation (and replacing X_1 with X), we obtain

$$
S(T'; X) - S(T; X) \leq \frac{U(T'; X) - U(T; X)}{T} \,.
\tag{6.19}
$$

Similarly, the work performed in the cycle

$$
(T'; X_1) \xrightarrow{\mathrm{qi}} (T'; X_3) \xrightarrow{\mathrm{qa}} (T; X_1) \xrightarrow{i'} (T'; X_1)
\tag{6.20}
$$

is

$$
W'_{\mathrm{cyc}} = -T'\{S(T'; X_1) - S(T'; X_3)\} + U(T'; X_1) - U(T; X_1) \,.
\tag{6.21}
$$

Again using (6.11) and Kelvin's principle, from this we obtain

$$
S(T'; X) - S(T; X) \geq \frac{U(T'; X) - U(T; X)}{T'} \,.
\tag{6.22}
$$

The two assertions follow readily from (6.19) and (6.22). First, applying Result 4.6 (p. 75) to (6.22), we immediately obtain $S(T'; X) - S(T; X) > 0$.

We have thus demonstrated (6.14). Next, combining (6.19) and (6.22) and introducing $\Delta T = T' - T$, we have

$$\frac{1}{T + \Delta T} \frac{U(T + \Delta T; X) - U(T; X)}{\Delta T} \leq \frac{S(T + \Delta T; X) - S(T; X)}{\Delta T}$$

$$\leq \frac{1}{T} \frac{U(T + \Delta T; X) - U(T; X)}{\Delta T} \, . \qquad (6.23)$$

Finally, assuming the differentiability of the energy and entropy with respect to T and taking the limit $\Delta T \searrow 0$, we arrive at (6.15). ∎

Derivation of Result 6.2 *With the initial state $(T_1; X_1)$, we carry out a quasi-static adiabatic operation under which the collective extensive variable is changed to X_2. This yields the transition $(T_1; X_1) \xrightarrow{\text{qa}} (T_3; X_2)$ for some temperature T_3, with the equality $S(T_3; X_2) = S(T_1; X_1)$ provided by Result 6.1. Thus, from the assumption (6.12) we obtain $S(T_3; X_2) = S(T_2; X_2)$. Combining this with (6.15), we conclude the equality $T_3 = T_2$.* ∎

6.1.3 Entropy and heat capacity

Comparing the definition of the heat capacity at constant volume (4.30) and the relation (6.15), we obtain the following:

$$C_{\mathrm{v}}(T; X) = T \frac{\partial S(T; X)}{\partial T} \, . \qquad (6.24)$$

Recall that in the derivation of (6.15), it was necessary to assume the differentiability of $S(T; X)$ and $U(T; X)$ with respect to T. If, on the other hand, these functions are not differentiable with respect to T at some point $(T; X)$, it is implied that $C_{\mathrm{v}}(T; X)$ diverges or changes discontinuously at this point. Actually, it is known that $C_{\mathrm{v}}(T; X)$ may diverge at the critical point of the liquid-gas phase transition.

Now, let us divide each side of (6.24) by T and integrate. This yields the following relation expressing the change in entropy in terms of the integral of the heat capacity at constant volume:

$$S(T; X) = S(T_0; X) + \int_{T_0}^{T} dT' \, \frac{C_{\mathrm{v}}(T'; X)}{T'} \, . \qquad (6.25)$$

In addition to the above, there exists a similar but more useful relation between the change in entropy and the heat capacity at constant pressure. This relation is given in (8.29).

As an example illustrating the use of (6.25), let us consider a system in which the collective extensive variable remains fixed at some value X_0 and the heat capacity, $C_{\mathrm{v}}(T; X_0)$, is a constant, which we write C_0. This situation represents the simplest treatment of a solid.[9] In this case, (6.25) is easily integrated, and we obtain the following temperature dependence of the entropy:

$$S(T; X_0) = S_0 + C_0 \log T \, . \tag{6.26}$$

6.1.4 Nernst-Planck postulate

In this section, we briefly discuss the Nernst-Planck postulate, which is sometimes called the *third law of thermodynamics*.

Let us reconsider (6.25). For given X, that equation allows us to determine the entropy at one temperature, T, in terms of that at another temperature, T_0. But, of course, this equation alone does not provide a specific value of $S(T; X)$, because $S(T_0; X)$ is arbitrary. However, according to the Nernst-Planck postulate, we can indeed determine a specific value of $S(T; X)$ from (6.25). This postulate establishes an absolute reference point for the entropy corresponding to the limit of vanishing temperature, asserting that for every X, $S(T; X)$ exhibits the limiting behavior $\lim_{T \searrow 0} S(T; X) = 0$. With this postulate, if we carry out experiments to measure the integral in (6.25), taking the limit $T_0 \to 0$, the value of $S(T; X)$ is determined explicitly.

Of course, confirmation of the Nernst-Planck postulate can only be accomplished through experiments on actual systems. To this time, there have been many experiments carried out to test this postulate, and its validity has indeed been confirmed by precise experimental data for various types of systems.[10]

Although the Nernst-Planck postulate is often called the *third law of thermodynamics*, it is important to realize that, unlike the first law (the law of energy conservation, stated in Postulate 4.5 (p. 69)) and the second law (as expressed by Kelvin's principle, stated in Postulate 3.1 (p. 47), the entropy principle, stated in Result 6.5 (p. 117), or Planck's principle, stated in Result 6.4 (p. 116), among other formulations), the third law does not contribute to the fundamental structure of thermodynamics. Whereas it is impossible (at least for us) to imagine a thermodynamic framework without the first or second law, the third law is unnecessary for the construction of this framework. In fact, there is evidence that casts doubt on its universal validity. A trivial counterexample is provided by the ideal gas entropy, appearing in (6.31) below. This entropy diverges to negative infinity in the $T \to 0$ limit. However, because for any actual system, behavior consistent with the ideal gas law is realized (with fixed V and N) only in the limit of large T, applying the $T \to 0$ limit to the ideal gas entropy is physically meaningless. But there do appear to be counterexamples among actual physical systems as well. For example, the existing experimental data on glass systems seem to indicate behavior inconsistent with the Nernst-Planck postulate. The problem of determining the types of systems for which this postulate holds is beyond the scope of this book.

---

[9] For many solids, this simple treatment yields results sufficiently accurate for many applications over a wide range of temperatures (including room temperature).

[10] Among such studies, there are many in which measurements of the heat capacity of a system ranging from extremely low temperatures to high temperatures are used to determine the entropy from (8.29) along with the Nernst-Planck postulate, and then this entropy is compared with that obtained from statistical mechanical computations at high temperature.

In application to chemistry, the third law can be combined with other experimental results to predict the conditions characterizing chemical equilibrium as well as the reaction direction in a system under given conditions. In many situations, such predictions are sufficiently accurate for practical applications.

6.1.5 Entropy of an ideal gas

We previously derived expressions for the free energy and energy of an ideal gas, (3.36) and (4.37). Substituting these into the definition of the entropy, (6.5), we obtain the following expression for the entropy of an ideal gas:

$$S(T; V, N) = cNR + \frac{Nu}{T} + NR \log \frac{V}{v(T)N} \, . \tag{6.27}$$

Recall that in (3.36) and (4.37), the function $v(T)$ was regarded as arbitrary. Below we determine its explicit form (up to an overall multiplicative constant) and thereby remove the indeterminacy from the above expression (up to an overall additive constant).

First, in accordance with (4.19), we choose the reference state as $(T^*; V^*, N)$, where $V^* = v^* N$, with constant v^*. Also, to simplify the discussion, let us choose the reference entropy as $S^* = cNR$. Next, we consider an arbitrary state $(T; V, N)$ for which the transitions $(T^*; V^*, N) \overset{\text{qa}}{\longleftrightarrow} (T; V, N)$ are realizable. Then from the Poisson relation (4.45), we obtain

$$V = \left(\frac{T^*}{T}\right)^c V^* = \left(\frac{T^*}{T}\right)^c v^* N \, . \tag{6.28}$$

Because $(T; V, N)$ can be reached from $(T^*; V^*, N)$ through a quasi-static adiabatic transition, we have $S(T; V, N) = S^* = cNR$. Therefore, from (6.27), we find

$$\frac{Nu}{T} + NR \log \frac{V}{v(T)N} = 0 \, . \tag{6.29}$$

Rewriting this equation slightly and using (6.28), we have

$$v(T) = \frac{V}{N} \exp\left(\frac{u}{RT}\right) = v^* \left(\frac{T^*}{T}\right)^c \exp\left(\frac{u}{RT}\right) \, . \tag{6.30}$$

The function $v(T)$ has thus been determined. Substituting this form into (6.27), the terms containing the energy constant u cancel, and we arrive at the final form of the entropy of an ideal gas:

$$S(T; V, N) = cNR + NR \log\left\{\left(\frac{T}{T^*}\right)^c \frac{V}{v^* N}\right\} \, . \tag{6.31}$$

6.2 Reversibility, irreversibility and entropy

In this section, we investigate the physical meaning of entropy, as defined in Section 6.1.1. As we discuss in Section 6.3, in the conventional treatment used in most textbooks, entropy is formulated in terms of the concept of heat. However, as demonstrated in this section, the essence of entropy becomes clearest through consideration of adiabatic operations, which involve no heat.

6.2.1 Reversibility and irreversibility in adiabatic operations

Here we formulate the concept of the reversibility/irreversibility of an adiabatic operation. The corresponding concept for isothermal operations is treated separately in Section 6.3.1.

It is not generally the case that the realizability of an adiabatic transition

$$(T_1; X_1) \xrightarrow{\text{a}} (T_2; X_2) \tag{6.32}$$

implies the realizability of the inverse adiabatic transition,

$$(T_2; X_2) \xrightarrow{\text{a}} (T_1; X_1) \ . \tag{6.33}$$

In the case that there exist operations under which both (6.32) and (6.33) are realizable, we say that these adiabatic operations are *reversible*.[11] In the case that there exists an operation that induces (6.32) but there exists no operation that induces (6.33), we say that the former operation is *irreversible*.

As noted in Section 4.1.1 and used repeatedly thereafter, quasi-static adiabatic operations can always be carried out in reverse, retracing the original path in the reverse direction. Thus, quasi-static adiabatic operations are necessarily reversible. Of course, operations in a purely mechanical system are also reversible, because for any dynamics exhibited by a mechanical system, the time-reversed dynamics are also realizable. The distinctive feature of thermodynamics is that there also exist adiabatic operations that are irreversible.

In Section 4.1.2, we examined adiabatic operations under which the extensive variables are unchanged. We then presented Postulate 4.1 (p. 66) and derived Result 4.3 (p. 67). Combining these two, we immediately obtain the following.

Result 6.4 (Planck's principle) *Consider an arbitrary thermodynamic system in an arbitrary state* $(T_1; X)$*. For any* $T_2 > T_1$*, any operation inducing the transition*

$$(T_1; X) \xrightarrow{\text{a}} (T_2; X) \tag{6.34}$$

is irreversible.

Recall that the existence of an operation of the type considered here, under which the temperature increases while the collective extensive variable is unchanged, follows from Postulate 4.1.

As noted below Result 4.3, the irreversibility of a temperature-increasing adiabatic operation of the type (6.34) is intuitively clear from our everyday experience.

---

[11] Note that reversibility does not require that the path followed by the system in the transition (6.33) be identically the direction-reversed path followed in (6.32). It is only necessary that the initial and final states of (6.32) be the final and initial states of (6.33).

Planck regarded such well-known phenomena as representing one of the essential aspects of thermodynamics and proposed it as the starting point of thermodynamic theory.[12] Although Planck is certainly best known for his contribution to quantum mechanics, as commemorated by the constant h that bears his name, he also made important contributions to the field of thermodynamics. In fact, his derivation of the equation describing blackbody radiation, which is known as Planck's law, relied on deep consideration of the thermodynamics of radiation (see Section 7.2.4).

Before moving on, let us point out that for adiabatic operations, although closely related, *reversible* and *quasi-static* are distinct characterizations, at least on a conceptual level. As discussed in Section 4.1.1, a quasi-static operation can be understood as an operation under which the system remains very close to equilibrium at all times. A quasi-static adiabatic operation is necessarily reversible. On the other hand, when we say that an adiabatic operation is reversible, we do not specify the nature of the operation. This only means that if we consider the initial and final states realized under this operation, there exists an operation under which these same states are realized, but with the roles reversed.

6.2.2 The entropy principle

The most essential aspect of entropy is that the realizability of an adiabatic transition is determined entirely by the relation between the values of the entropy for its initial and final states. We refer to this fact as the *entropy principle* (see Figure 6.2).

Result 6.5 (The entropy principle) *Consider an arbitrary thermodynamic system and two arbitrary values of its collective extensive variable, X_1 and X_2, that can be reached from each other operationally. Then, for any values of the temperature T_1 and T_2, there exists an adiabatic operation that realizes the transition*

$$(T_1; X_1) \xrightarrow{\text{a}} (T_2; X_2) \tag{6.35}$$

if and only if the inequality

$$S(T_1; X_1) \leq S(T_2; X_2) \tag{6.36}$$

holds.

Clearly, the entropy principle implies that the equality of the entropies of the initial and final states of an adiabatic operation is a necessary and sufficient condition for that operation to be reversible. The invariance of the entropy under quasi-static adiabatic operations, stated in Result 6.1 (p. 111), can be regarded as a special case of the entropy principle.

Derivation of Result 6.5 *We first demonstrate that the relation (6.36) implies the existence of an operation under which (6.35) is realized. Beginning in the*

---

[12] According to Ref. [9], this idea appears in Planck's doctoral thesis.

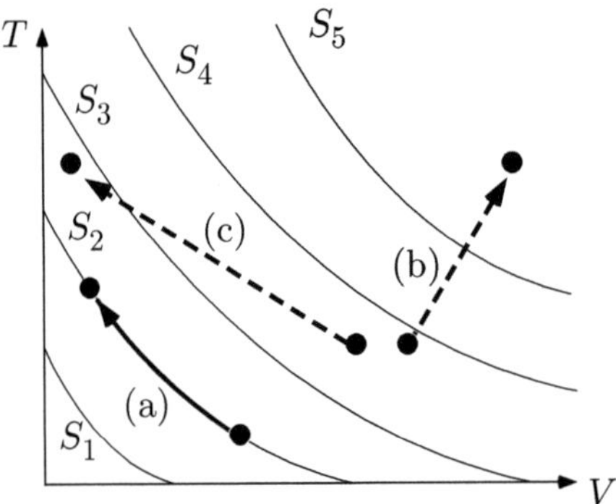

Figure 6.2 Schematic depiction of the entropy principle. The equilibrium states of an individual fluid system are plotted in the V-T plane. (The amount of substance is fixed.) Each curve in the figure consists of points representing states of equal entropy. The values of the entropy for these curves are related as $S_1 < S_2 < S_3 < S_4 < S_5$. Under a quasi-static adiabatic operation, the state of the system follows such a curve, as illustrated by (a). More generally, the entropy under an adiabatic operation can increase, as in the case of (b). This transition is plotted as a broken line to indicate that during the operation, the states of the system are not equilibrium states and hence cannot be represented by points in this plane. The type of adiabatic operation corresponding to (c), in which the entropy of the system decreases, cannot be realized in an isolated system (i.e., a system that interacts with no other system). However, if we consider two interacting systems and the possibility of one system undergoing the transition (b) and the other undergoing the transition (c), then from the additivity of entropy, the change in entropy for the composite system increases, and therefore a joint operation yielding these transitions can indeed be realized. In this sense, entropy can serve as a quantitative measure of realizability/unrealizability.

state $(T_1; X_1)$, we carry out a quasi-static adiabatic operation through which the system undergoes a transition to some state $(T_3; X_2)$. From Result 6.1, we have $S(T_3; X_2) = S(T_1; X_1)$. Therefore, the premise (6.36) implies $S(T_3; X_2) \leq S(T_2; X_2)$, and hence, from the fact that entropy is an increasing function of temperature (see Result 6.3), we conclude the relation $T_3 \leq T_2$. Thus, according to Postulate 4.1 (p. 66), there exists an adiabatic operation under which $(T_3; X_2) \xrightarrow{\text{a}} (T_2; X_2)$ is realized. Combining these operations to realize

$$(T_1; X_1) \xrightarrow{\text{qa}} (T_3; X_2) \xrightarrow{\text{a}} (T_2; X_2) , \tag{6.37}$$

we obtain the transition (6.35), as desired.

Next, we show that the existence of an operation yielding (6.35) implies the inequality (6.36). Beginning in the state $(T_1; X_1)$, we first carry out an adiabatic operation inducing the transition (6.35), and then carry out a quasi-static adiabatic operation inducing the transition $(T_2; X_2) \xrightarrow{\text{qa}} (T_4; X_1)$ for some temperature T_4. We thereby obtain the adiabatic transition $(T_1; X_1) \xrightarrow{\text{a}} (T_4; X_1)$. With Planck's principle, this implies the relation $T_1 \leq T_4$. Hence, from Result 6.3 we have $S(T_1; X_1) \leq S(T_4; X_1)$, and finally, applying Result 6.1, we arrive at (6.36). ∎

In this book, employing the universality of the maximum heat, we introduced the entropy through (6.5) as a state function that is invariant under quasi-static adiabatic operations. But we found that this quantity has a deeper meaning, as expressed by Result 6.5: Entropy provides a universally applicable means of determining whether or not a given transition is realizable under an adiabatic operation. Furthermore, the significance of Result 6.5 is not merely that it points out this useful role of entropy in the investigation of adiabatic transitions. The true importance of Result 6.5 lies in the fact that it expresses the essential characterization of the state function entropy, as entropy is the only additive and extensive state function that could satisfy this principle.[13] This important fact is demonstrated in Appendix B. Thus, the nature of entropy, which appeared somewhat mysterious upon first encounter, is found to be definitively elucidated by Result 6.5.

6.2.3 Entropy as a quantitative measure of realizability/unrealizability

As discussed above, for a single system, the values of the entropy realized in individual states determine the possibility for transitions among these states through adiabatic operations. However, if we consider combinations of multiple systems, entropy acquires a richer physical meaning, acting as a quantitative measure of realizability/unrealizability.[14]

Let us consider a system and two of its states $(T_1; X_1)$ and $(T_2; X_2)$ whose entropies are related as $S(T_1; X_1) > S(T_2; X_2)$. In this situation, it is not possible to reach the state $(T_2; X_2)$ from the state $(T_1; X_1)$ through an adiabatic operation carried out on this system in isolation. However, if we introduce a second system and consider two of its states $(T_1; Y_1)$ and $(T_2; Y_2)$ for which the relation

$$S(T_1; X_1) + S(T_1; Y_1) \leq S(T_2; X_2) + S(T_2; Y_2) \tag{6.38}$$

holds, then, from the additivity property (6.10), for the system consisting of the combination of these two systems, we have

$$S(T_1; X_1, Y_1) \leq S(T_2; X_2, Y_2) \, . \tag{6.39}$$

Therefore, because the entropy principle holds for any type of thermodynamic system, the above relation implies that there exists an adiabatic operation under which the transition

$$(T_1; X_1, Y_1) \xrightarrow{\text{a}} (T_2; X_2, Y_2) \tag{6.40}$$

is realized.[15] In rough terms, we can understand this as the unrealizability of the transition in the original system being overcome by "sufficiently strong" realizability of the transition in the second system. It is thus seen that a transition that

---

[13] Of course, any quantity that differs from the entropy by only an additive or (positive) multiplicative constant (or both) also satisfies this principle, but such a quantity can be regarded as equivalent to the entropy.

[14] The conventional expression is "quantitative measure of irreversibility," but we believe that our terminology is more accurate.

[15] Of course, we assume here that the collective extensive variables can be varied operationally from X_1 and Y_1 to X_2 and Y_2, respectively.

would be unrealizable for a system in isolation can become realizable when this system is combined with another system (see Figure 6.2).[16]

6.2.4 Examples

In this section, we consider two simple examples illustrating the entropy principle.

Adiabatic expansion of a gas into a vacuum

As an example illustrating an increase of entropy in an adiabatic operation, here we treat the expansion of a gas into a vacuum. For this purpose, we reconsider the Gay-Lussac experiment, which we first studied in Section 4.4.

Consider a container divided into two chambers, one of volume V' containing an amount N of a gas prepared with temperature T, and the other of volume $V - V'$ containing no material. The entire apparatus is enclosed within adiabatic walls. After the system reaches equilibrium, we remove the barrier separating the chambers. We thus realize a transition

$$(T; V', N) \xrightarrow{\text{a}} (T'; V, N) , \tag{6.41}$$

where the temperature T' depends on the specific properties of the system. For a transition of this type, in general, the entropies before and after are related as

$$S(T; V', N) < S(T'; V, N) . \tag{6.42}$$

Therefore, the adiabatic free expansion of a gas into a vacuum is necessarily irreversible.

Derivation of (6.42) *Suppose that after the free expansion and the system realizes the state $(T'; V, N)$, we next apply a quasi-static adiabatic operation through which the gas is compressed by a piston until the volume of the system returns to its original value, V', as depicted in Figure 6.3. This operation results in a transition*

$$(T'; V, N) \xrightarrow{\text{qa}} (T''; V', N) \tag{6.43}$$

for some temperature T''. In this operation, the mechanical world performs a positive amount of work on the system. From the law of energy conservation, we thus have $U(T; V', N) < U(T''; V', N)$ (as no work is performed in the removal of the barrier). Because energy is an increasing function of temperature, this implies the inequality $T < T''$. (Deriving an explicit expression for T'' in the case of an ideal gas is the task of Problem 1.1.) Because entropy is also an increasing function of temperature, the relation $S(T; V', N) < S(T''; V', N)$ follows. Finally, because the entropy is unchanged through the transition (6.43), we arrive at (6.42). ∎

---

[16] In the classic chemical thermodynamics textbook by Lewis and Randall [10], the concept of entropy is introduced on the basis of such considerations. Also, the definition of entropy given by Sasa [11] represents a somewhat more theoretically advanced form of this idea.

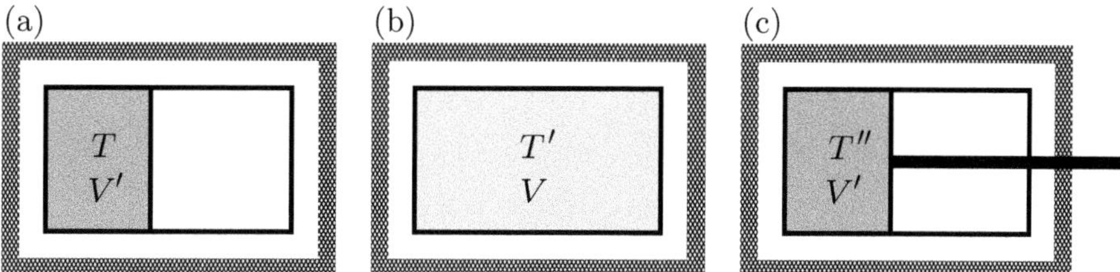

Figure 6.3 Demonstration that the adiabatic free expansion of a gas into a vacuum is irreversible. (a) The system begins in a state in which the gas is contained entirely in the left-hand chamber. (b) After the barrier between the chambers is removed, the gas expands into the right-hand side, and eventually comes to fill both sides. (c) The gas is subsequently compressed through a quasi-static adiabatic operation. At the end of this operation, the volume of the system is the same as before the barrier was removed. Because positive work is done on the system during the compression, the relation $T'' > T$ follows from the law of energy conservation. This demonstrates that entropy increases in the free expansion of a gas.

In the case of an ideal gas, an explicit result is easily obtained. First, recall that in this case, we have $T = T'$. Therefore, using the form of the entropy appearing in (6.31), for the change in entropy resulting from the transition (6.41), we obtain the following:

$$S(T; V, N) - S(T; V', N) = NR \log \left(\frac{V}{V'} \right) > 0 \,. \tag{6.44}$$

A system consisting of an ideal gas and a solid
From Result 6.3 (p. 111), we know that the entropy of a system is necessarily an increasing function of temperature. This and Result 6.5 (p. 117) together imply that for any system, if the value of the collective extensive variable is fixed, an adiabatic operation that increases the temperature is possible, but an adiabatic operation that decreases the temperature is not.[17] However, with the idea presented in the previous section of unrealizability in one system being "overcome" by realizability in another, it is possible to realize an operation through which the temperature of a system is decreased while its collective extensive variable is held fixed if we combine this system with a second system that is simultaneously undergoing an operation through which its entropy increases by a sufficient amount.

As an example, let us consider a system consisting of an ideal gas and a solid. Suppose that this is an idealized solid, whose entropy is given by (6.26), with constant heat capacity C_0. We write the extensive variables of the ideal gas as (V, N) and those of the solid as X_0, which is regarded as fixed. Then, using the expressions for the entropies of these two systems considered separately, (6.26)

---

[17] This is simply a restatement of Postulate 4.1 (p. 66) and Result 4.3 (p. 67).

and (6.31), from the additivity of the entropy, expressed by (6.10), we obtain the following form for the entropy of the combined system:

$$
\begin{aligned}
S(T; X_0, V, N) &= S(T; X_0) + S(T; V, N) \\
&= S_0 + C_0 \log T + NR \log(T^c V N^{-1}) \\
&= S_0 + NR \log \left\{ \frac{T^{c+c'} V}{N} \right\} .
\end{aligned}
\tag{6.45}
$$

Here, S_0 and c' are constants, with $c' = C_0/(NR)$. Then, we consider the possibility of performing an operation under which the volume of the ideal gas is changed from a value V_1 to a value $V_2 > V_1$ while the temperature of the combined system is changed from a value T_1 to a value $T_2 < T_1$. With values of V_1, V_2, T_1 and T_2 such that the condition $V_2 \geq (T_1/T_2)^{c+c'} V_1$ is satisfied,[18] the inequality $S(T_1; X_0, V_1, N) \leq S(T_2; X_0, V_2, N)$ holds, and hence such an operation does exist. Thus, in this case, we realize a transition

$$
(T_1; X_0, V_1, N) \xrightarrow{\ a\ } (T_2; X_0, V_2, N) ,
\tag{6.46}
$$

through which the solid experiences a temperature decrease while its extensive variables remain fixed. In particular, this is the case for a quasi-static operation, under which the equality $V_2 = (T_1/T_2)^{c+c'} V_1$ holds (see Problem 6.1).

6.3 Entropy and heat

In Section 6.1, we defined the entropy in terms of the difference between the energy and the free energy. Subsequently, in Section 6.2, we characterized the entropy as a quantitative measure of the realizability/unrealizability of adiabatic operations. As noted previously, however, this is not the conventional approach. In many textbooks, following Clausius, entropy is formulated in terms of the heat exchanged by a system and its environment. In this section, after introducing the concept of reversibility/irreversibility in an isothermal operation, we outline this traditional formulation as understood within the framework employed in this book.

6.3.1 Reversibility and irreversibility in isothermal operations

We begin our investigation of reversibility/irreversibility in isothermal systems by considering the Clausius inequality, which plays an essential role in the definition of entropy in the traditional formulation.

For an arbitrary thermodynamic system, consider an arbitrary isothermal operation that induces the transition

$$
(T; X_1) \xrightarrow{\ i\ } (T; X_2) .
\tag{6.47}
$$

---

[18] Obviously, for any given V_1, T_1 and (nonzero) $T_2 < T_1$, we can choose V_2 sufficiently large to satisfy this inequality.

Let Q denote the amount of heat that the system absorbs from the environment during this transition. Then, with the relation $Q \leq Q_{\max}(T; X_1 \to X_2)$ from the definition of the maximum heat, (6.7) implies

$$\frac{Q}{T} \leq S(T; X_2) - S(T; X_1) \,. \tag{6.48}$$

This is a special case of the Clausius inequality.[19]

We now formulate the concept of the reversibility/irreversibility of isothermal operations. In rough terms, we can think of a reversible operation—whether isothermal or adiabatic—to be an operation whose effect can be "undone" by another operation. For an adiabatic operation (see Section 6.2.1), because its only effect is the transition it causes, the reversibility of the operation is equivalent to the reversibility of this transition. For isothermal operations, however, the situation is different. As discussed in Section 3.1.1, for any transition $(T; X_1) \xrightarrow{\text{i}} (T; X_2)$ realizable under an isothermal operation, there exists an isothermal operation that realizes the inverse transition, $(T; X_2) \xrightarrow{\text{i}} (T; X_1)$. In other words, isothermal transitions are trivially reversible. But, of course, there is another effect of an isothermal operation, that which it has on the environment. We therefore define the reversibility of an isothermal operation with respect to this effect.

Definition 6.6 (Reversibility of an isothermal operation) *Consider two isothermal operations, one inducing the transition* $(T; X_1) \xrightarrow{\text{i}} (T; X_2)$*, and the other inducing the inverse transition,* $(T; X_2) \xrightarrow{\text{i}} (T; X_1)$*. Let Q and Q' denote the amounts of heat absorbed by the system from the environment in these two operations. These operations are said to be* reversible *if $Q = -Q'$ and* irreversible *if $Q \neq -Q'$.*

Every quasi-static isothermal operation is reversible, as may be intuitively clear. Formally, this follows from the sign rule of the maximum heat, (5.8), along with the fact that the heat absorbed by the system under a quasi-static isothermal operation is identically the maximum heat (see Section 5.1.3).

Next, we derive a necessary and sufficient condition for an isothermal operation to be reversible expressed in terms of the entropy. Consider two arbitrary isothermal operations yielding the transitions $(T; X_1) \xrightarrow{\text{i}} (T; X_2)$ and $(T; X_2) \xrightarrow{\text{i}} (T; X_1)$. Let Q and Q' denote the amounts of heat that the system absorbs from the environment during the former and latter transitions, respectively. For these transitions, we have the Clausius inequalities

$$\frac{Q}{T} \leq S(T; X_2) - S(T; X_1) \tag{6.49}$$

and

$$\frac{Q'}{T} \leq S(T; X_1) - S(T; X_2) \,. \tag{6.50}$$

---

[19] The inequality (6.48) was derived assuming isothermal conditions. The Clausius inequality is more general because it applies to the case of adiabatic conditions as well. In the latter case, it is simply $S(T; X_2) - S(T; X_1) \geq 0$.

It is seen that the equality

$$\frac{Q}{T} = S(T; X_2) - S(T; X_1) \qquad (6.51)$$

holds if and only if $Q' = -Q$ (i.e., if and only if the two operations are reversible). We have thus shown the following.

Result 6.7 (Reversibility of an isothermal operation) *An isothermal operation inducing a transition $(T; X_1) \xrightarrow{\mathrm{i}} (T; X_2)$ during which the system absorbs an amount of heat Q from the environment is reversible if and only if the equality (6.51) holds.*

It is illuminating to rewrite (6.51) as

$$S(T; X_1) + S_0 = S(T; X_2) + S_0 - \frac{Q}{T} . \qquad (6.52)$$

If we interpret S_0 and $S_0 - Q/T$ here as the entropies of the environment before and after the transition, respectively, then we see that (6.51) expresses the invariance of the total entropy. Of course, this confirms our expectation, because the invariance of the total entropy is the condition for the reversibility of an adiabatic operation, as follows from the entropy principle, Result 6.5 (p. 117).

6.3.2 Understanding of entropy as defined by Clausius

For an arbitrary system, consider a quasi-static isothermal operation causing a transition $(T; X_1) \xrightarrow{\mathrm{qi}} (T; X_2)$. Let $\Delta Q = Q_{\max}(T; X_1 \to X_2)$ and $\Delta S = S(T; X_2) - S(T; X_1)$ represent the heat absorbed from the environment and the change in the entropy of the system under this operation. Then, from (6.7) (or (6.51)), we have

$$\Delta S = \frac{\Delta Q}{T} . \qquad (6.53)$$

Note that the quantities ΔQ and ΔS here need not be infinitesimal. Next, let us consider a quasi-static adiabatic operation inducing a transition $(T_1; X_1) \xrightarrow{\mathrm{qa}} (T_2; X_2)$. Because the entropy is unchanged under a quasi-static adiabatic operation and because no heat is exchanged in any adiabatic operation, with analogously defined ΔS and ΔQ (and choosing T as, say, T_1), (6.53) holds in this case too, with both sides vanishing.

The relation (6.53) represents the central idea in the understanding of entropy as formulated by Clausius. The quantity ΔQ is the energy absorbed by the system from the environment in a particular operation, and thus it does not represent the change in any state function. However, this quantity divided by the temperature is equal to the difference between the values of the state function entropy in the two states connected by the operation. This meaning expressed by (6.53) constitutes one essential aspect of thermodynamics.

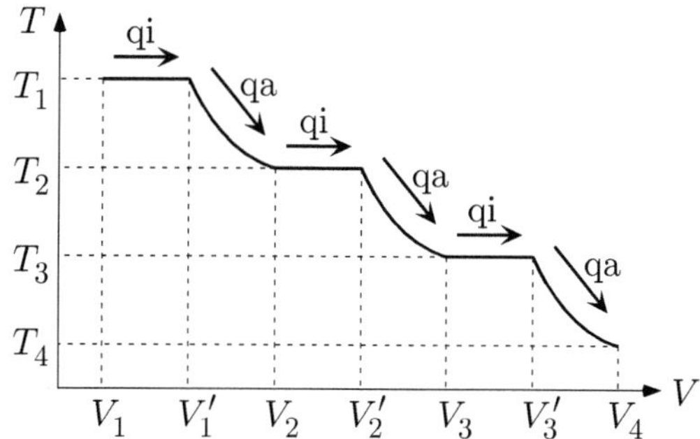

Figure 6.4 Through repeated isothermal and adiabatic operations, the system makes a transition from the state $(T_1; V_1, N)$ to the state $(T_4; V_4, N)$. The difference between the entropies in these two states can be obtained by applying (6.53) multiple times. This example demonstrates the manner of thinking on which the Clausius definition of entropy is based.

Now, let us consider a sequence of transitions resulting from repeated applications of quasi-static isothermal and quasi-static adiabatic operations carried out in turn:

$$(T_1; X_1) \xrightarrow{\text{qi}} (T_1; X_1') \xrightarrow{\text{qa}} (T_2; X_2) \xrightarrow{\text{qi}} (T_2; X_2') \xrightarrow{\text{qa}} \cdots$$

$$\cdots \xrightarrow{\text{qa}} (T_{n-1}; X_{n-1}) \xrightarrow{\text{qi}} (T_{n-1}; X_{n-1}') \xrightarrow{\text{qa}} (T_n; X_n) . \quad (6.54)$$

Together, these individual transitions form a transition from the state $(T_1; X_1)$ to the state $(T_n; X_n)$ (see Figure 6.4). The change in entropy resulting from each operation is given by (6.53). Summing these changes, we obtain the change caused by the entire sequence of operations as

$$S(T_n; X_n) - S(T_1; X_1) = \sum_{i=1}^{n-1} \frac{\Delta Q_i}{T_i} , \quad (6.55)$$

where ΔQ_i is the heat absorbed by the system during the quasi-static isothermal operation carried out at temperature T_i. In this way, we can obtain the difference between the entropies of the initial and final states. The relation (6.55) is the foundation of the Clausius definition of entropy.

6.3.3 Integral expression for the entropy

Consider the series of operations depicted in Figure 6.5(a). In this situation, making the changes undergone by T and X in each of the operations very small and the number of individual operations n very large, we obtain a smooth path in the state space like that displayed in Figure 6.5(b). In this limit, parameterizing the path followed by the system with a continuous variable τ as $(T(\tau); X(\tau))_{0 \leq \tau \leq 1}$, we obtain the continuous limit of (6.55):

$$S(T(1); X(1)) - S(T(0); X(0)) = \int_0^1 \frac{dQ(\tau)}{T(\tau)} . \quad (6.56)$$

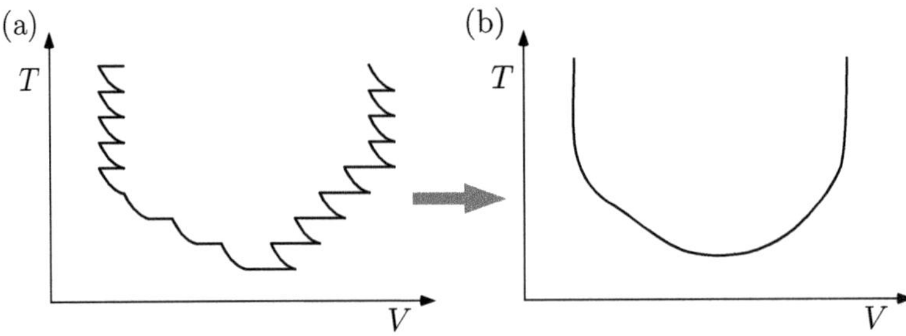

Figure 6.5 (a) A jagged path in state space resulting from repeated application of quasi-static isothermal and quasi-static adiabatic operations. (b) A smooth path obtained in the limit that the changes undergone by T and V in each operation become infinitesimal. It should be noted here that although the smooth path in (b) can be obtained in this way from alternating quasi-static isothermal and quasi-static adiabatic operations, this is just one possibility. In general, there does not exist a unique prescription for producing a given path in state space through such a limiting procedure.

This is the integral representation of the entropy. Here, $dQ(\tau)$ is the heat absorbed by the system in an infinitesimal operation.[20] In most thermodynamics textbooks, an equation of this type is used as the definition of the entropy.

The expression for the entropy appearing in (6.56) was obtained by considering the construction of a path in state space as the limit of small alternating quasi-static isothermal and quasi-static adiabatic operations, as depicted in Figure 6.5. We could also consider the reverse situation, in which we are given a path in state space like that in Figure 6.5(b) and attempt to determine the series of actual operations (perhaps expressed as a limit) from which it is constructed. However, in general, for a given path, there is not a unique way to accomplish this. In other words, the mathematical procedure through which a given path can be constructed is generally not unique. Taking the continuous limit of a series of alternating quasi-static isothermal and quasi-static adiabatic operations is just one special possibility. In general, there will exist many possible ways to construct a given path in state space from actual operations, and (6.56) will hold for only some of these. (This point is considered in Problems 6.2 and 6.3.)

Finally, let us note that although here the integral formula (6.56) was obtained as the continuous limit of (6.55), there are other ways in which it can be derived. For example, if we interpret $C_{\mathrm{v}}(T'; X)\, dT'$ in (6.25), which relates the entropy and the heat capacity at constant volume (or $C_{\mathrm{p}}(T', p; N)\, dT'$ in (8.29), which relates the entropy and the heat capacity at constant pressure), as the absorbed heat, dQ, then this equation takes precisely the form of (6.56).[21]

---

[20] More precisely, $dQ(\tau)$ should be interpreted as $Q'(\tau)d\tau$, where Q' is $dQ/d\tau$, the rate at which heat is absorbed by the system with the variation of τ.

[21] However, the equations (6.25) and (8.29) themselves cannot be derived through this kind of reasoning. (Some textbooks contain erroneous or misleading assertions related to this point.) Furthermore, (6.25) and (8.29) are exact equalities relating state functions, and thus whether the change in temperature involved in these equations results from the exchange of heat or from the performance of work is irrelevant. The validity of these equations is independent of the type of operations that may be considered in any particular situation.

6.4 Law of increasing entropy

According to the entropy principle, Result 6.5 (p. 117), whether or not it is possible for a transition between two states of a thermodynamic system to be realized through an adiabatic operation is determined entirely by the relationship between the values of the entropy in these two states: It is possible to reach a state $(T_2; X_2)$ from a state $(T_1; X_1)$ through an adiabatic operation if and only if the relation $S(T_1; X_1) \leq S(T_2; X_2)$ is satisfied. Also, carrying out experiments, one realizes that the equality $S(T_1; X_1) = S(T_2; X_2)$ represents only a limiting situation. We know that this equality holds in a quasi-static adiabatic operation, but the "quasi-static" condition is truly realized only in the physically unattainable limit of an infinitesimally slow operation. For this reason, in actual experiments, it is likely that we can only approach the equality $S(T_1; X_1) = S(T_2; X_2)$, without ever realizing it exactly. Thus, in general, the inequality

$$S(T_1; X_1) < S(T_2; X_2) \tag{6.57}$$

applies to every practically realizable adiabatic operation carried out on a thermodynamic system. In other words, applying manipulations to an adiabatic system inevitably causes its entropy to increase. This is the well-known *law of increasing entropy*.

As a special type of adiabatic operation, let us consider the removal of an interior wall separating two subsystems. The mechanical world does no work on the system in this operation, and in this sense, the change undergone by the system following the removal of the wall is more naturally regarded as a spontaneous process taking place within the system, rather than a process caused by the external world. This observation leads us to the following statement representing the law of increasing entropy within a narrow context: Any spontaneous evolution undergone by an adiabatic system is accompanied by an increase in entropy. Many textbooks consider only this narrow form of the law of increasing entropy. But the implications of the law of increasing entropy are much broader and deeper, because even if energy is added to the system as the mechanical world does work on it or if energy is extracted from the system as it does work on the mechanical world, there is no way to circumvent this law.[22]

The law of increasing entropy and the law of energy conservation are both fundamental physical laws, and they have a complementary relationship. Clausius described the essence of thermodynamics as follows: "Die Energie der Welt ist constant. Die Entropie der Welt strebt einem Maximum zu." ("The energy of the world is constant. The entropy of the world tends toward a maximum.") He seems to have used "Welt" ("world") here with a meaning of *everything*, and thus as a synonym of *universe*. Hence, the meaning of this statement is that if we regard the entire universe as a single adiabatic system, then because this system exchanges no energy with any other system, its total energy is conserved, while its entropy

---

[22] In many books, it is stated that the entropy can decrease in systems that are able to exchange energy with the external world. Although this statement is true, it is misleading, because it does not distinguish between mechanical and non-mechanical forms of energy: Although the entropy can decrease in a system that exchanges energy in the form of heat with another system, the entropy can never decrease in a system that exchanges no energy or exchanges energy only in the form of work. This is the essential meaning of the law of increasing entropy.

is continually increasing. Pursuing this idea to its logical conclusion, one would be led to the conception that with the passage of time, the entropy of the universe will increase and increase until, eventually, a complete state of thermal equilibrium—in which the entropy possesses its largest possible value—will be realized. Then, because in a state of thermal equilibrium there exists no macroscopic time evolution, the realization of such a state would seem to imply a subsequently changeless universe. For this reason, this kind of terminal state is sometimes referred to as the "heat death of the universe." Of course, viewed with a modern understanding of the universe, according to which the structure of space-time itself has been evolving since the Big Bang, this manner of thinking regarding the entropy of the universe is too simplistic. Furthermore, it is not at all clear whether the total entropy of the universe, a system that has never experienced equilibrium, can be treated and understood in the same manner as the entropy of thermodynamic systems undergoing transitions between equilibrium states. However, if we interpret the word "Welt" in Clausius's statement as referring not to the actual universe itself, but to some sufficiently isolated thermodynamic system, then this concise statement does accurately express the essence of thermodynamics.

As we have stated several times to this point, in thermodynamics, when considering an operation that connects two equilibrium states, there is generally no need to assume that the system remains in equilibrium during the performance of the operation. This is sometimes a point of confusion, and there is a common misconception that thermodynamics applies only to equilibrium states.[23] But in fact, for any operation, no matter how fast or slow, how violent or gentle, as long as the initial and final states are equilibrium states, the law of increasing entropy always holds exactly. However, it must be kept in mind that the concept of entropy itself only applies to equilibrium states. At least with the definition used in this book, entropy is not defined for non-equilibrium states, and thus it makes no sense to ask whether the entropy of a system is increasing during the time that it is evolving between equilibrium states. Discovering a thermodynamic quantity that generalizes entropy in such a manner that it can be defined even for non-equilibrium states and showing how the change in entropy that occurs in a transition between equilibrium states can be understood as the continuous variation of this quantity are topics in the field of non-equilibrium thermodynamics. Some progress has been made in this field (which is not treated in this book), but there does not yet exist a complete theoretical framework for non-equilibrium thermodynamic phenomena.

Because the law of increasing entropy has a deeply philosophical, perhaps even mysterious character, it has been a subject of discourse even outside of science. In particular, the idea that entropy is an ever-increasing quantity and that the inevitable fate of the universe is a "silent," unchanging state of absolute maximal entropy provides a kind of theoretical validation for the fatalistic ideas of ultimate destiny and loss that we all contemplate from time to time. However, because the world in which we live is essentially a non-equilibrium state, it is not possible to apply the law of increasing entropy to it in such a simple, direct manner. For this reason, conclusions derived from this kind of naive thinking should be considered with deep skepticism.

---

[23] As discussed in Section 1.2, equilibrium statistical mechanics does only apply to systems in equilibrium. This fact, along with the erroneous belief that thermodynamics can be derived entirely from statistical mechanics, may be the source of this confusion about thermodynamics.

Before ending this section, let us shift our attention from the universe to a much smaller system, Earth. For the time being ignoring its inherently non-equilibrium nature, if we attempt to treat Earth as a thermodynamic system, it clearly cannot be regarded as adiabatic. Earth continually gains energy from the sun, mainly through the absorption of infrared, visible and ultraviolet wavelengths of light, and loses energy to space, mainly through the emission of infrared wavelengths of light. Clearly, for a system of this kind, there is no reason for the law of increasing entropy to hold, since this law applies only to adiabatic systems. With this understanding, we recognize the speciousness of claims often made in popular literature that the appearance and evolution of life on Earth violate the law of increasing entropy. Such claims merely reflect a lack of elementary scientific knowledge.[24]

6.5 The entropy principle and the entropy of a compound state

To this point, our consideration of equilibrium states has been limited to states of single systems, which are characterized by a single temperature. However, it is also possible to define equilibrium states for a collection of multiple systems at different temperatures. States of this kind are referred to as *compound states*. Here, we study how the concept of entropy and the entropy principle are generalized for the treatment of compound states. In this book, consideration of compound states is limited to this section.[25]

6.5.1 Entropy of a compound state

Consider a set of n separate thermodynamic systems each surrounded by adiabatic walls, and suppose that these systems are in equilibrium states $(T_1; X_1)$, $(T_2; X_2)$, ..., $(T_n; X_n)$, where the temperatures T_i need not be the same. Now, we can imagine gathering these systems together and regarding them as collectively forming a kind of compound system. Then we can regard the state of this compound system as a new type of equilibrium state, extending the concept defined in Section 2.4.2. We express this state as $\{(T_1; X_1)|(T_2; X_2)|\cdots|(T_n; X_n)\}$ and call it a *compound state*. Here, the vertical bars indicate that the states $(T_i; X_i)$ are equilibrium states of distinct systems, separated by adiabatic walls.

Now, let us introduce the entropy of a compound state. In order to make this an additive quantity, we define the entropy of a state $\{(T_1; X_1)|(T_2; X_2)|\cdots|(T_n; X_n)\}$ as

$$S\left((T_1; X_1)|(T_2; X_2)|\cdots|(T_n; X_n)\right) = \sum_{i=1}^{n} S(T_i; X_i), \qquad (6.58)$$

where $S(T_i; X_i)$ is the entropy of the ith system as defined in Section 6.1.1. The entropy of a compound system as defined here satisfies the following generalized entropy principle.

Result 6.8 (Entropy principle for compound states) *Consider a compound system consisting of n systems described by n collective extensive variables.*

---

[24] See the caption of Figure 6.6 for related discussion.
[25] For a systematic treatment, see, for example, Ref. [9].

Then, suppose that there exist operations under which these n extensive variables with the values $(X_1, \ldots, X_n)$ and some set of ℓ extensive variables with the values $(X'_1, \ldots, X'_\ell)$ can be interchanged in either direction.[26] Also, let $T_1, \ldots, T_n, T'_1, \ldots, T'_\ell$ be arbitrary temperatures. Then there exists an adiabatic operation that realizes the transition

$$\{(T_1; X_1)| \cdots |(T_n; X_n)\} \xrightarrow{\text{a}} \{(T'_1; X'_1)| \cdots |(T'_\ell; X'_\ell)\} \tag{6.59}$$

if and only if the relation

$$S\left((T_1; X_1)| \cdots |(T_n; X_n)\right) \leq S\left((T'_1; X'_1)| \cdots |(T'_\ell; X'_\ell)\right) \tag{6.60}$$

holds.

In Section 6.2.3, it was shown how combining the additivity of the entropy with the entropy principle provides a useful method to quantify the degree of realizability/unrealizability of an adiabatic transition. With additivity and the entropy principle having now been generalized for application to states characterized by multiple temperatures, a richer physical content of the entropy principle is realized, and the usefulness of this quantity is greatly extended.

The generalized entropy principle stated above guarantees that any entropy-increasing transition is realizable (as long as the corresponding changes in the extensive variables can be carried out freely). In any given situation, devising an actual operation that realizes such a transition may be very difficult, but this is merely a "technical" problem. Doing this for the situation depicted in Figure 6.6 is the topic of Problem 6.5.

Derivation of Result 6.8 *The strategy of this derivation is to reduce the problem to one for a single system, characterized by a single temperature, in which case the result follows from Result 6.5 (p. 117).*

Consider a compound system in the compound state $\{(T_1; X_1)| \cdots |(T_n; X_n)\}$. Then, with each of the n constituent systems remaining within its adiabatic walls, we apply a quasi-static operation through which the 2nd, 3rd, ..., nth systems all undergo transitions to new states characterized by the temperature of the 1st system, T_1. Next, we replace the adiabatic walls separating the n systems with diathermal walls:

$$\{(T_1; X_1)|(T_2; X_2)| \cdots |(T_n; X_n)\} \xleftrightarrow{\text{qa}} \{(T_1; X_1)|(T_1; \widetilde{X}_2)| \cdots |(T_1; \widetilde{X}_n)\}$$

$$\xleftrightarrow{\text{qa}} (T_1; X_1, \widetilde{X}_2, \ldots, \widetilde{X}_n) \, . \tag{6.61}$$

In this way, we obtain a single composite system characterized by a single temperature. From (6.10), (6.11) and (6.58), it is seen that the operations under which the transitions in (6.61) are realized do not alter the total entropy of the

---

[26] Here, ℓ is understood as an arbitrary positive integer, which need not be the same as n. The reason for this is that the operation considered here may include the insertion or removal of any number of adiabatic walls.

system. Similarly, we can realize quasi-static adiabatic operations through which the compound system undergoes the transition

$$\{(T_1'; X_1')|(T_2'; X_2')|\cdots|(T_\ell'; X_\ell')\} \xleftrightarrow{\text{qa}} (T_1'; X_1', \widetilde{X}_2', \ldots, \widetilde{X}_\ell') , \qquad (6.62)$$

and the entropy is unchanged under these operations as well.

Next, note that the realizability of an adiabatic operation through which the single composite system makes a transition from $(T_1; X_1, \widetilde{X}_2, \ldots, \widetilde{X}_n)$ to $(T_1'; X_1', \widetilde{X}_2', \ldots, \widetilde{X}_\ell')$ is determined entirely by the entropies of these two states, as expressed by Result 6.5. Finally, applying the reversibility of the operations corresponding to (6.61) and (6.62), we arrive at the desired result. ∎

6.5.2 Entropy increase resulting from thermal contact

As an example illustrating the entropy principle for a compound system, here we study the phenomenon of thermal contact.

Let us consider two systems under adiabatic conditions that are identical except that they possess different temperatures. Suppose that these systems are composed of an idealized solid, and hence that their heat capacities are independent of temperature (see (6.26)). Initially the two systems are in equilibrium states $(T_1; X_0)$ and $(T_2; X_0)$, where $T_1 \neq T_2$, and separated by an adiabatic wall. We then establish thermal contact between the systems by replacing this partitioning wall with a diathermal wall. According to the "zeroth law of thermodynamics," Result 2.6 (p. 42), this composite system will subsequently settle into an equilibrium state in which the two individual systems have the same temperature. (See Problem 5.1 for the interpretation of this transition in terms of the "flow of heat.") This is an adiabatic operation under which the system does no work on the mechanical world. We write the resulting transition as

$$\{(T_1; X_0)|(T_2; X_0)\} \xrightarrow{\text{a}} (T_{\text{f}}; X_0, X_0) , \qquad (6.63)$$

where, because the heat capacity is constant, the final temperature is found to be $T_{\text{f}} = (T_1 + T_2)/2$ from the law of energy conservation. Then, using the additivity of the entropy, expressed by (6.58), and the expression for the entropy of a solid given in (6.26), we can evaluate the total change in entropy resulting from the transition (6.63) as

$$2S(T_{\text{f}}; X_0) - \{S(T_1; X_0) + S(T_2; X_0)\} = 2C_0 \log \frac{T_1 + T_2}{2} - C_0 \log T_1 - C_0 \log T_2$$

$$= C_0 \log \frac{(T_1 + T_2)^2}{4T_1 T_2} > 0 . \qquad (6.64)$$

We conclude that the adiabatic operation considered here is irreversible. (The general case is treated in Problem 6.4.)

Finally, we briefly illustrate how entropy realizes its full usefulness as a measure of the realizability/unrealizability of an adiabatic transition in the context of compound states. For this purpose, we note that the meaning of the irreversibility found above is that after the state $(T_{\text{f}}; X_0, X_0)$ has been realized, it is not possible to re-establish the initial states, $(T_1; X_0)$ and $(T_2; X_0)$, through an adiabatic

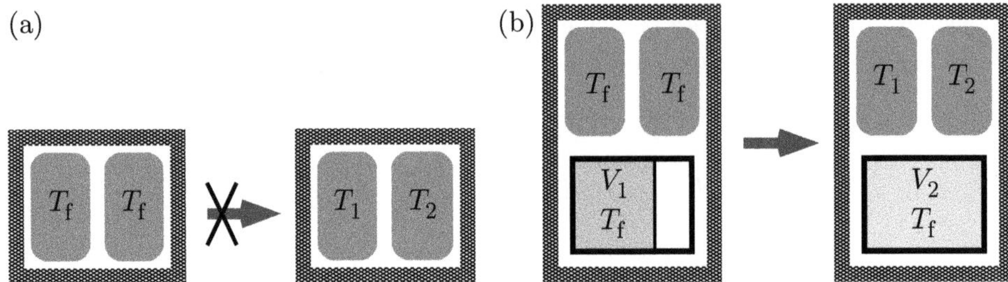

Figure 6.6 Example of the "cancellation of unrealizability." (a) The compound state $\{(T_f; X_0)|(T_f; X_0)\}$, with $T_f = (T_1 + T_2)/2$, has a greater entropy than the compound state $\{(T_1; X_0)|(T_2; X_0)\}$. Therefore, in accordance with the entropy principle, it is not possible to carry out an adiabatic operation through which this compound system makes a transition from the former to the latter. (b) However, if we introduce an additional system, then, again in accordance with the entropy principle, it is possible to realize an operation of this kind if it is combined with an operation on the additional system through which its entropy is increased by a sufficiently large amount. Such a situation is illustrated here, with the additional system being a gas system that undergoes expansion. (Designing an actual dual operation of this kind is the topic of Problem 6.5.) In this case, the increase in entropy of the gas system more than offsets the decrease in entropy of the compound solid system.

It is natural to regard the compound state in which the two solid systems possess different temperatures as having more "structure" than the state in which they have the same temperature. Comparing states that differ only with regard to the amount of structure that they possess, it is generally the case that the state with more structure will have a smaller entropy. Therefore, it is generally not possible to realize isolated adiabatic operations through which structure emerges in a system in which none existed previously. However, when two systems are combined, structure can emerge in one system under an adiabatic operation if the entropy of the other system increases sufficiently under the same operation. This fact has a deep connection to the many types of ordering processes that take place in nature. This illustrates how the true power of the entropy principle becomes apparent when it is applied to the study of multiple interacting systems.

operation, as long as these two systems interact with no other system. However, it is possible to do this if such an entropy-decreasing operation is carried out in conjunction with an entropy-increasing operation applied to another system. This situation is illustrated in Figure 6.6. That figure depicts a dual operation in which the two solid systems are combined with a gas system, and the latter undergoes an operation through which it expands. (This dual operation is also considered in Problem 6.5.)

Problems 6

6.1 (Section 6.2.4) Consider the transition (6.46) for the system consisting of a solid and an ideal gas studied there. Show explicitly that in the case of a quasi-static adiabatic operation, the relation $V' = (T/T')^{c+c'}V$ holds, and hence that this transition is realizable.

Hint: Follow the derivation of the Poisson relation given in Section 4.4.

6.2 (Sections 6.3.2 and 6.3.3) Give an example of a series of operations under which the state of the system changes along a "smooth path" in state space but for which the integral expression for the change in entropy given in (6.56) does not hold.

Hint: Consider the limit of infinitely many infinitesimally small repeated operations.

6.3 (Sections 6.3.2 and 6.3.3) Consider the series of generalized isothermal transitions $(T; X) = (T_1; X_1) \xrightarrow{\text{i}'} (T_2; X_2) \xrightarrow{\text{i}'} \cdots \xrightarrow{\text{i}'} (T_n; X_n) = (T'; X')$ (see Problem 5.2). Let us refer to the work done by the system on the mechanical world in the operation giving rise to the transition $(T_i; X_i) \xrightarrow{\text{i}'} (T_{i+1}; X_{i+1})$ as ΔW_i. Regarding n to be a large number, suppose that the quantities $|T_{i+1} - T_i|$, $|X_{i+1} - X_i|$ and ΔW_i are $O(n^{-1})$ for all i. We assume that there exists an operation through which the inverse transition $(T_{i+1}; X_{i+1}) \xrightarrow{\text{i}'} (T_i; X_i)$ is realized and that in this operation an amount of work $-\Delta W_i + O(n^{-2})$ is performed by the system.[27] Taking the limit $n \nearrow \infty$, the transitions undergone by the system in the individual operations come to form a smooth curve in state space connecting $(T; X)$ and $(T'; X')$. Show that in this limit, the integral expression for the change in entropy (6.56) holds for the series of operations as a whole. (Note that (6.25) and (8.29), which relate the heat capacity and entropy, are special cases of (6.56).)

Hint: Following Problem 5.2, evaluate the amounts of heat transferred in the transitions $(T_i; X_i) \xrightarrow{\text{i}'} (T_{i+1}; X_{i+1})$ and $(T_{i+1}; X_{i+1}) \xrightarrow{\text{i}'} (T_i; X_i)$.

6.4 (Section 6.5.2) For the type of adiabatic operation formalized in Problem 5.1 (under which two systems are placed into thermal contact), show that the entropy necessarily increases. (You may assume that the entropy is differentiable with respect to the temperature.)

6.5 (Section 6.5) Give a detailed explanation of the example considered in Figure 6.6 for the case in which the gas system consists of an ideal gas. In particular, for the operation depicted in Figure 6.6(b), with given V_1 and V_2, design a quasi-static adiabatic operation under which the total entropy is unchanged. (It is necessary to break up the operation into several steps.)

Hint: The operation constructed in Problem 6.1 is useful.

---

[27] For slow operations during which the pressure is defined at all times, this condition is nearly always satisfied.

7

Helmholtz Free Energy and Variational Principles

We introduced the Helmholtz free energy in Chapter 3, and we obtained its final form, including its temperature dependence, after the entropy was introduced, in Chapter 6. In this chapter, we carry out a detailed investigation of the Helmholtz free energy in this completed form. As we have seen in previous chapters, the Helmholtz free energy plays a fundamental role in the treatment of systems under isothermal conditions, representing the potential for a system to perform work through continuous extensive operations under such conditions. It is noteworthy that in this context, the external agent is always able to choose the state of the system, $(T; X)$. This is intimately connected to the fact that when regarded as a function of $(T; X)$, the Helmholtz free energy is a complete thermodynamic function, from which all other state functions can be derived. This special status of $F[T; X]$ is demonstrated explicitly in this chapter. We also study the Maxwell relations and related formulas, which embody the quantitative universality of thermodynamic theory. Next, we give a detailed treatment of variational principles, which play an essential role in thermodynamics, and derive balance conditions. As an application of the balance conditions, we study phase transitions and the coexistence of phases for a pure substance.

7.1 Deriving state functions from the Helmholtz free energy

In this section, we discuss the status of the Helmholtz free energy as a complete thermodynamic function and demonstrate how various state functions can be derived from it.

7.1.1 Helmholtz free energy as a complete thermodynamic function

As mentioned in Section 3.6, a complete thermodynamic function is a function from which all thermodynamically meaningful state functions can be derived.[1] The Helmholtz free energy $F[T; X]$ is one such function. In this section, we illustrate this point by showing how several important state functions, including the energy $U(T; X)$, the entropy $S(T; X)$, and the pressure $p(T; V, N)$, can be derived from $F[T; X]$.

Because the distinction between complete thermodynamic functions and other state functions is of fundamental importance, in any investigation it is necessary

---

[1] We can understand this capability of a complete thermodynamic function to result from the fact that it incorporates all information regarding the work performed in adiabatic operations and quasi-static isothermal operations.

Thermodynamics: A Modern Approach. Hal Tasaki and Glenn Paquette, Oxford University Press.
© Hal Tasaki and Glenn Paquette (2026). DOI: 10.1093/9780191878091.003.0007

to keep clear the status of each state function under consideration. For example, $U(T; X)$, $S(T; X)$ and $p(T; V, N)$ are not complete thermodynamic functions, as is clear from the fact that $F[T; X]$ cannot be derived from any of these functions.[2] The importance of clearly distinguishing between the two types of state functions was fully recognized by Gibbs,[3] but unfortunately, very few thermodynamics textbooks treat the concept of complete thermodynamic functions. In this book, complete thermodynamic functions are distinguished from other state functions by the use of square brackets to enclose their arguments.[4] A detailed treatment of complete thermodynamic functions is given in Appendix F.

7.1.2 Energy, entropy and heat capacity

We begin by rewriting the definition of the entropy, (6.5), to express the Helmholtz free energy as[5]

$$F[T; X] = U(T; X) - T\,S(T; X) .\tag{7.1}$$

We assume that $U(T; X)$ and $S(T; X)$ are differentiable with respect to T. Then, differentiating (7.1) with respect to T and using (6.15), we obtain

$$\frac{\partial F[T; X]}{\partial T} = \frac{\partial U(T; X)}{\partial T} - S(T; X) - T\frac{\partial S(T; X)}{\partial T}$$
$$= -S(T; X) .\tag{7.2}$$

It is thus shown that if we know the Helmholtz free energy, we can obtain the entropy. Further, using (7.1) and (7.2), we find

$$U(T; X) = F[T; X] + T\,S(T; X) = F[T; X] - T\frac{\partial F[T; X]}{\partial T}$$
$$= -T^2\frac{\partial}{\partial T}\left\{\frac{F[T; X]}{T}\right\} .\tag{7.3}$$

Thus, if we know the Helmholtz free energy, we can also obtain the energy.[6] The equation (7.3) is one member of a family of exact relations among thermodynamic

---

[2] As seen from (7.2) and (7.3), the T dependence of $F[T; X]$ can be determined from either $U(T; X)$ or $S(T; X)$, while as seen from (3.31), the V dependence of $F[T; V, N]$ can be determined from $p(T; V, N)$. However, the V dependence of $F[T; V, N]$ cannot be determined from either $U(T; V, N)$ or $S(T; V, N)$ alone, while the T dependence of $F[T; V, N]$ cannot be determined from $p(T; V, N)$ alone.

[3] Gibbs referred to equations expressing the functional dependence of complete thermodynamic functions as *fundamental equations*.

[4] It is important to understand that the status of a state function as a complete thermodynamic function depends on its arguments. For example, if we regard F as a function of T, p and N, it is not a complete thermodynamic function, and thus in this case, we express it as $F(T, p; N)$. (See Problem 8.4 for investigation of this point.) Conversely, U becomes a complete thermodynamic function when we consider it as a function of S, V and N, and S becomes a complete thermodynamic function when we consider it as a function of U, V and N.

[5] In many textbooks, (7.1) defines the Helmholtz free energy. Recall that we define this quantity (other than its temperature dependence) in terms of the maximum work, and use (7.1) as the definition of the entropy.

[6] The relations (7.2) and (7.3) were derived assuming the differentiability of $U(T; X)$ and $S(T; X)$ with respect to T. However, in the results, only the functions $U(T; X)$ and $S(T; X)$ appear, not their derivatives. Because $U(T; X)$ and $S(T; X)$ are continuous functions of T, these equations imply that $F[T; X]$ is always differentiable with respect to T. In fact, (7.2) and (7.3) hold even at critical points

functions known as the *Gibbs-Helmholtz equations*. That we can obtain the energy and the entropy from the Helmholtz free energy is of fundamental importance within the theory of thermodynamics, as discussed below.[7]

Next, substituting (7.2) into the expression for the heat capacity at constant volume given in (6.24), we obtain the following relation expressing it in terms of the Helmholtz free energy:

$$C_{\mathrm{v}}(T; X) = T\frac{\partial S(T; X)}{\partial T} = -T\frac{\partial^2 F[T; X]}{\partial T^2} . \tag{7.4}$$

7.1.3 Pressure and chemical potential

Deriving $p(T; V, N)$ and $\mu(T; V, N)$ from $F[T; V, N]$

For the remainder of this section, we consider an individual system parameterized by $(T; V, N)$ as the physical system under investigation. In this case, as seen in (3.31), the pressure is given by the derivative of the free energy with respect to the volume:

$$p(T; V, N) = -\frac{\partial F[T; V, N]}{\partial V} . \tag{7.5}$$

The right-hand side of this equation can be understood as representing the "response" experienced by an external agent when altering the volume of the system by a small amount.

We can also consider variation of the free energy with respect to the amount of substance. In this case, in analogy to (7.5), we have[8]

$$\mu(T; V, N) := \frac{\partial F[T; V, N]}{\partial N} . \tag{7.6}$$

It should be noted, however, that we cannot attribute any operational meaning to the quantity $\mu(T; V, N)$, because there is no operation under which the amount of substance is changed for a single-component system.[9] However, using an analogy to (7.5), we can think of $\mu(T; V, N)$ as representing the response that we would feel if we were to alter N by a small amount. The state function $\mu(T; V, N)$ is termed the *chemical potential*.[10] In (3.32), we demonstrated from the extensivity of the free energy that the pressure is an intensive quantity. From the same type of reasoning, it follows that the chemical potential is also an intensive quantity.

The full usefulness of the chemical potential defined here becomes clear when we treat multi-component systems in Chapter 9.

where $U(T; X)$ and $S(T; X)$ are not differentiable with respect to T. (**Advanced note**: At the triple point, however, $U(T; X)$ and $S(T; X)$ exhibit discontinuous change as functions of T, and hence $F[T; X]$ is not differentiable with respect to T at this point. A detailed treatment of this and related points is given in Appendix D.)

[7] In a more systematic treatment, the energy and entropy are obtained from the Helmholtz free energy through Legendre transformations (see Appendices F and H).

[8] The free energy $F[T; V, N]$ is differentiable with respect to V. From this fact, using (7.8), we conclude that $F[T; V, N]$ is also differentiable with respect to N.

[9] In Chapter 9, we consider operations on multi-component systems under which amounts of substance change through chemical reactions, but even under operations of that type, it is not possible to change the amount of substance of just a single species.

[10] This may seem to be a somewhat unfitting name. Its meaning is discussed in Footnote 15 on p. 176.

Euler equation for the Helmholtz free energy

Here, using the extensivity of the Helmholtz free energy, we derive an equation expressing it in terms of the pressure and chemical potential.

The extensivity of the Helmholtz free energy (see (3.24)) is expressed by

$$\lambda F[T; V, N] = F[T; \lambda V, \lambda N] , \tag{7.7}$$

which holds for arbitrary $\lambda > 0$. Let us take the derivative of both sides of this equation with respect to λ, and then set $\lambda = 1$. Then, using (7.5) and (7.6), we obtain

$$F[T; V, N] = V \frac{\partial F[T; V, N]}{\partial V} + N \frac{\partial F[T; V, N]}{\partial N}$$
$$= -V p(T; V, N) + N \mu(T; V, N) . \tag{7.8}$$

This equation corresponds to a particular case of Euler's general theory of homogeneous equations, and for this reason, it is referred to as an *Euler equation*.

As discussed previously, the Helmholtz free energy $F[T; V, N]$ and the pressure $p(T; V, N)$ can be determined from information obtained by measuring the work performed in quasi-static isothermal and quasi-static adiabatic operations. From (7.8), it is seen that the chemical potential $\mu(T; V, N)$ can therefore also be determined operationally.

Chemical potential of an ideal gas

In Section 3.7.2, we derived the expression (3.36) as the Helmholtz free energy of a single-component ideal gas. However, at that point, the function $v(T)$ was undetermined. Then, in Section 6.1.5, we derived the form (6.30) for this function through consideration of the entropy. Substituting this form for $v(T)$ into (3.36), we obtain

$$F[T; V, N] = -NRT \log \left\{ \left(\frac{T}{T^*} \right)^c \frac{V}{v^* N} \right\} + Nu . \tag{7.9}$$

This is the final form of the Helmholtz free energy for a single-component ideal gas. The only arbitrary values remaining here are T^* and v^*, which are determined by the reference point, and u, which is determined by the reference value of the energy.

With (7.9), we can derive the entropy using (7.2) and the energy using (7.3). Of course, the results obtained in this way are identical to the previously obtained results, (6.31) and (4.37). Finally, using (7.9) and (7.6), we find that the chemical potential for a single-component ideal gas is given by

$$\mu(T; V, N) = RT - RT \log \left\{ \left(\frac{T}{T^*} \right)^c \frac{V}{v^* N} \right\} + u . \tag{7.10}$$

7.1.4 Differential forms

From (7.2), (7.5) (or (3.31)) and (7.6), we have seen how differentiation of the Helmholtz free energy of an individual system yields the following expressions for the entropy, pressure and chemical potential:

$$\frac{\partial F[T; V, N]}{\partial T} = -S(T; V, N) \,, \qquad \frac{\partial F[T; V, N]}{\partial V} = -p(T; V, N) \,,$$

$$\frac{\partial F[T; V, N]}{\partial N} = \mu(T; V, N) \,. \tag{7.11}$$

Here, recall that $F[T; V, N]$ is essentially always differentiable with respect to T, V and N, even at critical points.[11] All of the relations in (7.11) can be expressed together in what is called a *differential form* as follows:

$$dF = -S\,dT - p\,dV + \mu\,dN \,. \tag{7.12}$$

It has become customary to express many of the relations in thermodynamics in such differential forms.

Although it is not necessary to use differential forms in thermodynamics,[12] they are quite convenient for the purpose of organizing (and memorizing) derivatives of complete thermodynamic functions and for computing Legendre transforms (see Section 8.2). In this book, we use differential forms in this manner, as merely a convenient tool, without carrying out a detailed mathematical treatment.[13] Below, we present the minimal explanation necessary to understand (7.12).

Among the variables determining the state of a system, T, V and N, let us first consider changing V by a small amount, ΔV. Writing the resulting change in $F[T; V, N]$ as $\Delta F = F[T; V + \Delta V, N] - F[T; V, N]$, from (7.11) we obtain

$$\Delta F = -p(T; V, N)\,\Delta V + O((\Delta V)^2) \,. \tag{7.13}$$

This relation connects small variations of the independent variable V with small variations of the dependent variable F in the particular case that T and N are held fixed. We can similarly derive analogous relations in the case that T is varied, with V and N held fixed, and in the case that N is varied, with V and T held fixed.

Now, in the case that the independent variables T, V and N are all varied (by ΔT, ΔV and ΔN),[14] the change in $F[T; V, N]$ is given by $\Delta F = F[T + \Delta T; V + \Delta V, N + \Delta N] - F[T; V, N]$. With (7.11), this expression can be written as

$$\Delta F = -S(T; V, N)\,\Delta T - p(T; V, N)\,\Delta V + \mu(T; V, N)\,\Delta N + O(\Delta^2) \,. \tag{7.14}$$

Here, $O(\Delta^2)$ represents a term of second order in ΔT, ΔV, ΔN, or any combination thereof. This equation describes the response of $F[T; V, N]$ to all possible small variations of its independent variables. Considering the case of infinitesimally small ΔT, ΔV, ΔN and ΔF (and rewriting these as dT, dV, dN and dF to express this meaning), we can ignore the second-order correction term in (7.14), and we obtain (7.12). In essence, the equations (7.12) and (7.14) express the entire behavior of $F[T; V, N]$, in contrast to relations like (7.13), which express the variation

---

[11] **Advanced note**: At the triple point, it is not differentiable with respect to T (see Appendix D).

[12] The axiomatic formulation of thermodynamics constructed by Carathéodory is perhaps an exception.

[13] For a comprehensive treatment of the mathematics of differential forms, see, for example, Ref. [4].

[14] These need not be regarded as changes imparted by physical operations. We are presently considering the situation from a strictly mathematical point of view and investigating how the function $F[T; V, N]$ changes if we alter its independent variables. Then, this information can be used, for example, to compare the free energies of different systems or different realizations of the same system.

in $F[T; V, N]$ with respect to the variation of only one independent variable. In (7.12), it is understood that T, V and N represent the complete set of independent variables of F, and with this understanding, there is no need to explicitly express F with its arguments, as we usually do in this book. In general, dF in this type of expression is referred to as the *total differential* of F.

We now give some general discussion concerning differential forms. Let f and g be arbitrary differentiable functions of T, V and N. Then, for the variation of the product of these functions with respect to small variations of T, V and N, we have

$$\begin{aligned}
\Delta(fg) &= f[T + \Delta T; V + \Delta V, N + \Delta N]\, g[T + \Delta T; V + \Delta V, N + \Delta N] \\
&\quad - f[T; V, N]\, g[T; V, N] \\
&= \{f[T + \Delta T; V + \Delta V, N + \Delta N] - f[T; V, N]\}\, g[T + \Delta T; V + \Delta V, N + \Delta N] \\
&\quad + f[T; V, N]\, \{g[T + \Delta T; V + \Delta V, N + \Delta N] - g[T; V, N]\} \\
&= g[T; V, N]\, \Delta f + f[T; V, N]\, \Delta g + O(\Delta^2)\,,
\end{aligned} \tag{7.15}$$

where we have used the fact that $g[T + \Delta T; V + \Delta V, N + \Delta N] = g[T; V, N] + O(\Delta)$. Taking the limit of infinitesimal variations and writing this in the form of (7.12), we obtain the following useful formula for the total differential of a product of functions:

$$d(fg) = g\, df + f\, dg\,. \tag{7.16}$$

Also, from the linearity of infinitesimal variations, for arbitrary constants α and β, we have

$$d(\alpha f + \beta g) = \alpha\, df + \beta\, dg\,. \tag{7.17}$$

With the basics of differential forms thus in hand, for the time being we can regard (7.12) as simply a convenient form for expressing the independent variables and derivatives of the Helmholtz free energy. Specifically, from the fact that dT, dV and dN appear on the right-hand side of the total differential of F, we know that we are regarding F as a function of the independent variables T, V and N. Also, from the terms appearing on the right-hand side, we can read off the partial derivatives of F with respect to each of these variables. For example, from the term $-S\, dT$, we infer the relation $\partial F/\partial T = -S$.

7.2 The Maxwell relations and their application

In this section, we derive several relations among state functions using derivatives of the Helmholtz free energy. Before proceeding, however, we wish to point out the significance of the universal nature of the relations derived here. It is remarkable that from only the general framework of thermodynamics, we are able to derive concrete, exact relations among thermodynamic quantities that hold rigorously for any thermodynamic system. Although some care is necessary when applying these equations to systems at phase transition points, in all other situations, they can be applied both universally and straightforwardly.

7.2.1 Maxwell relations

Let us consider an arbitrary individual system parameterized by $(T; V, N)$. We assume that $F[T; V, N]$ is twice continuously differentiable at some state $(T; V, N)$. This is generally true for any thermodynamic system, except at phase transition points. Then, because the order of multiple derivatives has no influence on the resulting function, for example, we have

$$\frac{\partial^2 F[T; V, N]}{\partial V \partial T} = \frac{\partial^2 F[T; V, N]}{\partial T \partial V} \, . \tag{7.18}$$

Applying (7.11) to this equation, we obtain the following relation:

$$\frac{\partial S(T; V, N)}{\partial V} = \frac{\partial p(T; V, N)}{\partial T} \, . \tag{7.19}$$

We have thus obtained an equation relating two quantities with no previously apparent connection, the rate of change of the entropy with respect to the volume and the rate of change of the pressure with respect to the temperature. Equation (7.19) is one of a collection of equations known as the *Maxwell relations*. Next, taking the derivatives of $F[T; V, N]$ with respect to T and N and with respect to V and N, and using the corresponding equalities akin to (7.18), we derive the other two Maxwell relations that can be obtained from the Helmholtz free energy:

$$\frac{\partial S(T; V, N)}{\partial N} = -\frac{\partial \mu(T; V, N)}{\partial T} \, , \quad \frac{\partial p(T; V, N)}{\partial N} = -\frac{\partial \mu(T; V, N)}{\partial V} \, . \tag{7.20}$$

7.2.2 The energy equation

Above, we saw how the Maxwell relations describe nonobvious connections among state functions. This is not the only reason that they are of interest, however. When combined with other relations, they also yield equations of great practical use. We now consider one such case.

First, let us rewrite (7.1) as $U(T; V, N) = T S(T; V, N) + F[T; V, N]$. Then, taking the derivative of both sides of this equation with respect to V and using (7.11) and the Maxwell relation (7.19), we obtain the following:

$$\frac{\partial U(T; V, N)}{\partial V} = T \frac{\partial S(T; V, N)}{\partial V} + \frac{\partial F[T; V, N]}{\partial V}$$
$$= T \frac{\partial p(T; V, N)}{\partial T} - p(T; V, N) \, . \tag{7.21}$$

This relation, which connects the volume dependence of the energy and the temperature dependence of the pressure, is known as the *energy equation*.

Below we consider the application of (7.21) to two very different types of systems.[15]

---

[15] For another application of the energy equation, see Problem 7.3.

7.2.3 The energy equation applied to ideal gases

First let us apply (7.21) to an ideal gas. As we know from the ideal gas equation of state, (3.35), the pressure of an ideal gas is proportional to its temperature. Hence the two terms on the right-hand side of (7.21) exactly cancel, and we have

$$\frac{\partial U(T; V, N)}{\partial V} = 0 \,. \tag{7.22}$$

In Section 4.4, we postulated that the energy of an ideal gas is independent of its volume on the basis of experimental results. With the above result, we now realize that this in fact follows from the general theoretical framework of thermodynamics along with the ideal gas equation of state.[16]

7.2.4 The energy equation applied to electromagnetic fields

Next, as another application of the energy equation, we treat the equilibrium state of an electromagnetic field in vacuum. We consider a container of volume V enclosing a perfect vacuum, and we suppose that this container is in thermal contact with an environment of temperature T. Within such a container, there will exist an electromagnetic field. To this point in the book, we have implicitly assumed that a vacuum can exchange no energy with the external world and hence that it plays no role in thermodynamics. However, electromagnetic fields do have the capacity to exchange energy with the external world, in the forms of both work and heat, and hence, in principle, they too must be included in the thermodynamic description. Indeed, when the temperature is sufficiently high, the role of electromagnetic fields in thermodynamic phenomena cannot be ignored.

We regard the degrees of freedom of an electromagnetic field as a thermodynamic system, and we carry out a thermodynamic analysis to elucidate the behavior of the electromagnetic field within the container.[17] We assume that the container is completely opaque to all wavelengths of radiation. In this case, after it is closed, the system will eventually reach an equilibrium state in which the container behaves as a *blackbody*. This is a body that emits electromagnetic radiation with a characteristic frequency spectrum that depends only on temperature, being entirely independent of the material composing the body. This phenomenon is known as *blackbody radiation*.

For a system consisting of an electromagnetic field, there is no conservation of particle number, and for this reason, there is just one independent extensive variable characterizing the system, the volume. We therefore represent an equilibrium

---

[16] It is important to realize that there is no circular logic here. In other words, (7.22) does not result from any assumption about the behavior of ideal gases that is somehow "built into" the general theory. Recall that, as discussed in Section 3.7, within the theoretical framework used in this book, the concept of an ideal gas is used only to set the temperature scale. It is not used in any way in the construction of this theoretical framework itself. The relation (7.22) is a statement about the nature of an ideal gas, as provided by this framework. This is independent of the temperature scale that we may employ. (With a different choice of the temperature scale, the equations from which we derive (7.22) would take different forms, but (7.22) itself must still hold.)

[17] This is the only instance within this book in which we consider the degrees of freedom of an electromagnetic field.

state by $(T; V)$. Thus, because the energy of the system is an extensive variable, we can write it in terms of the energy density, $u(T)$, as

$$U(T; V) = u(T)V .\tag{7.23}$$

Maxwell's equations of electromagnetism imply that the pressure and energy density of isotropic radiation are related by[18]

$$p(T; V) = \frac{u(T)}{3} .\tag{7.24}$$

Substituting (7.23) and (7.24) into (7.21) and simplifying, we obtain

$$\frac{\partial u(T)}{\partial T} = \frac{4u(T)}{T} .\tag{7.25}$$

This yields the form

$$u(T) = (\text{constant}) \times T^4 .\tag{7.26}$$

This relation is referred to as the *Stefan-Boltzmann law*. The value of the constant here cannot be determined from thermodynamic considerations alone.[19]

Note how simply we were able to derive the above relation using the theory of thermodynamics. This contrasts with the great trouble that was encountered when people first attempted to derive it using statistical mechanics. It is the resolution of this problem that famously led to the development of quantum mechanics.

7.3 Variational principles and the direction of spontaneous change

Variational principles provide the most important, elegant and useful point of view for thermodynamics. Employing variational principles, we are able to systematically determine balance conditions and the direction of spontaneous change in thermodynamic systems. In this section, we investigate the variational principle for the Helmholtz free energy.

7.3.1 The variational inequality for the Helmholtz free energy

For some fixed volume V and amount of substance N, choose arbitrary (positive) V_1, V_2, N_1 and N_2 satisfying the relations $V_1 + V_2 = V$ and $N_1 + N_2 = N$. Then, consider a fluid system under isothermal conditions consisting of two subsystems in the equilibrium states $(T; V_1, N_1)$ and $(T; V_2, N_2)$ that are separated by a thin wall. We

---

[18] It can be shown that for a plane electromagnetic wave incident on a perfectly reflecting wall, the radiation pressure is given by $p = u\cos^2(\theta)$, where θ is the angle of incidence and u is the energy density of the total field consisting of the incident and reflected waves. (We treat the simplest case with $\theta = 0$ in Problem 7.4.) Averaging this over all solid angles, we obtain $p = u/3$ in the case of isotropic radiation.

[19] Applying the theory of quantum statistical mechanics, this constant is computed as $(\pi^2 k_{\mathrm{B}}^4)/(15c^3\hbar^3) \simeq 7.57 \times 10^{-16}\,\mathrm{J\,m^{-3}K^{-4}}$, where k_{B} is the Boltzmann constant, c is the speed of light in vacuum, and $\hbar$ is the Planck constant divided by 2π.

represent the state of this composite system by $(T; (V_1, N_1), (V_2, N_2))$. Next, suppose that, maintaining thermal contact between the system and the environment, we remove the wall separating the subsystems. We thus obtain an unpartitioned system of volume V and amount of substance N, and if we wait a sufficiently long time, this system will reach the equilibrium state $(T; V, N)$. In this way, we have realized an isothermal operation causing the transition

$$(T; (V_1, N_1), (V_2, N_2)) \xrightarrow{\text{i}} (T; V, N) \ . \tag{7.27}$$

In general, this operation will not be quasi-static. Also, because it consists only of removing the partition between the two original subsystems, no work is done by the system while it is being carried out. Thus, from the definition of the maximum work, we have

$$W_{\max}(T; \{(V_1, N_1), (V_2, N_2)\} \to (V, N)) \geq 0 \ . \tag{7.28}$$

With the relation between the maximum work and the Helmholtz free energy, given in (3.27), the above can be rewritten as

$$F[T; (V_1, N_1), (V_2, N_2)] - F[T; V, N] \geq 0 \ . \tag{7.29}$$

Combining this inequality with the additivity of the Helmholtz free energy, expressed by (3.25), we find the following:

$$F[T; V, N] \leq F[T; V_1, N_1] + F[T; V_2, N_2] \ . \tag{7.30}$$

This inequality holds for any values of V, V_1, V_2, N, N_1 and N_2 satisfying the equalities $V_1 + V_2 = V$ and $N_1 + N_2 = N$. The above relation is termed the *variational inequality* for the Helmholtz free energy.

7.3.2 Variational principles

We now discuss the application of the inequality (7.30). Here again, we consider the state $(T; (V_1, N_1), (V_2, N_2))$ of a system under isothermal conditions consisting of two subsystems separated by a thin wall, with arbitrary V, V_1, V_2, N, N_1 and N_2 satisfying $V_1 + V_2 = V$ and $N_1 + N_2 = N$. Below we investigate two situations in which the wall undergoes two kinds of changes (see Figure 7.1).

First we consider the case in which a small hole is made in the wall, and as a result, fluid can flow between the two compartments. In this case, the volumes of the two compartments remain fixed, and the amounts of substance change in such a manner that the system approaches a new equilibrium state. Let us write the amounts of substance in the two compartments in this new equilibrium state as $\widetilde{N}_1$ and $\widetilde{N}_2$. Then, patching the hole, we have the state $\{T; (V_1, \widetilde{N}_1), (V_2, \widetilde{N}_2)\}$.[20] In this state, if we were to remove the wall, there would be no change in the system. This implies the equality $F[T; (V_1, \widetilde{N}_1), (V_2, \widetilde{N}_2)] = F[T; V, N]$. Applying to this

---

[20] While the procedures of creating and patching the hole are clearly not operations, they are equivalent to the operations of removing and replacing the wall entirely.

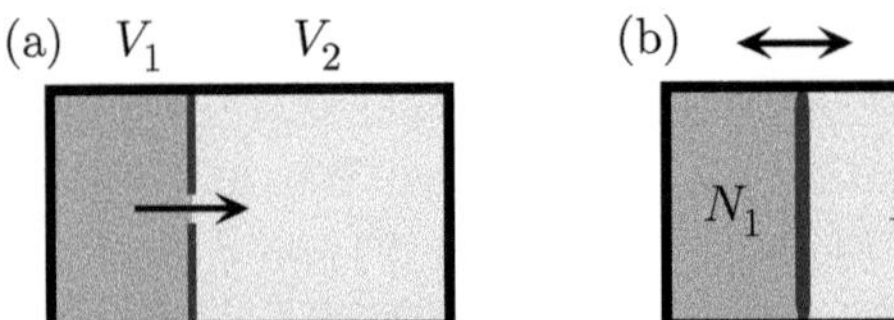

Figure 7.1 Two situations that illustrate the physical meaning of variational principles. (a) A hole is opened in the wall, allowing fluid to move between the two compartments. Hence, with V_1 and V_2 fixed, the distribution of the amounts of substance will evolve toward that which minimizes the right-hand side of (7.31), eventually realizing equilibrium with $\widetilde{N}_1$ in the left compartment and $\widetilde{N}_2$ in the right compartment. (b) The wall is allowed to move. In this case, the amounts of substance in the compartments remain fixed at N_1 and N_2, while the position of the wall moves by such an amount to minimize the right-hand side of (7.32). This yields an equilibrium state in which the left compartment has a volume $\widetilde{V}_1$ and the right compartment has a volume $\widetilde{V}_2$. In either case, a change in the nature of the wall results in a change in the properties of the system, as it undergoes a transition to a new equilibrium state that minimizes the Helmholtz free energy under the new conditions.

the additivity of the Helmholtz free energy, we obtain $F[T; V_1, \widetilde{N}_1] + F[T; V_2, \widetilde{N}_2] = F[T; V, N]$. Combining this with the inequality (7.30), we arrive at the following variational principle regarding the transport of substance:

$$F[T; V_1, \widetilde{N}_1] + F[T; V_2, \widetilde{N}_2] = \min_{\substack{N_1, N_2 \\ (N_1 + N_2 = N)}} \left\{ F[T; V_1, N_1] + F[T; V_2, N_2] \right\}. \quad (7.31)$$

The "min" operator on the right-hand side represents the procedure by which, maintaining the condition $N_1 + N_2 = N$, we vary N_1 and N_2 to find the values at which the sum of the two free energies is minimized.[21]

Next, we consider the case in which the wall itself remains impermeable, but it is no longer fixed in place and can move laterally. In this case, the amounts of substance remain fixed at N_1 and N_2, but the volumes of the two subsystems will generally change. Under these conditions, proceeding analogously to the case considered above, we obtain the following variational principle with regard to the movement of the wall:

$$F[T; \widetilde{V}_1, N_1] + F[T; \widetilde{V}_2, N_2] = \min_{\substack{V_1, V_2 \\ (V_1 + V_2 = V)}} \left\{ F[T; V_1, N_1] + F[T; V_2, N_2] \right\}. \quad (7.32)$$

Here, $\widetilde{V}_1$ and $\widetilde{V}_2$ are the volumes of the two compartments in the new equilibrium state.

In the situations described by both the variational principle (7.31), regarding changes in the amounts of substance, and the variational principle (7.32), regarding changes in the volumes, the system evolves in such a manner that the Helmholtz free energy for the total system decreases. In either case, the final state of the system is that in which the Helmholtz free energy realizes its minimum value,

---

[21] Obviously, this minimum is realized with $N_1 = \widetilde{N}_1$ and $N_2 = \widetilde{N}_2$, but there may be another pair of values of N_1 and N_2 at which it is also realized. In fact, at a phase transition point, there are infinitely many such pairs. The same is true for $\widetilde{V}_1$ and $\widetilde{V}_2$ in (7.32).

subject to the new conditions. The results found here can be generalized as follows: For an isothermal system consisting of two subsystems separated by a wall, if the conditions of the wall are changed, and as a result a new extensive degree of freedom emerges, the system will change spontaneously in such a manner that it eventually realizes a new equilibrium state minimizing the free energy of the total system under its new conditions.[22] This is a statement specific to the presently considered situation of an important law that in fact holds much more generally.

7.3.3 Convexity of the Helmholtz free energy

We next present a mathematical result that is intimately related to variational principles.

Using the inequality (7.30) and the extensivity of the Helmholtz free energy, for arbitrary (positive) V_1, V_2, N_1 and N_2 and λ satisfying $0 \leq \lambda \leq 1$, we obtain the following inequality:

$$F[T; \lambda V_1 + (1 - \lambda)V_2, \lambda N_1 + (1 - \lambda)N_2] \leq \lambda F[T; V_1, N_1] + (1 - \lambda)F[T; V_2, N_2] .$$
$$(7.33)$$

This inequality expresses the fact that $F[T; V, N]$ is *convex* (sometimes called *convex downward*) with respect to V and N (see Figure 7.2). The convexity of thermodynamic functions is an essential property, fundamental to the framework of thermodynamic theory. A detailed mathematical treatment of convex functions is given in Appendix G.

The following important result can be derived from the convexity of the Helmholtz free energy.

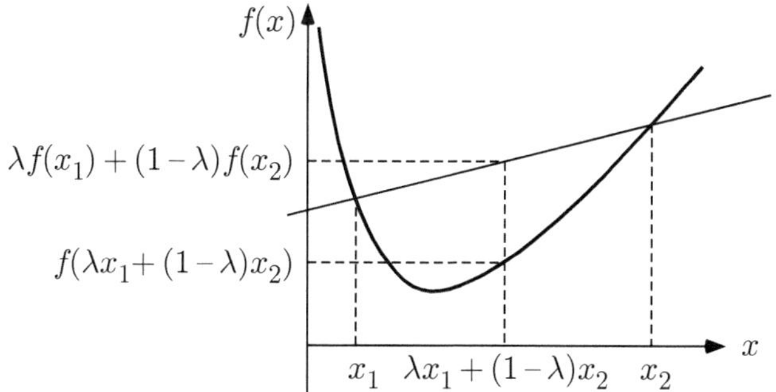

Figure 7.2 A function $f(x)$ for which the relation $f(\lambda x_1 + (1 - \lambda)x_2) \leq \lambda f(x_1) + (1 - \lambda)f(x_2)$ holds for arbitrary x_1 and x_2 and for λ satisfying $0 \leq \lambda \leq 1$ is said to be a *convex* function of x. The defining property of a convex function can be understood from the figure. There, it is seen that the line connecting the points $(x_1, f(x_1))$ and $(x_2, f(x_2))$ is above $f(x)$ for all values of x satisfying $x_1 < x < x_2$. The above definition implies that the derivative of a convex function is a non-decreasing function over any interval on which it is defined. (For detailed discussion, see Appendix G.)

---

[22] Of course, within (equilibrium) thermodynamics, we are only able to study the original and final equilibrium states and their relation. Describing the evolution that occurs between these states is beyond the scope of this theory.

Result 7.1 (Dependence of the pressure on the volume) *The pressure* $p(T;V,N)$ *is a non-increasing function of* V.

As stated in the caption of Figure 7.2, for a convex function $f(x)$, the derivative is a non-decreasing function on any interval over which it is defined. The pressure $p(T;V,N)$ is by definition the opposite of the derivative of $F[T;V,N]$ with respect to V, and hence it is seen that Result 7.1 represents a special case of this general property of convex functions. In Appendix G, we prove a general theorem in this regard (Theorem G.5), but in order to obtain an understanding of the role played by the convexity of thermodynamic functions, here we give a complete proof for the presently considered case.[23]

Derivation of Result 7.1 *In the following derivation, we assume only that* $F[T;V,N]$ *is once differentiable with respect to* V*. (Recall that this assumption is valid even at phase transition points.)*

Choose the quantities V*,* V' *and* ε *in accordance with the relations* $0 < V < V + \varepsilon < V' < V' + \varepsilon$*. Then, setting* $\lambda = (V' - V)/(V' - V + \varepsilon)$*, from the convexity inequality (7.33) we obtain*

$$F[T;V',N] = F[T;\lambda(V'+\varepsilon) + (1-\lambda)V, \lambda N + (1-\lambda)N]$$
$$\leq \lambda F[T;V'+\varepsilon,N] + (1-\lambda)F[T;V,N] \,. \tag{7.34}$$

In a similar manner, we find

$$F[T;V+\varepsilon,N] = F[T;(1-\lambda)(V'+\varepsilon) + \lambda V, (1-\lambda)N + \lambda N]$$
$$\leq (1-\lambda)F[T;V'+\varepsilon,N] + \lambda F[T;V,N] \,. \tag{7.35}$$

Adding these two equations yields

$$F[T;V+\varepsilon,N] + F[T;V',N] \leq F[T;V'+\varepsilon,N] + F[T;V,N] \,. \tag{7.36}$$

Combining this inequality with the definition of the pressure, we have

$$p(T;V',N) - p(T;V,N)$$
$$= \lim_{\varepsilon \searrow 0}\left\{ -\frac{F[T;V'+\varepsilon,N] - F[T;V',N]}{\varepsilon} + \frac{F[T;V+\varepsilon,N] - F[T;V,N]}{\varepsilon} \right\}$$
$$= \lim_{\varepsilon \searrow 0}\frac{(F[T;V+\varepsilon,N] + F[T;V',N]) - (F[T;V'+\varepsilon,N] + F[T;V,N])}{\varepsilon} \leq 0 \,,$$
$$\tag{7.37}$$

which holds for arbitrary V *and* V' *satisfying* $V' > V$*.* ∎

The fact that the pressure is a non-increasing function of the volume has a deep connection to the stability of thermodynamic systems. To understand this, let us suppose that $p(T;V,N)$ were an increasing function of V. Then, consider a system like that in Figure 3.3, consisting of a fluid contained in a vessel whose

---

[23] With the same approach, it can also be shown that the chemical potential $\mu(T;V,N)$ is a non-decreasing function of N.

volume is controlled by a piston on which the external world exerts a force $\mathcal{F}$, exactly cancelling the force exerted by the fluid. Writing the cross-sectional area of the piston as A, we thus have $\mathcal{F} = p(T; V, N)A$. Now, suppose that we carry out an operation through which the volume of the fluid increases slightly to $V + \Delta V$. After this operation, the force exerted by the fluid on the piston is $p(T; V + \Delta V, N)A$. By assumption, this is larger than the original force, $\mathcal{F} = p(T; V, N)A$. For this reason, the imbalance of the forces will be such that the volume will increase further. In this way, the fluid will spontaneously continue to expand with ever-growing explosiveness. It is thus seen that, in this counterfactual scenario, the original force balance was unstable.

As we discuss in detail in the next section, variational principles with respect to extensive variables guarantee many kinds of stability in thermodynamic systems.[24]

7.4 Balance conditions

Consider a fluid system under isothermal conditions in an equilibrium state $(T; V, N)$. Suppose that a partition is gently inserted into this system, creating two subsystems in the states $(T; \widetilde{V}_1, \widetilde{N}_1)$ and $(T; \widetilde{V}_2, \widetilde{N}_2)$. Because the original system was in equilibrium before the partition was inserted, subsequently (gently) removing it and reuniting the two subsystems will have no observable effect. This reflects the fact that the two subsystems in the states $(T; \widetilde{V}_1, \widetilde{N}_1)$ and $(T; \widetilde{V}_2, \widetilde{N}_2)$ are *balanced*.[25] In this section, we investigate the general conditions necessary for thermodynamic balance.

Let us study the situation described above in some detail. Using additivity (and the fact that inserting the partition imparts no measurable change), we have $F[T; V, N] = F[T; (\widetilde{V}_1, \widetilde{N}_1), (\widetilde{V}_2, \widetilde{N}_2)] = F[T; \widetilde{V}_1, \widetilde{N}_1] + F[T; \widetilde{V}_2, \widetilde{N}_2]$. Substituting this into the variational inequality (7.30), we obtain the following form of the variational principle, which is applicable to situations in which the volumes and amounts of substance are both varied:

$$F[T; \widetilde{V}_1, \widetilde{N}_1] + F[T; \widetilde{V}_2, \widetilde{N}_2] = \min_{v,n} \left\{ F[T; \widetilde{V}_1 + v, \widetilde{N}_1 + n] + F[T; \widetilde{V}_2 - v, \widetilde{N}_2 - n] \right\} .$$

$$(7.38)$$

In evaluating the minimum here, it is understood that v and n can take any values as long as the two volumes and amounts of substance are positive. Obviously, the minimum is realized at $v = 0$, $n = 0$, and hence we have

$$\frac{\partial}{\partial v} \left(F[T; \widetilde{V}_1 + v, \widetilde{N}_1 + n] + F[T; \widetilde{V}_2 - v, \widetilde{N}_2 - n] \right) \Bigg|_{v=0, n=0} = 0 \qquad (7.39)$$

and

$$\frac{\partial}{\partial n} \left(F[T; \widetilde{V}_1 + v, \widetilde{N}_1 + n] + F[T; \widetilde{V}_2 - v, \widetilde{N}_2 - n] \right) \Bigg|_{v=0, n=0} = 0 . \qquad (7.40)$$

---

[24] The Le Chatelier-Braun principle, which is closely related to this fact, is treated in Problem 7.7.

[25] In this and the following section, we use tildes to indicate that the volumes $\widetilde{V}_1$ and $\widetilde{V}_2$ and amounts of substance $\widetilde{N}_1$ and $\widetilde{N}_2$ characterize two systems that are balanced.

Next, rewriting the derivatives of the Helmholtz free energy above in terms of the pressure and chemical potential, in accordance with (3.31) and (7.6), from the above equations we obtain

$$p(T; \widetilde{V}_1, \widetilde{N}_1) = p(T; \widetilde{V}_2, \widetilde{N}_2) \, , \quad \mu(T; \widetilde{V}_1, \widetilde{N}_1) = \mu(T; \widetilde{V}_2, \widetilde{N}_2) \, . \tag{7.41}$$

Thus, we find that in the presently considered situation, the equality of the pressures and chemical potentials is a necessary condition for the balance of the subsystems.

It should be understood here that the condition expressed by (7.38)—that the sum of the two free energies be minimal with respect to variation of v and n—is a *global* condition, while the condition expressed by (7.39) and (7.40)—that this sum realize a critical value—is merely a *local* condition. It may thus seem that these latter conditions could yield not only the desired solution, corresponding to the global minimum, but also other solutions, corresponding to other types of critical points. However, the following powerful result allows us to dismiss this possibility.

Result 7.2 (Balance conditions) *The local balance expressed by (7.41) is a necessary and sufficient condition for the global balance expressed by (7.38).*

This result follows from the fact that the function of v and n acting as the operand of the min operator in (7.38) (below referred to as $f(v, n)$) is a convex function of v and n. As can be understood from the graph in Figure 7.2, a convex function can have at most one critical point, and this is a minimum. This fact is proven in Appendix G in the general case (see Theorems G.6 and G.10), but below we give a complete proof for the present case, because it is instructive.

Derivation of Result 7.2 *The following is a mathematically rigorous proof of the above result. The only assumption employed is that $F[T; V, N]$ is once differentiable with respect to both V and N. (Therefore, the proof holds even in the case of phase transitions.)*

It is obvious that local balance is a necessary condition for global balance, and thus we only demonstrate that it is also a sufficient condition. First, let us introduce the following function:

$$f(v, n) = F[T; \widetilde{V}_1 + v, \widetilde{N}_1 + n] + F[T; \widetilde{V}_2 - v, \widetilde{N}_2 - n] \, . \tag{7.42}$$

We wish to show that the equalities

$$\left. \frac{\partial f(v, 0)}{\partial v} \right|_{v=0} = 0 \, , \quad \left. \frac{\partial f(0, n)}{\partial n} \right|_{n=0} = 0 \tag{7.43}$$

imply the inequality $f(v, n) \geq f(0, 0)$ for arbitrary v and n.

From the convexity inequality (7.33), we know that $f(v, n)$ is a convex function of both v and n. Hence, the relation

$$f(\lambda v + (1 - \lambda)\,0, \lambda n + (1 - \lambda)\,0) \leq \lambda\, f(v, n) + (1 - \lambda)\, f(0, 0) \qquad (7.44)$$

holds. Through some straightforward manipulations, this expression can be rewritten as follows:

$$f(v, n) \geq f(0, 0) + \frac{f(\lambda v, \lambda n) - f(0, 0)}{\lambda}\,. \qquad (7.45)$$

Next, note that by virtue of (7.43), in the limit $\lambda \to 0$, we have

$$\frac{f(\lambda v, \lambda n) - f(0, 0)}{\lambda} \to v\, \left.\frac{\partial f(v, 0)}{\partial v}\right|_{v=0} + n\, \left.\frac{\partial f(0, n)}{\partial n}\right|_{n=0} = 0\,. \qquad (7.46)$$

Because (7.45) holds for arbitrary $\lambda \in [0, 1]$, it must also hold in this limit. We thus obtain the relation $f(v, n) \geq f(0, 0)$. ∎

The fact that this kind of local balance condition implies global stability is a remarkable property of thermodynamic systems. (The stability discussed near the end of Section 7.3 is a manifestation of the same property.) Let us compare thermodynamic systems with mechanical systems in this regard. In the case of mechanical systems, there are both unstable balanced states (e.g., billiard balls balanced one upon another) and locally stable but globally unstable balanced states (e.g., a ball sitting in a small depression at the top of a hill). Within thermodynamics, however, such states do not exist.[26] The non-existence of unstable balanced states is a simplifying aspect of the theory, allowing powerful conclusions to be obtained with great ease in many situations. However, it can also be said that it introduces a limitation on the applicability of thermodynamics. It is known that actual thermodynamic systems can exhibit *metastable states*, as in the case of supersaturated solutions and supercooled liquids. Although metastable states are not actual equilibrium states, a system in a metastable state will remain there indefinitely unless it is subject to some kind of external perturbation, and in this sense, they do possess a kind of stability. Intuitively, metastable states can be regarded as locally stable but globally unstable. We are thus led to conclude that the equivalence of local and global stability within the theory of thermodynamics implies that the theoretical framework of thermodynamics (at least in its ordinary form presented in this book) cannot be applied to metastable states.[27]

---

[26] Despite their non-existence within the theory of thermodynamics, systems that exhibit unstable balance can be constructed by properly combining thermodynamic and mechanical systems. This point is considered in Problem 7.6.

[27] There are many works in which a quantity referred to as a "free energy" that possesses one or more local minima in addition to the global minimum is used to describe a thermodynamic system. In such works, generally, the global and local minima of this "free energy" are identified as the equilibrium state and metastable states, respectively. However, theories of this kind are not based on a firm theoretical foundation. As far as the authors are aware, among works of this kind, there are none that give serious consideration to defining the "free energy" or the quantities acting as its independent variables. Metastable states are quite common in nature, and thus the fact that we are not yet able to properly treat them theoretically is frustrating. However, the authors believe that formulating a proper treatment of metastable states is an extremely difficult problem. For related discussion, see Footnotes 11, 23 and 24 on pp. 245, 257 and 258. Also, see Problem 7.6.

7.5 Phase transitions and the coexistence of phases

7.5.1 Introduction

In this section, we treat the fascinating topic of phase transitions. This treatment is limited to the case of single-component individual systems parameterized by $(T; V, N)$. But even simple systems of this kind can sometimes exhibit very complicated phase transitions and phase structures. A notable example is water, whose phase diagram is indeed quite complicated.

Within the regime of sufficiently high temperatures and sufficiently low pressures, all substances exhibit transitions between gas and liquid phases as well as states characterized by the coexistence of gas and liquid phases (see Figure 7.3). We study such transitions and states in this section. Before doing so, however, we present a brief discussion explaining the concepts of the *phase* of a substance and *phase equilibrium* from a general point of view.

We can understand a system to be in a single phase when all macroscopic local regions are identical in the sense that if we were to isolate any two regions of the same size from the rest of the system (and from each other) by introducing walls, all macroscopically measurable properties of these two regions would be the same. For example, water at room temperature and atmospheric pressure exists in a single phase. Contrastingly, a mixture of water and oil does not typically exist in a single phase. For such a system, through the introduction of walls, we could isolate parts of the system in different phases, one consisting mainly of water with a small amount of oil diffused into it, and one consisting mainly of oil with a small amount of water diffused into it. This kind of coexistence of phases is also possible in the case of a single substance. For example, at a temperature of $0°C$ and a pressure of 1 atm, liquid and solid water can coexist.

In general, an equilibrium state that is realized as a state of balance with respect to the transfer of matter among subsystems is said to be in *phase equilibrium*.[28] In

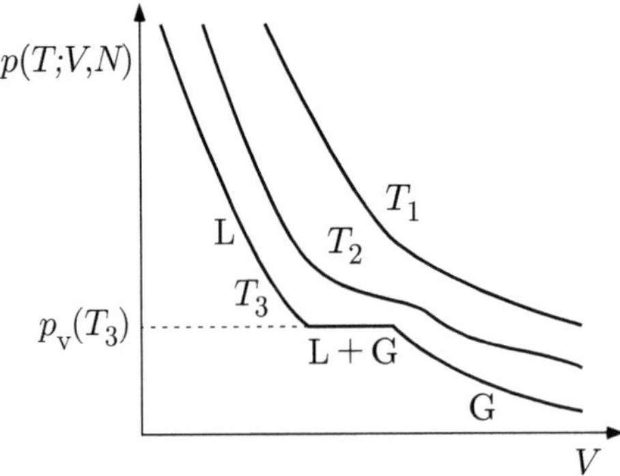

Figure 7.3 A rough depiction of the volume dependence of the pressure for a real fluid at three temperatures. (The amount of substance has the same value in all cases.) The temperatures satisfy $T_1 > T_2 > T_3$. In the figure, "G" indicates the gas phase, "L" indicates the liquid phase, and "L + G" indicates the coexistence of the two phases.

---

[28] This should be compared with *chemical equilibrium*, which is realized as a state of balance with respect to chemical reactions (see Sections 9.4 and 9.5).

this section, we treat phase equilibrium for systems consisting of a single substance. Phase equilibrium in systems consisting of multiple substances is treated Chapter 9.

7.5.2 Transition between gas and liquid phases

In Figure 7.3, the pressure, $p(T; V, N)$, is plotted as a function of the volume for three values of the temperature, related as $T_1 > T_2 > T_3$. (In all of the figures presented in this section, N is regarded as fixed.) It is seen that the dependence of the pressure on the volume at the highest temperature, T_1, is similar to that for an ideal gas, $p \propto 1/V$. However, at $T = T_2$, the functional form of the pressure differs significantly from that of an ideal gas, and in fact there are points at which its curvature changes sign. At the lowest temperature, T_3, there is a range of values of V for which the pressure is independent of the volume, remaining fixed at the value $p_{\mathrm{v}}(T_3)$ throughout. The pressure $p_{\mathrm{v}}(T)$ is referred to as the *saturated vapor pressure* or, simply, *vapor pressure* at temperature T. This distinctive behavior of the pressure reflects the occurrence of a phase transition: The fluid is a liquid to the left of the constant pressure region and a gas to the right. In the constant pressure region itself, the gas and liquid phases coexist. As the volume is increased within this region, the proportion of the fluid existing in the gas phase increases, while the pressure remains constant.

Now, as a simple introduction to our treatment of phase equilibrium, we study the volume dependence of the pressure, Helmholtz free energy, and chemical potential on the basis of the behavior described by Figure 7.3.

First we consider a temperature at which $p(T; V, N)$ is a decreasing function of V, as in the cases $T = T_1$ and $T = T_2$ in Figure 7.3. With this behavior of $p(T; V, N)$, from (3.33) we know that $F[T; V, N]$ is a decreasing function of V and $\partial F[T; V, N]/\partial V$ is an increasing function of V, while from (7.8) we know that $\mu(T; V, N)$ is a decreasing function of V. This behavior is depicted in Figure 7.4.

Next, let us consider the situation corresponding to the case $T = T_3$ in Figure 7.3, in which the fluid can exist in both the gas and liquid phases. As seen in Figure 7.5, in the region $V_{\mathrm{L}}(T; N) \leq V \leq V_{\mathrm{G}}(T; N)$, the pressure is fixed at the value $p_{\mathrm{v}}(T)$.[29]

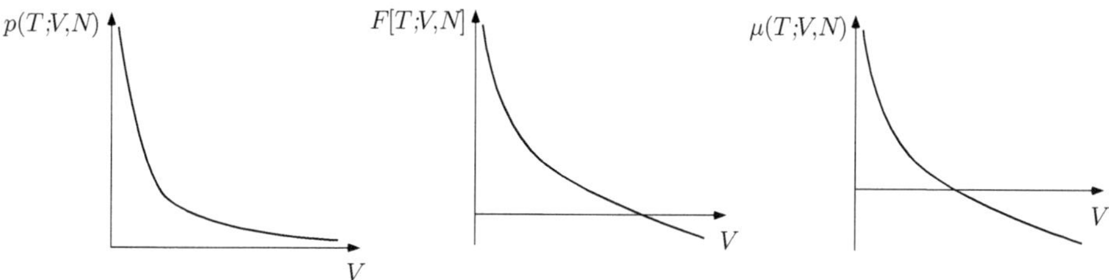

Figure 7.4 Approximate dependences of the pressure, Helmholtz free energy and chemical potential on the volume for a temperature at which the fluid is in the gas phase. In each case, the quantity is a decreasing function of V. (The values of T and N are fixed.)

---

[29] Note that from the extensivity of the volume, it follows that the volume per unit amount of substance, given by $v_{\mathrm{L}}(T) = V_{\mathrm{L}}(T; N)/N$ and $v_{\mathrm{G}}(T) = V_{\mathrm{G}}(T; N)/N$ in the two phases, respectively, depends only on T.

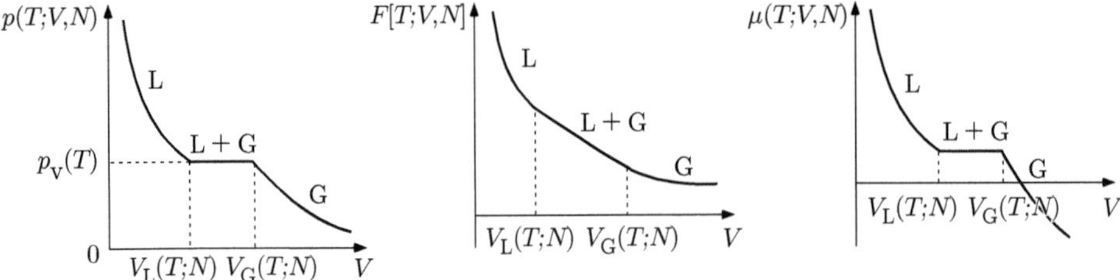

Figure 7.5 Approximate dependences of the pressure, Helmholtz free energy and chemical potential on the volume for a temperature at which the liquid and gas phases can coexist. In the region $V_L(T; N) \leq V \leq V_G(T; N)$, the pressure and chemical potential are constant, and the free energy decreases with a constant slope of $-p_v(T)$. In the figure, "G" indicates the gas phase, "L" indicates the liquid phase, and "L + G" indicates the coexistence of these two phases. (The values of T and N are fixed.)

Therefore, as found from (3.31), in this region $F[T; V, N]$ decreases with a constant slope of $-p_v(T)$. Also, considering (7.8), we find that in this region the V dependence of $F[T; V, N]$ on the left-hand side exactly cancels $-Vp$ on the right-hand side, and hence $\mu(T; V, N)$ is independent of V. This behavior is seen in Figure 7.5.

We next study the coexistence of phases from a general point of view.

7.5.3 Coexistence of liquid and gas phases

Here, we investigate the coexistence of separate phases on the basis of the balance conditions derived in Section 7.4.

Let us consider the case in which a partition is introduced into a system originally in the equilibrium state $(T; V, N)$, and two subsystems in the states $(T; \widetilde{V}_1, \widetilde{N}_1)$ and $(T; \widetilde{V}_2, \widetilde{N}_2)$ are thereby produced. Obviously, we have $V = \widetilde{V}_1 + \widetilde{V}_2$ and $N = \widetilde{N}_1 + \widetilde{N}_2$. Because these states are balanced, from (7.41) we know that the following equalities hold:

$$p(T; \widetilde{V}_1, \widetilde{N}_1) = p(T; \widetilde{V}_2, \widetilde{N}_2) , \quad \mu(T; \widetilde{V}_1, \widetilde{N}_1) = \mu(T; \widetilde{V}_2, \widetilde{N}_2) . \tag{7.47}$$

Intuitively, we understand that these conditions should be satisfied in the case that the two subsystems have equal concentrations of material, corresponding to the relation

$$\frac{\widetilde{N}_1}{\widetilde{V}_1} = \frac{\widetilde{N}_2}{\widetilde{V}_2} . \tag{7.48}$$

Indeed, it is easily demonstrated that this intuitive understanding is correct: Setting $\lambda = \widetilde{V}_1/\widetilde{V}_2$ in the relation expressing the intensivity of the pressure, (3.32), we obtain

$$p(T; \widetilde{V}_1, \widetilde{N}_1) = p(T; \frac{\widetilde{V}_1}{\widetilde{V}_2}\widetilde{V}_2, \frac{\widetilde{V}_1}{\widetilde{V}_2}\widetilde{N}_2) = p(T; \widetilde{V}_2, \widetilde{N}_2) . \tag{7.49}$$

Then, because the chemical potential is also intensive, similar reasoning yields its balance condition as well.

The above result is just what one would expect from the simplest considerations, and therefore not particularly interesting. What is more interesting is whether (and when) the balance conditions in (7.47) can be satisfied even if (7.48) does not hold. Here we seek a precise answer to this question.

First, let us consider the situation described by Figure 7.4, in which the fluid behaves as a gas for all volumes under consideration, and both $p(T; V, N)$ and $\mu(T; V, N)$ are decreasing functions of V/N. In this situation, in order to satisfy the balance conditions in (7.47), clearly it is necessary to satisfy (7.48). We thus conclude that for any temperature at which the pressure is a strictly decreasing function of V, a state of balance requires that the concentrations of material in the two subsystems be equal. This can hold only if the system consists of a single phase.

Next, let us consider the case of lower temperature, with the system exhibiting the behavior described by Figure 7.5. For such a temperature, because there exists a region in which the pressure is independent of the volume, it is possible to realize balance for subsystems with different concentrations. As is clear from the graphs of the pressure and chemical potential, the balance conditions in (7.47) are satisfied as long as both $\widetilde{V}_1/\widetilde{N}_1$ and $\widetilde{V}_2/\widetilde{N}_2$ take values between $v_{\mathrm{L}}(T)$ and $v_{\mathrm{G}}(T)$. Thus, in this regime, it is possible to realize the situation in which, for example, the insertion of the partition separates a subsystem consisting of only liquid and a subsystem consisting of only gas.

It is important to understand that the coexistence of the liquid and gas phases considered here is different from the coexistence of liquid water (liquid H_2O) and air containing water vapor (gas H_2O), which we experience in our daily lives. In this familiar case, because the air contains substances in addition to H_2O, this is a multi-component system. As we are all well aware, this kind of coexistence is stable for broad ranges of the temperature and pressure. (A problem related to this point is treated in Section 9.3.2.) The type of coexistence that can be treated with the theory described in this section is that in which, for example, a vessel contains pure H_2O only. Phase coexistence in a system of this kind can be prepared as follows. First, we draw up some amount of pure liquid water into a syringe, seal its end, and place it in thermal contact with a fixed-temperature environment. Then, after allowing the water to reach equilibrium, we pull the piston, increasing the volume of the system. When we do this, we will observe the vaporization of water beginning to take place. Continuing to increase the volume, we can realize a system in which finite amounts of the liquid and gas phases coexist. Under such conditions, if we move the piston in either direction over a certain range, the pressure will remain fixed while the proportions of the liquid and gas phases vary (see Figure 7.5). At a temperature of $100°C$, this coexistence of phases will exist at a pressure of 1 atm.

7.5.4 Phase diagrams and the Clausius-Clapeyron relation

In this subsection, we carry out a more thorough treatment of phase transitions between gas and liquid phases. In particular, we take our first look at graphical representations of phase equilibrium states known as *phase diagrams*, introduce the state function *enthalpy*, and derive the Clausius-Clapeyron relation, which demonstrates the universal, quantitatively exact predictive power of thermodynamics in this context.

Phase diagram of a pure substance

Figure 7.6, which is a more detailed version of Figure 7.3, qualitatively depicts the volume dependence of the pressure of a fluid at various temperatures, with phase regions indicated. To the left of the bell-shaped region, the fluid is in the liquid phase, and to the right, it is in the gas phase. Within the bell-shaped region, the pressure is independent of the volume, and the liquid and gas phases coexist. The temperature at which the pressure curve intersects the coexistence region at a single point is called the *critical temperature*, denoted by T_c.[30] The pressure corresponding to this unique point of intersection is called the *critical pressure*, denoted by p_c.

At temperatures below T_c, the fluid exhibits a phase transition between liquid and gas phases as the volume is varied, while above this temperature, no such phase transition exists. For a fluid that is initially in a liquid state with a temperature below T_c, if the volume is increased with the temperature fixed, the system will eventually enter the coexistence region. As the system moves through this region (from left to right in Figure 7.6), more and more of the fluid will change from liquid to gas, and the coexisting phases will generally form an inhomogeneous mixture. If the volume is increased further, the system will eventually leave the coexistence region and be in a gas state. Comparing the purely liquid state that exists just

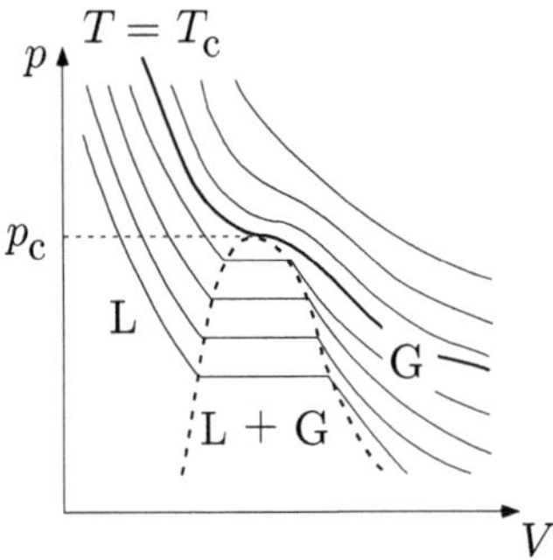

Figure 7.6 A simple phase diagram for a fluid in the V-p plane. Here, the volume dependence of the pressure is plotted for several temperatures. (These pressure curves should not be regarded as quantitatively realistic. In particular, because actual liquids are nearly incompressible, quantitatively accurate curves would generally be much steeper in the liquid region.) The amount of substance is fixed and equal in all cases. In the left-hand region (indicated by "L"), the fluid is a liquid, and in the right-hand region (indicated by "G"), the fluid is a gas. In the bell-shaped region below the broken curve, between the L and G regions, the liquid and gas phases coexist. In this region, the pressure is independent of the volume. The critical temperature, T_c, is that temperature for which the pressure curve intersects the coexistence region at a single point, and the critical pressure, p_c, is the value of the pressure at this intersection point. When the pressure and temperature are both above their critical values, there do not exist distinct liquid and gas phases. For such values of p and T (but below the solid phase region), the substance exists as a *supercritical fluid*.

---

[30] There are various types of interesting *critical phenomena* that occur in the neighborhood of the critical point, including the divergence of the specific heat. (See Section 10.4 for detailed discussion.)

before entering the coexistence region and the purely gas state that exists just after leaving it, we find that the densities differ by a finite amount (as long as $T_c - T$ is finitely large).

The phase diagram in Figure 7.7 depicts the phase regions for a single-component system in the T-p plane.[31] The curves represent the boundaries between the solid, liquid and gas phase regions. The points (T_3, p_3) and (T_c, p_c) are the triple point and critical point, respectively. The curve separating the liquid and gas regions, which connects these points, is the *vaporization curve*.[32] The temperature at each point on the vaporization curve is the boiling point for the corresponding value of the pressure.

As seen in Figure 7.7, if we begin in the liquid region, below p_c, and raise the temperature while keeping p fixed, the system will eventually reach the boiling point, $T_b(p)$, at which the fluid changes from a liquid to a gas. The density of the fluid changes discontinuously at $T_b(p)$, as the vaporization curve is crossed. Note that this curve corresponds to the entire bell-shaped coexistence region in Figure 7.6. In other words, the two-dimensional coexistence region in the V-p plane collapses into a one-dimensional curve in the T-p plane. Thus, while in the case that the temperature is fixed and the volume is varied, the system travels through the coexistence region, and hence the density changes continuously when moving between the liquid and gas phases, in the case that the pressure is fixed and the

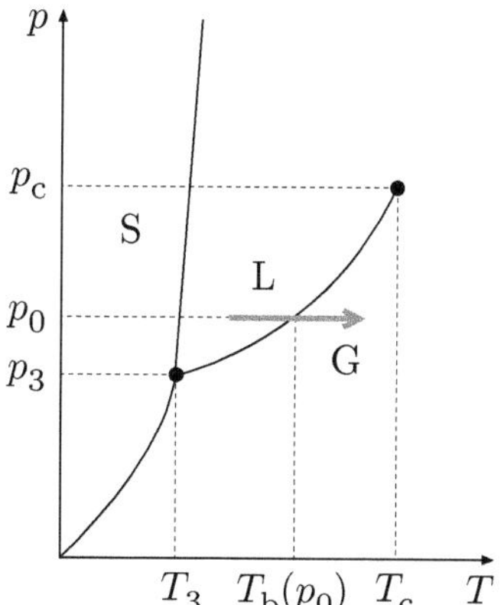

Figure 7.7 Typical phase diagram for a single-component system in the T-p plane, with solid ("S"), liquid ("L"), and gas ("G") phase regions indicated. The point at which the three phases coexist, (T_3, p_3), is called the *triple point*. The point (T_c, p_c) is called the *critical point*. The curve separating the gas and liquid phases, extending from the triple point to the critical point, is the *vaporization curve*. The temperature $T_b(p_0)$ is the *boiling point* at pressure p_0. Following the path indicated by the arrow, a fluid will undergo a phase transition from a liquid to a gas at $T_b(p_0)$, as it crosses the vaporization curve. At this curve, the density of the fluid changes discontinuously as a function of T. Unlike the gas and liquid phase regions, the solid phase region is separated entirely from the other two by a phase boundary.

---

[31] Like Figure 7.6, this should be regarded as a simplified, generic phase diagram. It roughly approximates phase diagrams for actual substances, which may be much more complicated.

[32] The curve separating the liquid and solid phases is referred to as the *fusion curve*, and that separating the gas and solid phases is referred to as the *sublimation curve*.

temperature is varied, the system effectively jumps over the coexistence region, and hence the density changes discontinuously.

From Figure 7.7, it is seen that liquid and gas states can also be reached from one another without crossing the vaporization curve, by going above the critical temperature. (In Figure 7.6, this corresponds to going above the bell-shaped region.) Along such a path, the density varies continuously while the system remains homogeneous (i.e., of a single phase) throughout the transition. The liquid and gas phases are thus seen to merge continuously into one another above the vaporization curve. Contrastingly, because the solid phase region is separated entirely from the other phase regions by a phase boundary, across which the density varies discontinuously, it is not possible to make a transition between a solid state and either a liquid or gas state during which the system remains homogeneous and never undergoes discontinuous change. This distinction is consistent with the view that while the difference between the liquid and gas phases is merely quantitative, the difference between the solid phase and the other phases is also qualitative.[33]

Finally, we briefly discuss the significance of the triple point, (T_3, p_3). At this point, all three phases can coexist, and with the temperature and pressure of the system held fixed, it is possible to vary their amounts within certain ranges. Thus, while the triple point is represented by a point in the T-p plane, it in fact consists of many equilibrium states. (For this reason, in a broader context, it is more accurate to refer to it as the "triple-point region.") Indeed, for a single-component system at $(T, p) = (T_3, p_3)$, equilibrium states are not uniquely specified by just T, V and N, as noted in Footnote 11 on p. 30. In this case (and only in this case), it is necessary to introduce a different parameterization in order to fully specify equilibrium states. For example, representing the amount of substance in the solid phase by N_S, equilibrium states in the triple-point region are fully specified by the parameterization $(T_3; V, N, N_S)$. Nevertheless, as we show in Appendix D, the Helmholtz free energy within the triple-point region is independent of the value of N_S, and hence it can be written simply as $F[T; V, N]$.

Absorbed heat and entropy change accompanying vaporization

Let us return to the phase transition between liquid and gas phases. We consider a quasi-static isothermal operation at temperature T (below T_c) through which the volume is increased from the value $V_L(T; N)$ at the left edge of the coexistence region, where the system is entirely in the liquid phase, to the value $V_G(T; N)$ at the right edge of the coexistence region, where the system is entirely in the gas phase. Thus, we have the transition

$$(T; V_L(T; N), N) \xrightarrow{\text{qi}} (T; V_G(T; N), N) \,. \tag{7.50}$$

In the initial state, the fluid has the smallest density that can be realized in the liquid phase at temperature T, while in the final state, it has the largest density that can be realized in the gas phase at temperature T. During the course of the

---

[33] On the basis of theoretical considerations, it is conjectured that with the proper use of a particular type of externally applied electric field (which cannot be realized with current technology), it would be possible to realize a transition between solid and liquid states without encountering a singularity (see, e.g., Section 3.3 of Ref. [13]).

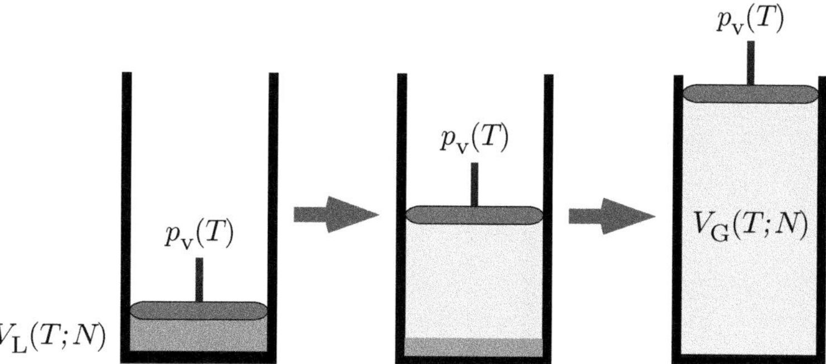

Figure 7.8 A quasi-static isothermal operation in which the volume of a fluid is increased from $V_L(T; N)$ to $V_G(T; N)$ with the temperature and pressure held fixed at T and $p_v(T)$. During the course of this operation, the liquid and gas phases necessarily coexist, and their proportions change as the volume is changed.

operation, the liquid and gas phases coexist, as depicted in Figure 7.8. Note that for all values of the volume between $V_L(T; N)$ and $V_G(T; N)$, the pressure remains fixed at $p_v(T)$. Hence, the amount of work done by the system on the mechanical world through the operation is given by

$$W_{\max}(T; (V_L(T; N), N) \to (V_G(T; N), N)) = p_v(T)\{V_G(T; N) - V_L(T; N)\} \ . \tag{7.51}$$

Using this expression for the maximum work, we obtain the following for the maximum heat, which is the energy absorbed from the environment by the system during the operation:

$$\begin{aligned}
Q_{\max}&(T; (V_L(T; N), N) \to (V_G(T; N), N)) \\
&= W_{\max}(T; (V_L(T; N), N) \to (V_G(T; N), N)) + U(T; V_G(T; N), N) \\
&\quad - U(T; V_L(T; N), N) \\
&= \{U(T; V_G(T; N), N) + p_v(T)V_G(T; N)\} - \{U(T; V_L(T; N), N) \\
&\quad + p_v(T)V_L(T; N)\} \ . \tag{7.52}
\end{aligned}$$

We next define a state function termed the *enthalpy*:[34]

$$H(T; V, N) := U(T; V, N) + p(T; V, N)V \ . \tag{7.53}$$

The enthalpy is an extensive quantity that plays a particularly important role in applications of thermodynamics to chemistry. In terms of the enthalpy, the heat absorbed in the transition from a liquid state to a gas state given in (7.52) can be written as

$$Q_{\max}(T; (V_L(T; N), N) \to (V_G(T; N), N)) = H(T; V_G(T; N), N) - H(T; V_L(T; N), N)$$

$$= H_{\text{vap}}(T; N) \ . \tag{7.54}$$

---

[34] The enthalpy is a complete thermodynamic function when regarded as a function of the pressure, entropy and amount of substance. In this case, we express it with square brackets, writing $H[p; S, N]$ (see Appendix F). For applications of the enthalpy, see Problems 7.10 and 8.6.

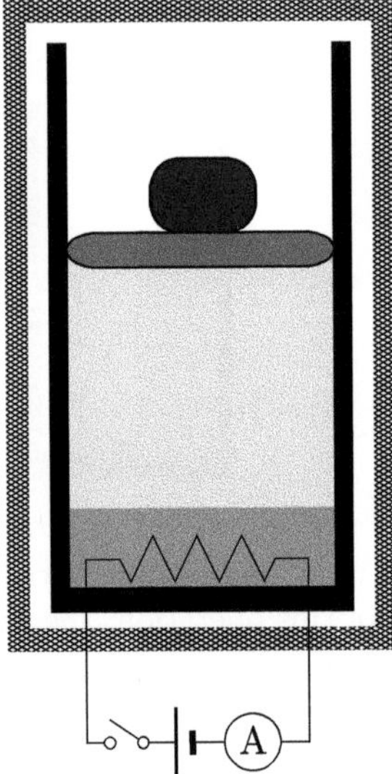

Figure 7.9 Apparatus for measuring the enthalpy of vaporization, $H_{\text{vap}}(T; N)$. Energy is supplied by an electric heating element to a fluid in the liquid state $(T; V_{\text{L}}(T; N), N)$ contained in a vessel surrounded by adiabatic walls. With the pressure held fixed at $p_{\text{v}}(T)$, energy is added until the fluid is entirely in the gas state $(T; V_{\text{G}}(T; N), N)$. In this situation, $H_{\text{vap}}(T; N)$ is equal to the work performed by the electric power source, W_{el}.

The quantity $H_{\text{vap}}(T; N)$, referred to as the *enthalpy of vaporization*, is an important quantity characterizing the transition from a liquid to a gas.[35] Next, using (7.54), along with the relation between the entropy and maximum heat given in (6.7), we can express the enthalpy of vaporization in terms of the difference between the entropy of the system in the gas phase at $V = V_{\text{G}}(T; N)$ and in the liquid phase at $V = V_{\text{L}}(T; N)$ as follows:

$$H_{\text{vap}}(T; N) = T\{S(T; V_{\text{G}}(T; N), N) - S(T; V_{\text{L}}(T; N), N)\} \,. \tag{7.55}$$

The quantity $H_{\text{vap}}(T; N)$ can be measured using the kind of experimental setup depicted in Figure 7.9. In the experiment considered, a system enclosed within adiabatic walls begins in the liquid state $(T; V_{\text{L}}(T; N), N)$. Then, an amount of work W_{el} is performed by the mechanical world on the system,[36] while the volume is allowed to increase with the pressure maintained at the value $p_{\text{v}}(T)$. Through this operation, the system makes a transition to the gas state $(T; V_{\text{G}}(T; N), N)$. In essence, the situation here is equivalent to that in the case of the quasi-static

---

[35] The enlthalpy of vaporization is also referred to as the *heat of vaporization, latent heat of vaporization* and *heat of evaporation*. The term *latent heat* is obviously a remnant of the caloric theory of heat. It hints at the outdated understanding that heat, in the form of the conserved substance "caloric," is absorbed by the liquid as it changes to a gas and that in this form it remains hidden within the gas. However, this is inconsistent with the understanding provided by the law of energy conservation, from which we know that part of the energy absorbed as heat from the environment is used by the system to perform work as its volume increases. The inability of the caloric theory to account for this fact reflects one limitation of the understanding that it provides. This failure played a large role in the eventual abandonment of this theory [15].

[36] In the situation considered in the figure, this work is done by an electric power source, but any means of adding energy while the pressure is held fixed could be used.

isothermal transition represented by (7.50), with the energy in the form of heat absorbed from the environment replaced by energy in the form of work done by the electric power source. Thus, we conclude that W_{el} is identical to the maximum heat in the case of (7.50), $Q_{\max}(T; (V_{\text{L}}(T; N), N) \to (V_{\text{G}}(T; N), N))$, and hence, also to the enthalpy of vaporization, $H_{\text{vap}}(T; N)$.[37]

The Clausius-Clapeyron relation

Taking the derivative of the vapor pressure, $p_{\text{v}}(T)$, with respect to T, we obtain the slope of the phase boundary in the T-p phase diagram appearing in Figure 7.7. Here we derive a relation in which this slope is expressed in terms of the enthalpy of vaporization. This is known as the *Clausius-Clapeyron relation*.[38]

Rewriting the left-hand side of (7.51) through use of (3.27), which relates the maximum work and the Helmholtz free energy, we have

$$F[T; V_{\text{L}}(T; N), N] - F[T; V_{\text{G}}(T; N), N] = p_{\text{v}}(T)\{V_{\text{G}}(T; N) - V_{\text{L}}(T; N)\} . \tag{7.56}$$

Then, taking the derivative of both sides with respect to T and employing (7.11), we obtain

$$-S(T; V_{\text{L}}(T; N), N) - \frac{\partial V_{\text{L}}(T; N)}{\partial T} p(T; V_{\text{L}}(T; N), N)$$

$$+S(T; V_{\text{G}}(T; N), N) + \frac{\partial V_{\text{G}}(T; N)}{\partial T} p(T; V_{\text{G}}(T; N), N)$$

$$= \frac{dp_{\text{v}}(T)}{dT}\{V_{\text{G}}(T; N) - V_{\text{L}}(T; N)\} + p_{\text{v}}(T) \left\{ \frac{\partial V_{\text{G}}(T; N)}{\partial T} - \frac{\partial V_{\text{L}}(T; N)}{\partial T} \right\} .$$
$$\tag{7.57}$$

With $p(T; V_{\text{L}}(T; N), N) = p(T; V_{\text{G}}(T; N), N) = p_{\text{v}}(T)$, this becomes

$$\frac{dp_{\text{v}}(T)}{dT} = \frac{S(T; V_{\text{G}}(T; N), N) - S(T; V_{\text{L}}(T; N), N)}{V_{\text{G}}(T; N) - V_{\text{L}}(T; N)} . \tag{7.58}$$

Finally, using (7.55), we arrive at the Clausius-Clapeyron relation:

$$\frac{dp_{\text{v}}(T)}{dT} = \frac{H_{\text{vap}}(T; N)}{T\{V_{\text{G}}(T; N) - V_{\text{L}}(T; N)\}} = \frac{h_{\text{vap}}(T)}{T\{v_{\text{G}}(T) - v_{\text{L}}(T)\}} . \tag{7.59}$$

Here, $h_{\text{vap}}(T) = H_{\text{vap}}(T; N)/N$ is the molar enthalpy of vaporization, and $v_{\text{G}}(T) = V_{\text{G}}(T; N)/N$ and $v_{\text{L}}(T) = V_{\text{L}}(T; N)/N$ are the molar volumes in the two states considered. These are all functions of T alone.

We stress that (7.59) is an exact relation that holds universally for all types of materials. This conclusion follows from the observation that if there were any discrepancy between actual physical phenomena and the predictions derived from this relation, then it would be possible to construct a perpetual motion machine of the second kind (see Problem 7.9).

---

[37] Of course, the same result can also be derived from the law of energy conservation for adiabatic operations, but the above derivation is simpler.

[38] The Clausius-Clapeyron relation is usually derived using the Gibbs free energy (see Problem 8.11). Its derivation using the Helmholtz free energy is somewhat longer, but this derivation provides a clearer understanding of the phenomena involved, and hence it is more instructive.

We can also derive a relation similar to the Clausius-Clapeyron relation that describes the phase boundary between the solid and liquid phases, like that depicted in Figure 7.7. In this case, with the solid-liquid phase boundary represented by $(T, p_{\mathrm{fus}}(T))$, using the molar enthalpy of fusion $h_{\mathrm{fus}}(T)$, defined similarly to $h_{\mathrm{vap}}(T)$, we obtain the following relation:

$$\frac{dp_{\mathrm{fus}}(T)}{dT} = \frac{h_{\mathrm{fus}}(T)}{T\{v_{\mathrm{L}}(T) - v_{\mathrm{S}}(T)\}} \,. \tag{7.60}$$

For the vast majority of substances, we have $h_{\mathrm{fus}}(T) > 0$ and $v_{\mathrm{L}}(T) > v_{\mathrm{S}}(T)$. Therefore, in accordance with (7.60), $p_{\mathrm{fus}}(T)$ is an increasing function of T. In other words, the plot of the solid-liquid phase boundary in the T-p plane is slanted upward to the right, as in Figure 7.7. However, for the solid-liquid phase transition with which we are most familiar, that between ice and water, at ordinary pressures, we have $h_{\mathrm{fus}}(T) > 0$ and $v_{\mathrm{L}}(T) < v_{\mathrm{S}}(T)$. For this reason, when water freezes, its volume increases. Due to this extremely familiar yet chemically atypical property of water, in a phase diagram like that in Figure 7.7, the solid-liquid phase boundary is slanted upward to the left at ordinary pressures. Thus, as the pressure increases, the melting point of water decreases.[39]

Problems 7

7.1 (Section 7.1.3) Using the Euler equation given in (7.8), derive the following:

$$V\frac{\partial p(T; V, N)}{\partial V} = N\frac{\partial \mu(T; V, N)}{\partial V}, \quad V\frac{\partial p(T; V, N)}{\partial N} = N\frac{\partial \mu(T; V, N)}{\partial N} \,. \tag{7.61}$$

7.2 (Sections 7.1.3, 7.1.4 and 8.2) Take the total differential of the Euler equation (7.8). Then, using (7.12) to rewrite the left-hand side and applying the formula (7.16) to the right-hand side, show that the following equation can be derived:

$$S\,dT - V\,dp + N\,d\mu = 0 \,. \tag{7.62}$$

This is the *Gibbs-Duhem equation*.

Show that the relations in (7.61) can be derived from (7.62) by first varying V while holding T and N fixed and then varying N while holding

---

[39] In many textbooks, it is stated that this is the phenomenon that makes ice skating possible. The standard argument here is something like the following. The skate blade presses on the ice, raising the local pressure. As a result, the melting point is lowered (in accordance with (7.60)), and hence the ice melts. In this way, a thin layer of water is created between the blade and the ice, and as a consequence, the friction experienced at the blade-ice interface is lowered. However, it is now believed that this explanation in fact does not account for the slipperiness of ice. This can be easily understood from the empirical fact that ice skating is possible at temperatures at least as low as $-30°\mathrm{C}$, while the melting point of ice can be decreased by only approximately $3.5°\mathrm{C}$ with a pressure of 500 atm. Although it is still believed that the presence of a thin layer of water on the surface of the ice plays a fundamental role in its slipperiness, experimental evidence strongly suggests that the existence of this layer is due to a more complicated phenomenon that is unrelated to pressure-induced melting. (For detailed discussion, see R. Rosenberg, "Why Is Ice Slippery?," Physics Today, vol. 58, iss. 12, 2005, 50–54 (https://doi.org/10.1063/1.2169444.)

T and V fixed. Next, show that (7.62) can also be derived using the Euler equation for the Gibbs free energy given in (8.19) and the total differential of the Gibbs free energy given in (8.22).

7.3 (Section 7.2) Assuming that $U(T; V, N)$ and $p(T; V, N)$ are twice differentiable, from (7.21) demonstrate the equality

$$\frac{\partial^2 U(T; V, N)}{\partial V \partial T} = T \frac{\partial^2 p(T; V, N)}{\partial T^2} . \tag{7.63}$$

Discuss what this equation implies with regard to the energy of a gas that behaves in accordance with the van der Waals equation of state, given in (3.37).

7.4 (Section 7.2.4) This is a problem from electromagnetism, in which we study the pressure exerted by an electromagnetic field in a simple situation. (We use MKS units.)

Consider electric and magnetic fields $\boldsymbol{E}(x, y, z, t)$ and $\boldsymbol{B}(x, y, z, t)$, where (x, y, z) represents position in Cartesian coordinates and t is time. Suppose that the region corresponding to $x \leq 0$ consists of a perfectly conducting material, and the region corresponding to $x > 0$ is vacuum. In the $x = 0$ plane, we impose the boundary conditions $E_x = E_y = 0$, where E_x and E_y are the x and y components of $\boldsymbol{E}$.

Let us study a solution to Maxwell's equations in the vacuum region. Specifically, we consider the situation in which a plane wave traveling in the negative x direction is perfectly reflected at the $x = 0$ boundary, and the incident and reflected waves form a standing wave. Show that as such a standing wave, the electromagnetic field consisting of the components

$$\boldsymbol{E}(x, y, z, t) = (0, 0, E_0 \sin kx \, \cos \omega t) ,$$

$$\boldsymbol{B}(x, y, z, t) = (0, B_0 \cos kx \, \sin \omega t, 0) , \tag{7.64}$$

where $B_0 = E_0/c$ and $\omega = ck$, with $c = (\varepsilon_0 \, \mu_0)^{-1/2}$, satisfies Maxwell's equations. (Here, c is the speed of light in vacuum, ε_0 is the vacuum permittivity, and μ_0 is the vacuum permeability.) Derive the energy density of this electromagnetic field and show that its average over space and time is $u = \varepsilon_0 E_0^2/4$.

In the $x < 0$ region, the magnetic field vanishes, but at $x = +0$, it has a finite magnitude. For this reason, electric current flows in the $x = 0$ plane. Derive the direction and density j of this flow. This current is subject to a force exerted by the magnetic field. Derive the direction and magnitude per unit area of this force. (This force is the same as that in the case that, rather than there existing a magnetic field on only one side of the current, there exists a magnetic field on both sides, with half the magnitude of the original.[40]) From this computation, it is found that a pressure $p = \varepsilon_0 E_0^2/4$ is exerted in the negative x direction. This demonstrates the relation $p = u$.

---

[40] This is clear from the fact that identical magnetic fields at $x = -0$ and $x = +0$ exert identical forces on the current in the $x = 0$ plane. (An analogous factor of $1/2$ arises in, for example, the computation of the force on the electrode plates of a parallel plate capacitor.) It can also be confirmed using the more systematic approach in which we first compute the magnetic field and force assuming that the region containing the current has a finite width and then take the limit of 0 width. In this computation, the factor of $1/2$ naturally appears.

7.5 (Section 7.4) Recall the function $f(v,n)$ given in (7.42), which plays an important role in the balance conditions. Show that if this function is sufficiently differentiable with respect to v and n, then both eigenvalues of the 2×2 matrix

$$D = \begin{pmatrix} \dfrac{\partial^2 f(v,n)}{\partial v^2}\Big|_{v=0,n=0} & \dfrac{\partial^2 f(v,n)}{\partial v \partial n}\Big|_{v=0,n=0} \\ \dfrac{\partial^2 f(v,n)}{\partial v \partial n}\Big|_{v=0,n=0} & \dfrac{\partial^2 f(v,n)}{\partial n^2}\Big|_{v=0,n=0} \end{pmatrix} \quad (7.65)$$

are positive.

7.6 (Section 7.4) In the main text, we saw that for a thermodynamic system, the global equilibrium state can be obtained from the local balance conditions. Here we study a simple system (which is a combination of thermodynamic and mechanical systems) for which this does not hold.[41]

Let us consider a tube like that depicted in Figure 7.10. The tube is in the form of a half-torus, and the radius of the semicircle passing through the center of each circular cross section is r. The tube is oriented such that this semicircle is in a plane parallel to the direction of gravity, and its highest point is its midpoint. The volume of the tube is V. Both ends of the tube are closed, and it contains a thin, circular piston of mass m. On each side of the piston there is an amount $N/2$ of gas. The entire system is in contact with an environment at temperature T.

Let θ represent the angular position of the piston, as shown in the right-hand panel of Figure 7.10. The piston is free to move in the range $0 < \theta < \pi$. The volumes on the left and right of the piston are $(\theta/\pi)V$ and $\{1 - (\theta/\pi)\}V$, respectively. Thus, with the piston fixed at a given value

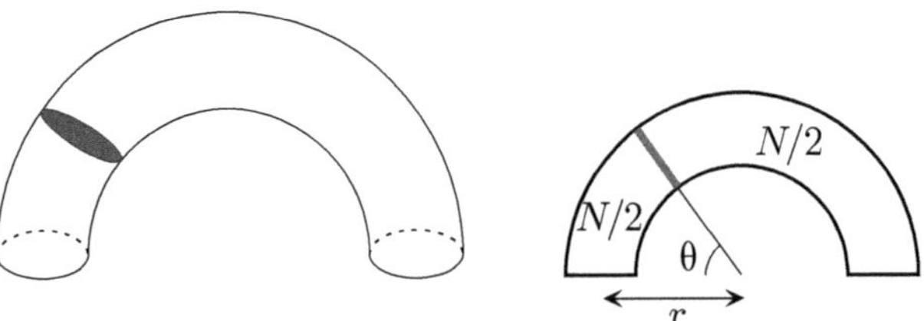

Figure 7.10 A closed tube in the form of a half-torus. A movable piston with nonzero mass separates the tube into two parts. The tube contains a gas, and the amounts on either side of the piston are equal. For any given mass of the piston, there exists a critical value of the temperature above which the pressure will be sufficiently large that in the equilibrium state, the piston will be positioned in the middle of the tube, and thus the volumes on its two sides will be equal. However, for temperatures below the critical value, the potential energy of the piston will play a role, and as a result, the state in which the piston is at the highest point will be unstable. In this case, in equilibrium, the piston will be positioned on one side or the other, some finite angle away from the middle of the tube.

---

[41] This example is due to Callen [2]. He attempts to connect this example to phase transitions in thermodynamic systems, but we do not believe that this connection can actually be made.

of θ, the Helmholtz free energy of the system is

$$F[T;\theta] = F\left[T; \frac{\theta}{\pi}V, \frac{N}{2}\right] + F\left[T; \left(1 - \frac{\theta}{\pi}\right)V, \frac{N}{2}\right] + mgr\sin\theta \ . \quad (7.66)$$

The first two terms on the right-hand side of this equation are the free energies of the parts of the tube to the left and right of the piston, respectively, and the third term is the potential energy of the piston.

Suppose that, beginning at some initial value, θ changes spontaneously and eventually comes to rest at an equilibrium value θ^*, at which no net force is exerted on the piston. Show that θ^* is such that $F[T;\theta]$ (with fixed T) realizes a local minimum at $\theta = \theta^*$.

From this point, let us assume that the gas is ideal. Show that there exists a value of the temperature, $T_c(m)$, such that for $T \geq T_c(m)$, $F[T;\theta]$ is minimized only at $\theta^* = \pi/2$, while for $T < T_c(m)$, $F[T;\theta]$ is minimized at two points on either side of $\theta = \pi/2$, while $\theta = \pi/2$ itself corresponds to a local maximum of $F[T;\theta]$. This result contradicts Result 7.2 (p. 148), which asserts that states of unstable and metastable balance do not exist. Find that part of the derivation of Result 7.2 that is incorrect for the present example. Finally, show that by changing the form of the tube, a metastable state can also be realized.[42]

7.7 (Section 7.3) Consider a system characterized by collective extensive variables X and Y. Let us assume that X is a single-component quantity and that $F[T; X, Y]$ is twice differentiable and convex with respect to X. Suppose that initially X and Y are both fixed, but then the conditions of the system are changed, and in response to this change, X is allowed to vary freely, while Y remains fixed. Show that under the new conditions, the value of X in equilibrium, $X^*(T;Y)$, is determined by

$$\left.\frac{\partial F[T; X, Y]}{\partial X}\right|_{X=X^*(T;Y)} = 0 \ . \quad (7.67)$$

Use this to derive the following:

$$\frac{\partial X^*(T;Y)}{\partial T} = \left[\left\{\frac{\partial^2 F[T; X, Y]}{\partial X^2}\right\}^{-1} \frac{\partial S(T; X, Y)}{\partial X}\right]_{X=X^*(T;Y)} \ . \quad (7.68)$$

From the convexity of $F[T; X, Y]$ with respect to X, we have $\partial^2 F/\partial X^2 \geq 0$. Also, from the relation between the maximum heat and the entropy given in (6.7), it is seen that if X is changed by a small amount, whether the system absorbs or expels heat is determined by the sign of $\partial S/\partial X$. Show using (7.68) that if the temperature increases, X^* changes in such a manner that the system absorbs heat, while if the temperature decreases,

---

[42] It might be thought that through the investigation and extension of models like this, we may be able to obtain an understanding of such metastable phenomena as supercooling and magnetic hysteresis. However, for phenomena of these types, there is no imaginable parameter that could play a role analogous to that of θ in the present example. We believe that any metastable state exhibited by the simple system considered here is fundamentally different from those of supercooling, magnetic hysteresis, and related phenomena exhibited by thermodynamic systems.

X^* changes in such a manner that the system expels heat. Confirm this explicitly for the example considered in Figure 8.1.

Next, regarding Y_i as one component of Y, apply similar considerations to $\partial X^*(T; Y)/\partial Y_i$ and give a simple example.

The examples considered here, in which a thermodynamic system responds to a small change in the conditions governing it by evolving in such a manner to partly counteract that change, are illustrative of a widely observed class of behavior. That such behavior is universal for thermodynamic systems is the assertion of Le Chatelier's principle, or the Le Chatelier-Braun principle (see Section 9.4).[43]

7.8 (Section 7.5) Here we investigate the van der Waals equation of state, given in (3.37). At sufficiently low temperatures, the "pressure" $\tilde{p}(V)$ in this equation of state first decreases, then increases, and then decreases again as a function of V, as shown in Figure 7.11(a). (In this problem, T and N are regarded as fixed and therefore omitted.) From this behavior, we conclude that $\tilde{p}(V)$ cannot be regarded as a physical pressure, because according to Result 7.1 (p. 146), pressure is a non-increasing function of V. To avoid this contradiction, we wish to construct a more well-behaved pressure that is independent of the volume over a certain interval, as in an actual liquid-gas phase transition (corresponding to the case $T = T_3$ in Figure 7.3). We propose to construct this pressure from the pathological pressure $\tilde{p}(V)$ by applying to it the smallest possible reasonable modification. The *Maxwell equal-area rule* provides a method for this purpose.

The Maxwell equal-area rule is implemented through the following procedure, known as the *Maxwell construction*. First, as shown in Figure 7.11(b), we draw a line parallel to the V axis that crosses $\tilde{p}(V)$ at three points. Second, we adjust the position of this horizontal line so that the areas of the two regions enclosed between it and the graph of $\tilde{p}(V)$, A_1 and A_2, are equal. Third, as shown in Figure 7.11(c), we replace the parts of the graph of $\tilde{p}(V)$ that border these two regions by the horizontal line. The resulting graph is regarded as representing the physical pressure, $p(V)$.

The original derivation of the Maxwell equal-area rule utilizes an operation under which the system passes through unstable states, and it is

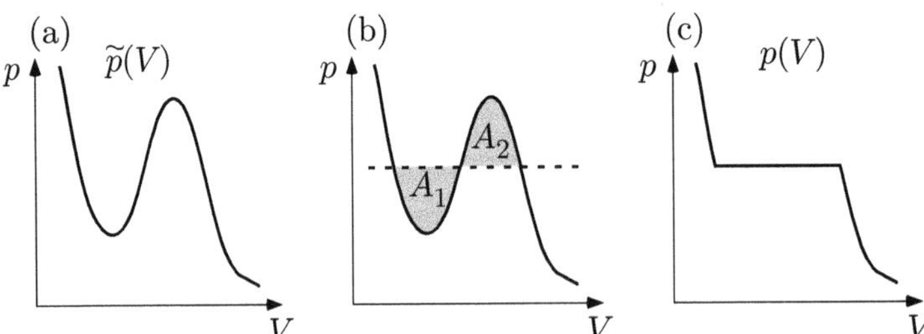

Figure 7.11 Method to obtain the physical pressure $p(V)$ in (c) from the non-physical pressure $\tilde{p}(V)$ in (a) using the Maxwell equal-area rule. The horizontal line in (b) is positioned such that the areas A_1 and A_2 are equal.

---

[43] Usually, in situations that the system responds directly to a change, the former name is used, while in situations that the system responds indirectly, the latter name is used.

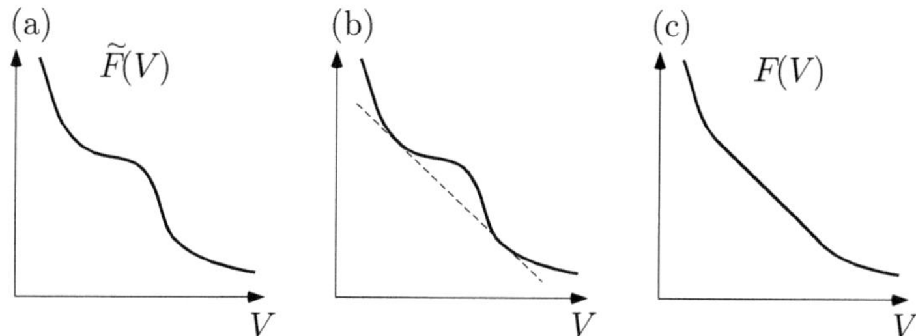

Figure 7.12 Constructing the convex function $F[V]$ "closest" to the pseudo free energy $\widetilde{F}[V]$ that corresponds to the non-physical pressure $\tilde{p}(V)$. The Maxwell equal-area rule can be obtained through comparison of $F[V]$ and $\widetilde{F}[V]$. Interestingly, the function $F[V]$ can also be obtained by applying the Legendre transformation and then its inverse to $\widetilde{F}[V]$.

therefore unconvincing on physical grounds. For this reason, here we do not present this derivation. Instead, below we show that this rule can be naturally derived from a fundamental postulate in thermodynamics, the convexity of the Helmholtz free energy.

If we formally substitute the non-physical pressure $\tilde{p}(V)$ for $p(T; V', N)$ in (3.33) and carry out the integral there to derive the Helmholtz free energy, we obtain a non-convex function $\widetilde{F}[V]$ like that plotted in Figure 7.12(a). Because this cannot be regarded as a physical free energy, let us call it a "pseudo free energy." (See Section 10.3 for related discussion.) The procedure depicted in Figure 7.12 allows us to obtain a convex free energy from this pseudo free energy with the smallest possible modification. In this procedure, first, the line tangent to $\widetilde{F}[V]$ at two points is drawn. Next, the part of the graph of $\widetilde{F}[V]$ between these two points is replaced with this tangent line. The function obtained in this way is plotted in (c). We refer to this function as $F[V]$.[44] This function is convex and can be regarded as a physical Helmholtz free energy.[45] Show that the

---

[44] As shown in Appendix H.4, $F[V]$ can also be obtained by applying the Legendre transformation and then its inverse to $\widetilde{F}[V]$. Mathematically, $F[V]$ is referred to as the *convex hull* of $\widetilde{F}[V]$.

[45] The reader may question the validity of this procedure for producing a physically meaningful free energy. Indeed, there is no rigorous proof that the free energy obtained in this manner is the "correct" one, and some skepticism is certainly warranted. However, we can offer the following argument in support of this procedure.

First, note that the van der Waals pressure, given in (3.37), is the simplest extension of the ideal gas pressure and, as such, it possesses an intuitively appealing universal form. It is thus reasonable to conjecture that this pressure and the free energy it yields would be physically meaningful in situations that the behavior of the system is insensitive to small variations in the system parameters. In general, we expect such insensitivity to exist sufficiently far from phase transitions, while we know that near phase transitions, systems in fact become quite sensitive to variations in the system parameters. Hence, there is good reason to believe that the form of the free energy obtained from the van der Waals pressure would be valid sufficiently far from the phase transition region, while the appearance of unphysical behavior (i.e., non-convexity) in the phase transition region is also understandable. It is thus reasonable that if we could find a trustworthy method to "fix" the unphysical behavior in the phase transition region, we could obtain a generally valid free energy. As a minimal requirement for this type of method, we know that it must yield a free energy that is a convex function of V. Although there are indefinitely many possible ways of accomplishing this, that demonstrated in Figure 7.12 is clearly the simplest. (Although, obviously, in science simplicity does not necessarily imply validity, experience tells us that there is some connection between the two.) Also, with the function $F(V)$ thus obtained, the region of constant slope corresponds to a coexistence region connecting the liquid and gas phases over which the pressure given by $p(V) = -dF[V]/dV$ is identical to that obtained from the Maxwell construction.

7.9 (Section 7.5.4) Consider a system consisting of an amount N of a pure substance, and suppose that it undergoes the following Carnot cycle:

$$(T'; V_0', N) \xrightarrow{\text{qi}} (T'; V_1', N) \xrightarrow{\text{qa}} (T; V_1, N) \xrightarrow{\text{qi}} (T; V_0, N) \xrightarrow{\text{qa}} (T'; V_0', N) \,. \tag{7.69}$$

Let us assume that the system remains in the liquid-gas coexistence region during the entire cycle. For simplicity, let us also assume the conditions $V_0' < V_1' < V_1 > V_0 > V_0'$ and $T' > T$. Derive the Clausius-Clapeyron relation by applying Carnot's theorem to this cycle.

Hint: It may be useful to examine the change undergone by the enthalpy in the case that $\Delta T = T' - T$ is infinitesimally small.

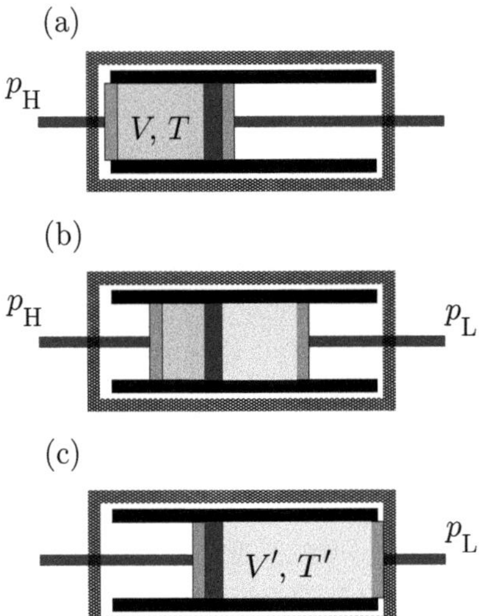

Figure 7.13 Schematic depiction of the Joule-Thomson effect. (a) Initially, the gas is entirely on the left side, in a state $(T; V, N)$ with pressure p_{H}. (b) Under the operation, gas is caused to slowly pass through the porous material to the right side. At all times during this operation, pressures p_{H} and p_{L} are maintained on the left and right sides, respectively, through the application of external forces on the pistons that exactly cancel those exerted by the gas. The entire system is surrounded by adiabatic walls. (c) At the completion of the operation, the gas is entirely on the right side, in a state $(T'; V', N)$ with pressure p_{L}. The enthalpies before and after the operation, $H(T; V, N)$ and $H(T'; V', N)$, are identical.

chemical potential is constant. We know that this is precisely the behavior seen in actual liquid-gas phase transitions (see Section 7.5).

On the basis of the above simple considerations, it can be said that it is at least reasonable to conjecture that the free energy $F(V)$ obtained in the manner demonstrated in Figure 7.12 is physically valid. Indeed, this conjecture is supported by experimental results, which are consistent with $F(V)$ in a wide variety of situations. (See Section 10.3.4 for a more detailed treatment of the same type of unphysical behavior in the context of ferromagnetic systems.)

7.10 (Section 7.5.4) In this problem we consider the famous Joule-Thomson experiment (although we study a somewhat altered experimental setup that allows for a simpler analysis).

The system that we consider is shown in Figure 7.13. A cylindrical container sealed on each side by a piston contains an amount N of a gas. A disk of porous material filling the cross section of the cylinder is fixed at a position between the pistons. Let us suppose that this porous material allows the gas to pass through it, but only very slowly. An adiabatic operation of the type described by the figure causes the transition $(T; V, N) \xrightarrow{\text{a}} (T'; V', N)$. As shown there, before the operation, all of the gas is to the left of the porous disk, and after, all of it is to the right.[46] We write the initial and final pressures of the system as $p_{\mathrm{H}} = p(T; V, N)$ and $p_{\mathrm{L}} = p(T'; V', N)$, assuming the relation $p_{\mathrm{H}} > p_{\mathrm{L}}$. During the operation, as depicted in (b), forces are applied externally to the pistons to exactly maintain the pressures p_{H} on the left side and p_{L} on the right. Note that although this is a slow adiabatic operation, it is not quasi-static, as seen from the fact that it cannot be carried out in reverse.

Show that the enthalpy of the system is unchanged by this operation. (For further consideration of the Joule-Thomson experiment, see Problems 8.6 and 8.7.)

---

[46] The original experiment was carried out not on gas contained in such a closed system, but instead on gas continuously flowing through an open pipe.

8

Gibbs Free Energy

In this chapter, we introduce and investigate the *Gibbs free energy*. Just as the Helmholtz free energy is of fundamental importance for the treatment of isothermal systems when the volume is a controlled quantity, the Gibbs free energy is of fundamental importance for the treatment of isothermal systems when the pressure is a controlled quantity, providing a complete thermodynamic description in this context.[1] The true descriptive power of the Gibbs free energy is realized in application to the treatment of multi-component systems, including systems exhibiting chemical reactions. However, in this chapter, we consider only single-component systems, in order to thoroughly elucidate the essential nature of the Gibbs free energy. Mathematically, this treatment provides the opportunity to introduce the Legendre transformation. The application of the Legendre transformation to convex functions lies at the foundation of an important method used extensively in the investigation of thermodynamic behavior.

8.1 Derivation of the Gibbs free energy

In this section, we define the Gibbs free energy in the case of an individual fluid system. We then investigate the fundamental properties of this quantity.

To this point in the book, we have considered systems for which the volume is directly controlled by the external agent and the pressure is determined by the state of the system. Now, we change our point of view, and instead regard the pressure as a controlled parameter. In this case, the volume is no longer controlled externally, but instead determined autonomously, through the evolution of the system itself.

8.1.1 Definition of the Gibbs free energy

When a thermodynamic system, say water in a beaker, is open to the atmosphere, its pressure is maintained at the ambient value. As discussed in Footnote 1 on p. 24, in such a situation, we model this mechanical interaction with the surrounding air by a purely mechanical device, as illustrated in Figure 8.1. For the purpose of describing the thermodynamic behavior of the system, this modeling method is entirely faithful, as long as the ambient air pressure can be treated as constant.

---

[1] As stated previously, the Helmholtz free energy is a complete thermodynamic function only when regarded as a function of its natural independent variables, T, V and N. As seen below, from this fact and the definition of the Gibbs free energy, it follows that the Gibbs free energy is a complete thermodynamic function only when regarded as a function of its natural independent variables, T, p and N. The entropy, energy and enthalpy too (with the proper independent variables) can act as complete thermodynamic functions, but in the construction of the theory given to this point, they have not been treated as such. (See Appendix F for detailed discussion.)

Thermodynamics: A Modern Approach. Hal Tasaki and Glenn Paquette, Oxford University Press.
© Hal Tasaki and Glenn Paquette (2026). DOI: 10.1093/9780191878091.003.0008

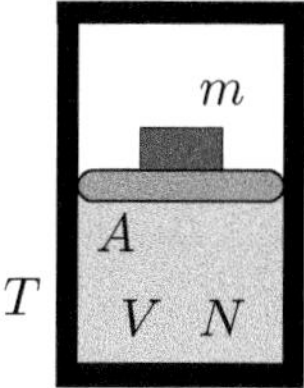

Figure 8.1 A fluid system whose pressure is fixed by a mechanical device. Fluid is contained in a vessel with a cross section of area A. A freely moveable massless piston acts as the lid of the vessel. A weight of mass m is positioned on top of the piston. In this situation, the pressure of the system is maintained at the value $p = mg/A$. (We have assumed that the space above the piston is a vacuum.)

In the situation described by Figure 8.1, a closed vessel of cross-sectional area A contains a fluid of amount N. A piston acts as the lid of the vessel, and a weight of mass m is positioned on top of the piston. For the sake of simplicity, we assume that the space above the piston is a vacuum and that the piston itself is massless. The entire system is in contact with an environment of temperature T. We assume that the piston is free to move spontaneously and that it eventually comes to rest at an equilibrium position at which the total force acting on it vanishes. Assuming that this system is located on Earth, a force of magnitude mg presses down on the piston (identifying the downward direction in the figure as the direction of gravity), and hence the pressure of the gas will be $p = mg/A$ in equilibrium.[2] According to the classification presented in Section 2.1.1, it is most natural to regard the weight maintaining the constant pressure as a part of the mechanical world. However, as seen below, to describe the thermodynamic behavior of this system, it is actually more useful to regard it as part of the system itself.

To begin our analysis of the system described by Figure 8.1, we recall (3.31), according to which the volume V of the fluid in equilibrium for given T, N and p can be determined by solving

$$-\frac{\partial}{\partial V}F[T;V,N] = p \,, \tag{8.1}$$

which here is regarded as an equation for V, with p fixed at the value mg/A.[3] With this equation and the convexity of $F[T;V,N]$, it is seen that the equilibrium value of V minimizes the quantity

$$F[T;V,N] + pV \,. \tag{8.2}$$

Now, if we write the volume as $V = Ah$, where h is the distance from the bottom of the container to the bottom of the piston, we find that the quantity pV is simply mgh, which is the familiar (mechanical) potential energy associated with the weight. Then, noting that a purely mechanical system can also be regarded

---

[2] Here, g is the acceleration due to gravity. Although the weight is subject to the gravitational force, we assume that gravity has no direct effect on the fluid (as done throughout the book, except in Appendix E). For an "ordinary," laboratory-sized system, this assumption is valid as long as the total mass of the fluid is sufficiently small.

[3] Under conditions characterized by phase coexistence, this equation has multiple solutions (see Section 8.4.1).

as a thermodynamic system for which both the energy and free energy are identified with the potential energy,[4] the quantity in (8.2) is interpreted as the total Helmholtz free energy of the composite system consisting of the fluid and the weight. Thus, we can also understand the fact that the equilibrium value of the volume is determined by minimizing (8.2) as being a result of a variational principle, as discussed in Section 7.3. Recall that, as demonstrated there, variational principles can be used to determine the equilibrium state of a system realized after a constraint has been removed. In the present case, this constraint is that of fixing the volume. From this point of view, (8.2) is regarded as the Helmholtz free energy of the composite system with the volume fixed at some arbitrary value V. Then, comparing the values of this Helmholtz free energy for all values of V, the variational principle (7.32) asserts that the value of V realized in equilibrium when the constraint is released, and the volume is able to vary freely, is that which minimizes (8.2).[5]

From the above discussion, we understand the minimum value of (8.2), which is a function of T, p and N, to be the Helmholtz free energy of the composite system when the volume is free to vary. But this is very different from the Helmholtz free energy studied to this point, and for this reason, we formally define it as a new quantity, $G[T, p; N]$:

$$G[T, p; N] := \min_{V} \{F[T; V, N] + pV\} \ . \tag{8.3}$$

We refer to this quantity as the *Gibbs free energy*.

Although the definition (8.3) was obtained through consideration of a particular system, because the treatment given here can be similarly applied to any system under conditions of fixed pressure, this definition can be regarded as entirely general. Also, because, unlike the Helmholtz free energy appearing in (8.2), the above Gibbs free energy has no explicit dependence on the potential energy of the pressure-fixing device, it is regarded as a quantity characterizing the fluid system alone.

As shown below, the Gibbs free energy is a complete thermodynamic function when regarded as a function of the intensive variables T and p and the extensive variable N. From the extensivity of the Helmholtz free energy, expressed by (3.24), for arbitrary $\lambda > 0$ we find that $G[T, p; \lambda N]$ behaves as

$$G[T, p; \lambda N] = \min_{\lambda V} \{F[T; \lambda V, \lambda N] + p\lambda V\}$$

$$= \lambda \min_{V} \{F[T; V, N] + pV\} = \lambda G[T, p; N] \ . \tag{8.4}$$

It is thus seen that $G[T, p; N]$ is also an extensive quantity.

---

[4] This follows from the facts that a mechanical system can only exchange energy through mechanical work and that its state does not depend on temperature.

[5] The physical situation considered here appears to be quite different from that for which the variational principle (7.32) was derived, but with the parts of the system identified appropriately, we see that in fact (7.32) applies to the present situation as well. First, let us compare Figures 7.1(b) and 8.1. We identify the fluid system to the left of the wall in Figure 7.1(b) with the fluid system of volume V in Figure 8.1, and the fluid system to the right of the wall in Figure 7.1(b) with the weight and the region of vacuum above the piston in Figure 8.1. With this rather formal identification, the total free energy $F[T; V_1, N_1] + F[T; V_2, N_2]$ on the right-hand side of (7.32) precisely corresponds term-by-term to $F[T; V, N] + pV$.

The procedure represented by (8.3), through which the function $G[T, p; N]$ is obtained from the original function $F[T; V, N]$, is known as the *Legendre transformation*. In Appendix H, we present a systematic treatment of the Legendre transformation from a mathematical point of view. Here we investigate it through consideration of $F[T; V, N]$ and $G[T, p; N]$.[6]

8.1.2 Geometrical meaning of the Legendre transformation and its inverse

In order to obtain a deep understanding of the Legendre transformation, as represented by (8.3), here we consider it from a geometrical point of view. In the following, in order to simplify the notation, we regard T and N to be fixed and write $F[T; V, N]$ as $f(V)$ and $G[T, p; N]$ as $g(p)$.

In Figure 8.2, a typical free energy $f(V)$ is plotted as a function of V. Choosing some $p > 0$ and V_0, we draw the line of slope $-p$ that passes through the point $(V_0, f(V_0))$. The equation for this line is $y = -p(V - V_0) + f(V_0)$, and hence its y-intercept is given by $y_0 = f(V_0) + pV_0$. Then, keeping p fixed, if we consider all the lines constructed in this manner, to each value of V_0, there corresponds a value of y_0. From (8.3), it is clear that the minimum among all such y_0 is the value of the Gibbs free energy at p. By repeating this procedure for all p, we obtain the Gibbs free energy $g(p)$ corresponding to $f(V)$.

Next, let us consider whether it is possible to recover the Helmholtz free energy, $f(V)$, from a given Gibbs free energy, $g(p)$. Mathematically, this is the problem of constructing the inverse Legendre transformation. This is done as follows. First, choosing a value of $p > 0$, we draw the line in the V-y plane of slope $-p$ whose y-intercept is $g(p)$. The equation for this line is

$$y = -pV + g(p) \ . \tag{8.5}$$

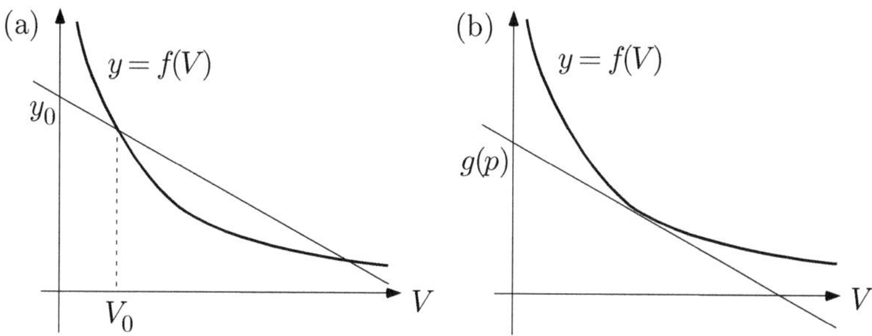

Figure 8.2 Geometrical interpretation of the Legendre transformation. The same Helmholtz free energy $y = f(V)$ is plotted in each graph. (a) For a given value of p, we draw the line of slope $-p$ that passes through the point $(V_0, f(V_0))$. The y-intercept of this line is written y_0. (b) Fixing the value of p, if we vary V_0, y_0 will also vary. The minimum value of y_0 obtained in this manner is $g(p)$, the value of the Gibbs free energy at p.

---

[6] As explained in Appendix H, the forms of the Legendre transformation conventionally used in thermodynamics and mathematics employ different sign conventions. While these forms are entirely equivalent, they do differ with regard to certain specific characteristics, as pointed out in Footnote 1 on p. 301.

This one line alone gives us very little information about $f(V)$. However, considering the method of obtaining $g(p)$ from $f(V)$ described above, we know that the graph of $y = f(V)$ is entirely above this line, except at one point (or on an interval), where they are tangent (see Figure 8.3(a)). Then, choosing a second value of p, we draw the line (8.5) corresponding to it. Again, we know that the graph of $y = f(V)$ must lie entirely above this line, except at the point(s) of tangency (see Figure 8.3(b)). Repeating the same procedure for many values of p, the form of $y = f(V)$ will gradually emerge (see Figure 8.3(c)). More precisely, the graph of $y = f(V)$ is reconstructed as the envelope of the set of lines defined by (8.5) for all values of $p > 0$.

Let us now express the above geometrical understanding of the inverse Legendre transformation in the form of an equation. Suppose that we wish to know the value of $f(V)$ for one particular value of V. Because the graph of $y = f(V)$ is never below the line (8.5), we have

$$f(V) \geq -pV + g(p) . \tag{8.6}$$

This inequality must hold for all values of $p > 0$, while equality is realized for at least one value of p. Hence, the value of $f(V)$ that we seek is obtained as

$$f(V) = \max_{p} \{g(p) - pV\} . \tag{8.7}$$

The right-hand side here is understood as the maximum value of the quantity in the brackets when regarded as a function of p in the domain $p > 0$.

Finally, making the T and N dependences explicit in (8.7), we have

$$F[T; V, N] = \max_{p} \{G[T, p; N] - pV\} . \tag{8.8}$$

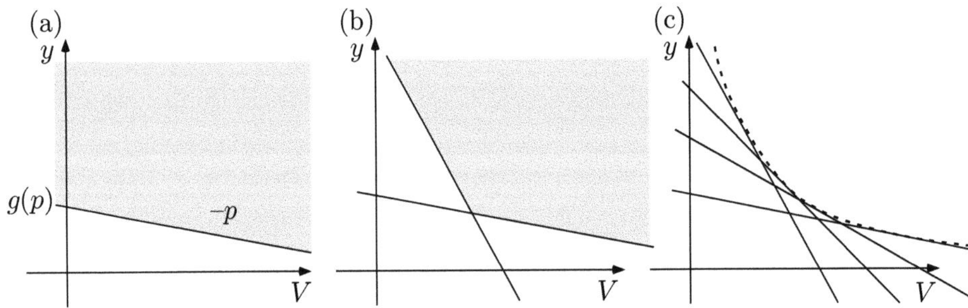

Figure 8.3 Intuitive understanding of the reconstruction of $f(V)$ from $g(p)$.
(a) Choosing an arbitrary value of $p > 0$, we draw the line of slope $-p$ and
y-intercept $g(p)$. The graph of $f(V)$ exists entirely inside the gray region above this
line, except at the point(s) where it is tangent to the line. (b) Choosing another value
of p, we draw another line according to the same prescription. In this way, the region
in which $f(V)$ resides is further narrowed. (c) With this process repeated for many
values of p, the graph of $y = f(V)$ emerges as the envelope of the set of lines defined
in this manner.

This equation allows us to derive the Helmholtz free energy $F[T; V, N]$ from a given Gibbs free energy $G[T, p; N]$.[7]

The fact that the Helmholtz free energy and Gibbs free energy can be derived from one another is of fundamental importance. This implies that we possess precisely the same information if we know either one. Thus, recalling that the Helmholtz free energy is a complete thermodynamic function, we conclude that the Gibbs free energy is also a complete thermodynamic function.[8]

8.1.3 Simplified definition of the Gibbs free energy

Here we introduce an alternative definition of the Gibbs free energy. This definition is less general than that given above, but it is often more useful in practice.

The right-hand side of (8.3) represents the procedure of varying V over all possible values and finding the minimum of the quantity in the brackets over this domain. In view of this aim, it is noteworthy that $F[T; V, N]$ is always differentiable with respect to V. For this reason, to find the minimum that we seek, we need only take the derivative of the quantity $F[T; V, N] + pV$ with respect to V and set the result equal to 0. Using (3.31), we obtain

$$\frac{\partial}{\partial V} \{F[T; V, N] + pV\} = -p(T; V, N) + p \, . \tag{8.9}$$

Hence, the condition determining the minimum can be written

$$p(T; V, N) = p \, . \tag{8.10}$$

This equation may appear strange, but note that the two quantities here have different interpretations: $p(T; V, N)$ is a state function, determined by T, V and N, while p represents the pressure as a controllable parameter that is fixed externally at the outset (for example, to mg/A in the case of the experimental setup described by Figure 8.1).[9] In the presently considered situation, T, p and N are given, and thus (8.10) should be regarded as an equation to determine V. Let us write the solution to this equation as $V(T, p; N)$. Rewriting (8.10) accordingly, we have

$$p(T; V(T, p; N), N) = p \, . \tag{8.11}$$

---

[7] The treatment here is not rigorous, but from the general theory of the Legendre transformation, it is possible to rigorously prove that for any $F[T; V, N]$ defined on an open interval of V and with $G[T, p; N]$ defined by (8.3), the inverse transformation (8.8) does yield $F[T; V, N]$. For related considerations, see Appendix H and Problem 8.5.

[8] It is now clear why the fact that F is a complete thermodynamic function only when regarded as a function of T, V and N and the definition of G together imply that G is a complete thermodynamic function only when regarded as a function of T, p and N. (For related discussion, see Problem 8.4 and Footnote 16 on p. 177.)

[9] Clearly, using the same letter to represent both of these quantities is poor notation. However, such notation is customary and accepted without question in scientific papers, and thus it is necessary to become accustomed to it. Generally, notation of this type does not cause any problem, because if the physical situation under consideration is clear, there should be no danger of confusing two quantities playing two such roles.

The quantity $V(T, p; N)$ is the volume that the system itself selects in the equilibrium state when T, p and N are given. This is identical to the value of V at which the quantity in the brackets on the right-hand side of (8.3) is minimized. In other words, $V(T, p; N)$ is the value of V at the point of tangency between the line and the graph of $y = f(V)$ shown in Figure 8.2(b).

Substituting $V(T, p; N)$ into (8.3), we immediately obtain

$$G[T, p; N] = F[T; V(T, p; N), N] + p\, V(T, p; N) \,. \tag{8.12}$$

In many textbooks, this equation is used as the definition of the Gibbs free energy.[10]

Now, note that in the above derivation, we have assumed that (8.10) determines a unique value of V. However, there are situations in which there exist multiple values of V that satisfy this equation. For example, this is true for systems in which the liquid and gas phases coexist. This is because, as we found in Section 7.5, in the liquid-gas coexistence region, it is possible to vary the volume without varying the pressure. In this case, (8.12) is invalid as a definition. Contrastingly, because the minimization with respect to V in (8.3) is meaningful even in this case, it remains a proper definition. It is thus seen why we must regard the rigorous definition of the Gibbs free energy to be that given in (8.3); (8.12) should be regarded as merely a convenient formula that holds in the case that (8.10) possesses a unique solution $V(T, p; N)$.[11]

With our shift from considering the Helmholtz free energy as a function of T, V and N to the Gibbs free energy as a function of T, p and N, the reader might expect that we will now regard an equilibrium state to be specified by $(T, p; N)$ rather than $(T; V, N)$. However, because the values of T, p and N do not determine the value of V in the liquid-gas coexistence region, these quantities do not provide a complete parameterization of the equilibrium states of a fluid system.[12] For this reason, in this book, we do not use $(T, p; N)$ to specify equilibrium states. However, we stress that the Gibbs free energy is well defined as a function of T, p and N,[13] and as such is a complete thermodynamic function.

8.1.4 Gibbs free energy of an ideal gas

Before ending this section, we derive the Gibbs free energy of an ideal gas. Because an ideal gas exhibits no phase transition, in this case we can use the simplified definition, (8.12). From the equation of state (3.35), we have $V(T, p; N) = NRT/p$,

---

[10] There are even some textbooks that omit the arguments of the functions appearing here, simply defining G as $F + pV$. With this definition, the functional dependence of G is left undetermined. (See Footnote 16 on p. 177 for related discussion.)

[11] However, it should be noted that even when there is phase coexistence, (8.12) yields the correct Gibbs free energy if we choose $V(T, p; N)$ to be one of the solutions of (8.10).

[12] In connection to this point, recall that even T, V and N themselves do not uniquely specify all equilibrium states at the triple point, where the volume or amount of substance of one of the three phases is necessary to provide a complete description. (For detailed discussion, see Appendix D.)

[13] In other words, if we consider two states with the same T, p and N but different V, the Gibbs free energy will necessarily take the same value in both states.

and thus from (8.12) and the form of the Helmholtz free energy given in (7.9), we immediately obtain

$$
\begin{aligned}
G[T,p;N] &= -NRT\log\left\{\left(\frac{T}{T^*}\right)^c \frac{V(T,p;N)}{v^*N}\right\} + Nu + pV(T,p;N)\\
&= NRT - NRT\log\left\{\left(\frac{T}{T^*}\right)^c \frac{RT}{v^*p}\right\} + Nu\\
&= NRT + NRT\log\left\{\left(\frac{T^*}{T}\right)^{c+1} \frac{p}{p^*}\right\} + Nu\,.
\end{aligned}
\tag{8.13}
$$

Here, we have introduced the reference pressure defined by $p^* = RT^*/v^*$.

8.2 Derivatives of the Gibbs free energy

8.2.1 Relations obtained through differentiation of $G[T,p;N]$

Here we compute derivatives of the Gibbs free energy and consider several resulting relations. Before beginning, however, let us point out that some care must be taken in the application of the results obtained here, because at phase transition points, $G[T,p;N]$ is not even once differentiable with respect to p or T. (This can be understood from Figure 8.6.) In the following, we exclude phase transition points, considering only regimes in which $V(T,p;N)$ is a well-defined function differentiable with respect to T, p and N, and hence $G[T,p;N]$ too is differentiable with respect to T, p and N.

Derivative with respect to T
Taking the derivative of (8.12) with respect to T and applying the equations expressing the derivatives of the Helmholtz free energy given in (7.11), we derive the following:

$$
\begin{aligned}
\frac{\partial G[T,p;N]}{\partial T} &= \frac{\partial}{\partial T}\left\{F[T;V(T,p;N),N] + pV(T,p;N)\right\}\\
&= -S(T;V(T,p;N),N) - \frac{\partial V(T,p;N)}{\partial T}p(T;V(T,p;N),N)\\
&\quad + p\frac{\partial V(T,p;N)}{\partial T}\\
&= -S(T,p;N)\,.
\end{aligned}
\tag{8.14}
$$

Here, we have used (8.11) and defined the entropy as a function of T, p and N as[14]

$$
S(T,p;N) = S(T;V(T,p;N),N)\,.
\tag{8.15}
$$

---

[14] Because we define the entropy in this way, when $V(T,p;N)$ is undefined or discontinuous, the same holds for $S(T,p;N)$. It may seem that the situation is similar for $\mu(T,p;N)$, but as seen from (8.19), $\mu(T,p;N)$ is continuous at all T and p.

Derivative with respect to p

Taking the derivative of (8.12) with respect to p and proceeding similarly, we obtain

$$\frac{\partial}{\partial p} G[T, p; N] = V(T, p; N) \,. \tag{8.16}$$

In analogy to the expression for the Helmholtz free energy given in (3.33), written as an integral of the pressure over the volume, integrating (8.16) over p, we obtain the following integral expression for the Gibbs free energy:

$$G[T, p; N] = \int_{p(T)}^{p} dp' \, V(T, p'; N) \,. \tag{8.17}$$

Because the reference value $p(T)$ is not specified by the equation of state alone, (8.17) does not determine $G[T, p; N]$ entirely, but instead only determines how $G[T, p; N]$ varies with p. The volume $V(T, p; N)$ changes discontinuously as p is varied across a phase transition point, but even if the integral contains such a point, (8.17) holds exactly.

Derivative with respect to N

Finally, taking the derivative of (8.12) with respect to N yields

$$\frac{\partial}{\partial N} G[T, p; N] = \mu(T, p; N) = \mu(T, p) \,. \tag{8.18}$$

Here, $\mu(T, p; N)$ is defined as $\mu(T; V(T, p; N), N)$, in analogy to $S(T, p; N)$. As indicated, the function $\mu(T, p; N)$ is often written simply as $\mu(T, p)$. The reason for this is that because the chemical potential is an intensive quantity, it satisfies the relation $\mu(T, p; \lambda N) = \mu(T, p; N)$, and hence for a single-component system, $\mu(T, p; N)$ is independent of N.

As noted above, the Gibbs free energy is an extensive quantity, and therefore it satisfies the relation $\lambda G[T, p; N] = G[T, p; \lambda N]$. If we take the derivative of both sides of this equation with respect to λ and then set $\lambda = 1$, using (8.18), we obtain

$$G[T, p; N] = N \frac{\partial}{\partial N} G[T, p; N] = N\mu(T, p) \,. \tag{8.19}$$

This is the Euler equation for the Gibbs free energy. (See (7.8) for comparison.) We thus see that for a single-component system, the chemical potential is simply the Gibbs free energy normalized by N.[15]

---

[15] The relation (8.19) may clarify the reason that μ is called the *chemical potential*. Let us compare the situation here with that for an electric potential. If an object of charge q is placed at a position with electric potential φ, its electric potential energy is $q\varphi$. Hence, because $G = N\mu$ is a type of potential energy, there is a direct correspondence between $q\varphi$ and $N\mu$. Thus, with the understanding that q and N both represent amounts of certain physical attributes, we can interpret μ and φ as being quantities of the same type. The validity of this interpretation becomes even clearer when we consider the definition of the electrochemical potential, given in (9.133).

Combining (8.19) and (8.13), we immediately obtain the chemical potential of an ideal gas as a function of T and p:

$$\mu(T,p) = RT + RT \log \left\{ \left(\frac{T^*}{T} \right)^{c+1} \frac{p}{p^*} \right\} + u \ . \tag{8.20}$$

Of course, this is identically what we would obtain if we applied the appropriate change of variables to the previously derived expression, (7.10).

8.2.2 Differential form for the Gibbs free energy

Using the differential form for the Helmholtz free energy given in (7.12) and the useful formulas appearing in (7.16) and (7.17), the expressions (8.14), (8.16) and (8.18) for the derivatives of the Gibbs free energy can be derived quite efficiently. This is the main method used for deriving these expressions in most thermodynamics textbooks, and it is indeed useful. For this reason, it is worth considering.

First, let us rewrite (8.12), simply omitting the arguments of the functions appearing there:

$$G = F + pV \ . \tag{8.21}$$

Taking the total differential of each side of this equation and using (7.12) and (7.16), we obtain

$$\begin{aligned} dG &= dF + d(pV) \\ &= -S\,dT - p\,dV + \mu\,dN + V\,dp + p\,dV \\ &= -S\,dT + V\,dp + \mu\,dN \ . \end{aligned} \tag{8.22}$$

Following the interpretation of differential forms explained at the end of Section 7.1.4, from (8.22) we conclude that the independent variables of G are T, p and N.[16] Then, from (8.22) we can simply read off the derivatives of G with respect to these variables, reproducing the results found above. Of course, what we have done here is equivalent to the computations carried out above to obtain (8.14), (8.16) and (8.18), which involve derivatives of composite functions, but clearly, the present approach is extremely simple and efficient. As long as the meaning of this approach is understood, it is a very useful computational method.

8.3 Heat capacity at constant pressure

The heat capacity at constant volume, $C_\mathrm{v}(T; V.N)$, introduced in Section 4.3, is an important quantity characterizing thermodynamic systems. However, in many experimental situations, the pressure is fixed, and the volume is free to change. In situations of this type, the relevant heat capacity is that at constant pressure, $C_\mathrm{p}(T, p; N)$. In this section, we study this quantity.

---

[16] We could define the Gibbs free energy as a function of T, V and N, writing $G(T, V; N) = F[T; V, N] + p(T; V, N)V$. (Although, in this case it would not be a complete thermodynamic function.) However, as seen above, applying the rule for interpreting differential forms, the "natural" independent variables of G are automatically obtained as T, p and N.

8.3.1 Expression for the heat capacity at constant pressure

Let us consider a system under adiabatic conditions whose pressure is maintained at a constant value, p. Then, suppose that an amount of energy ΔW is transferred to the system through some means (for example through work or an electric heating element). As a result, generally, the temperature of the system will increase, and simultaneously the volume of the system will change.[17] Thus, we understand that the energy added to the system, ΔW, goes into both temperature and volume changes.[18]

In the situation described above, let ΔT be the change in temperature resulting from the input of energy ΔW. Then, the heat capacity at constant pressure of this system is defined as $C_\mathrm{p}(T, p; N) := \lim_{\Delta W \to 0} \Delta W / \Delta T$. This definition reflects the experimental situation considered here, in which ΔW is controlled, and ΔT is a function of ΔW. However, because ΔW does not represent the change of a state variable, while ΔT does, we wish to regard ΔW as a function of ΔT. In "ordinary" situations, this role reversal is possible.[19] With this understanding, we rewrite the above definition in its standard form, $C_\mathrm{p}(T, p; N) := \lim_{\Delta T \to 0} \Delta W / \Delta T$.

We next derive an explicit form for the heat capacity at constant pressure. Before doing so, however, we point out a common misunderstanding regarding this quantity. Noting that the internal energy as a function of T, p and N is $U(T, p; N) = U(T; V(T, p; N), N)$, and considering the expression for the heat capacity at constant volume appearing in (4.30), one might be tempted to conclude that the heat capacity of a system with fixed pressure is given simply by

$$C_\mathrm{p}(T, p; N) = \frac{\partial U(T, p; N)}{\partial T} \quad \text{(incorrect)} . \tag{8.23}$$

However, considering the physical situation under investigation, we realize that this expression does not include the work done by the system on the mechanical world as it changes volume (see Figure 8.4), and hence it is incorrect.

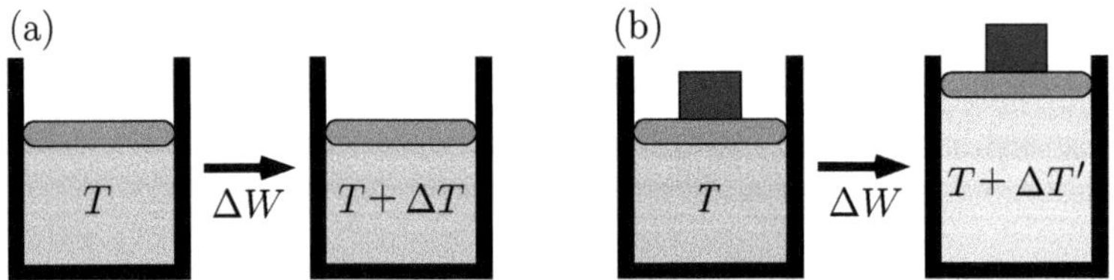

Figure 8.4 (a) In the case of a system at constant volume, the energy input from the external world, ΔW, acts solely to increase the temperature of the fluid. (b) In the case of a system at constant pressure, ΔW acts not only to increase the temperature of the fluid but also to increase its volume, thereby raising the weight.

---

[17] For most types of systems, in such a situation, the volume is an increasing function of temperature. However, there are exceptional cases. For example, for liquid water at atmospheric pressure below 4°C, the volume is a decreasing function of temperature.

[18] The energy that goes into the change of volume consists of both the work done against external forces and the change in the internal energy, $U(T; V, N)$, due to its volume dependence.

[19] Ordinarily, ΔT is an increasing function of ΔW, and hence the $\Delta W \to 0$ and $\Delta T \to 0$ limits are equivalent. The only exception is in the liquid-gas coexistence region. In this region, ΔT vanishes even for finite ΔW, and thus the quantity $\Delta W / \Delta T$ diverges, and $C_\mathrm{p}(T, p; N)$ is ill defined.

To obtain the correct expression for $C_\mathrm{p}(T, p; N)$, we consider the situation depicted in Figures 8.1 and 8.4(b), where a constant pressure is maintained by the weight situated on top of the piston with cross-sectional area A. Suppose that, as in Figure 8.4(b), the temperature of the fluid increases from T to $T + \Delta T$ with the pressure fixed at p, and as a result, the volume of the fluid increases by some amount $\Delta V = V(T + \Delta T, p; N) - V(T, p; N)$. This implies that the piston rises by an amount $\Delta h = \Delta V / A$, and the potential energy of the weight thus increases by $mg\,\Delta h = p\,\Delta V$, where $p = mg/A$. Then, using the law of energy conservation, we have the equality

$$\Delta W = U(T + \Delta T; V_0 + \Delta V, N) - U(T; V_0, N) + p\,\Delta V \,, \tag{8.24}$$

where $V_0 = V(T, p; N)$ is the initial volume.[20]

Applying a Taylor expansion in ΔT and ΔV to (8.24), we obtain

$$
\begin{aligned}
C_\mathrm{p}(T, p; N) &= \lim_{\Delta T \to 0} \frac{\Delta W}{\Delta T} \\
&= \left.\frac{\partial U(T; V, N)}{\partial T}\right|_{V=V(T,p;N)} + \frac{\partial V(T, p; N)}{\partial T} \left.\frac{\partial U(T; V, N)}{\partial V}\right|_{V=V(T,p;N)} \\
&\quad + p\,\frac{\partial V(T, p; N)}{\partial T} \,.
\end{aligned}
\tag{8.25}
$$

It is important here to clearly distinguish between the independent variable V and the function $V(T, p; N)$. Now, from the definition of the entropy, (6.5), we have

$$
\begin{aligned}
\frac{\partial U(T; V, N)}{\partial V} &= \frac{\partial F[T; V, N]}{\partial V} + T\,\frac{\partial S(T; V, N)}{\partial V} \\
&= -p(T; V, N) + T\,\frac{\partial S(T; V, N)}{\partial V} \,,
\end{aligned}
\tag{8.26}
$$

where we have used the definition of the pressure given in (3.31). Using (6.15) and (8.26), we can rewrite (8.25) as

$$
\begin{aligned}
C_\mathrm{p}(T, p; N) &= T\left.\frac{\partial S(T; V, N)}{\partial T}\right|_{V=V(T,p;N)} + T\,\frac{\partial V(T, p; N)}{\partial T}\left.\frac{\partial S(T; V, N)}{\partial V}\right|_{V=V(T,p;N)} \\
&= T\,\frac{\partial S(T; V(T, p; N), N)}{\partial T} \,,
\end{aligned}
\tag{8.27}
$$

where we have applied (8.11). Then, recalling (8.15) and (8.14), we arrive at the following:

$$C_\mathrm{p}(T, p; N) = T\frac{\partial S(T, p; N)}{\partial T} = -T\frac{\partial^2 G[T, p; N]}{\partial T^2} \,. \tag{8.28}$$

Note the similarity between (8.28) and the expression for the heat capacity at constant volume given in (7.4).[21] Finally, it should be noted that although the

---

[20] The "weight" considered here is not to be regarded as a proper thermodynamic system. Rather, this is a (hypothetical) purely mechanical system possessing only mass, with 0 intrinsic heat capacity. Thus there is no contribution from the internal energy of the weight in (8.24).

[21] This similarity is not a mere coincidence. This can be understood from the following shortcut for obtaining (8.28). Consider the situation depicted in Figure 8.1, but now regard the entire system

above derivation of (8.28) was carried out in reference to the situation depicted in Figure 8.1, in fact, the arguments used here and (8.28) itself are entirely general.

Integrating both sides of (8.28), we obtain

$$S(T, p; N) = S(T_0, p; N) + \int_{T_0}^{T} dT' \, \frac{C_{\mathrm{p}}(T', p; N)}{T'} \, . \tag{8.29}$$

We have thus derived a useful relation from which we can determine the change in entropy due to a change in temperature using the heat capacity at constant pressure.[22] Recall that a relation similar to (8.29) involving the heat capacity at constant volume is given in (6.25). (See Section 6.3 for related discussion.)

The heat capacity at constant pressure is an extensive state function, and hence it varies in proportion to the size of the system. Similarly to the case of the heat capacity at constant volume, in actual applications, it is common to use the *specific heat at constant pressure*, which generically is the heat capacity at constant pressure divided by an extensive parameter. For a single-component fluid, the molar specific heat at constant pressure, $c_{\mathrm{p}}(T, p) = C_{\mathrm{p}}(T, p; N)/N$, is a function of T and p alone.

The heat capacity at constant pressure is an experimentally measurable quantity. Therefore, applying the Nernst-Planck postulate to fix $S(T_0, p; N)$ in (8.29), we can determine $S(T, p; N)$ explicitly from experimental data. We can then derive the T dependence of $G[T, p; N]$ from (8.14). Furthermore, using experimental data for the pressure dependence of the volume, we can determine the p dependence of $G[T, p; N]$ from (8.17). Combining these, for given N, we can entirely determine $G[T, p; N]$ up to an overall additive constant. (From (8.19), we know that this constant is proportional to N.) Therefore, because the Gibbs free energy is a complete thermodynamic function, in this way, we can obtain complete thermodynamic information about the system in question. Today, with a growing body of experimental data and progress in computer-aided data analysis (and data sharing) technology, we are gradually compiling accurate quantitative representations of the Gibbs free energy and other complete thermodynamic functions for actual physical systems.

8.3.2 Heat capacity at constant pressure for an ideal gas

From (8.28) and the expression for the Gibbs free energy given in (8.13), we immediately obtain the following expression for the heat capacity at constant pressure in the case of an ideal gas:

$$C_{\mathrm{p}}(T, p; N) = (c + 1)NR \, . \tag{8.30}$$

(including the weight) as our thermodynamic system. This system has only one extensive variable, N. Writing the energy and the Helmholtz free energy of this system as $U_{\mathrm{whole}}(T; N)$ and $F_{\mathrm{whole}}[T; N]$, from (7.4) we find $C_p(T, p; N) = \partial U_{\mathrm{whole}}(T; N)/\partial T = -T\partial^2 F_{\mathrm{whole}}[T; N]/\partial T^2$. This is identical to (8.28), because (8.3) implies the equality $F_{\mathrm{whole}}[T, N] = G[T, \mu, N]$.

[22] In analogy to $S(T_0; X)$ in (6.25), the constant of integration $S(T_0, p; N)$ here cannot be determined from the heat capacity alone. But, as discussed in Section 6.1.4, according to the Nernst-Planck postulate, $S(T_0, p; N)$ is determined by the condition $\lim_{T \to 0} S(T, p; N) \searrow 0$.

Recall that we postulated the constant value $C_v(T; V, N) = cNR$ for the heat capacity at constant volume of an ideal gas. With this, writing the heat capacity at constant pressure as

$$C_p(T, p; N) = C_v(T; V(T, p; N), N) + NR\,, \tag{8.31}$$

we can understand the first term as corresponding to the energy needed to increase the temperature of the gas and the second term as corresponding to the work done on the mechanical world as the gas expands.[23]

8.4 Properties of the Gibbs free energy

In this section, we investigate the general behavior of the Gibbs free energy for a single-component fluid, $G[T, p; N] = N\mu(T, p)$.

8.4.1 Pressure dependence of the Gibbs free energy

Let us consider a system under isothermal conditions at a temperature T for which the system does not undergo a phase transition, such as T_1 or T_2 in the situation described by Figure 7.3. In this case, the Helmholtz free energy is a smoothly decreasing convex function of V, as shown in Figure 7.4. In Figure 8.5(a), the value of $G[T, p; N]$ at each of the three values of the pressure considered, p_1, p_2 and p_3, is determined using the geometrical method described in Section 8.1. Carrying out the procedure depicted in Figure 8.5(a) for all values of $p > 0$, we obtain $G[T, p; N]$ as the smoothly increasing function graphed in Figure 8.5(b).

Next, let us consider a temperature (like T_3 in the case of Figure 7.3) at which the system does exhibit a gas-liquid phase transition. The Helmholtz free energy for such a system is plotted in Figure 7.5. As seen there, over the range of values of the volume satisfying $V_L(T; N) \le V \le V_G(T; N)$, this graph has a constant slope of $-p_v(T)$. Now let us carry out the construction of the Gibbs free energy in this case using the same geometrical method as above. This procedure is depicted in Figure 8.6(a). First, for a value of the pressure p_1 below $p_v(T)$, the line of slope $-p_1$ determining the value of $G[T, p_1; N]$ is tangent to the graph of $F[T; V, N]$ at one point in the gas region. Next, for $p = p_v(T)$, the line of slope $-p_v(T)$ determining the value of $G[T, p_v(T); N]$ coincides with the graph of $F[T; V, N]$ over the entire coexistence region. Finally, for a value of the pressure p_2 above $p_v(T)$, the line of slope $-p_2$ determining the value of $G[T, p_2; N]$ is tangent to the graph of $F[T; V, N]$ at one point in the liquid region. We thus find that due to the existence of the region in which the slope of $F[T; V, N]$ is constant, the slope of $G[T, p; N]$ is undefined at $p = p_v(T)$, as shown in Figure 8.6(b). This behavior reflects the fact that the single point on the graph at $p = p_v(T)$ represents the entire liquid-gas coexistence region. For this reason, many actual states are "compressed" into this one point.

---

[23] As here, it is generally the case that the heat capacities of a system satisfy the inequality $C_p(T, p; N) \ge C_v(T; V(T, p; N), N)$ (see Problem 8.9).

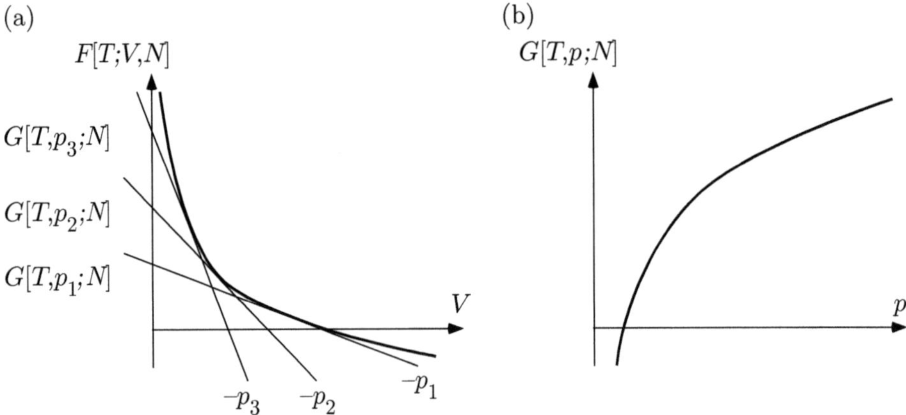

Figure 8.5 Construction of the Gibbs free energy from the Helmholtz free energy for a system exhibiting no phase transition. (a) The y-intercept of the line with slope $-p$ tangent to the graph of $F[T; V, N]$ is $G[T, p; N]$. (b) The graph of $G[T, p; N]$ constructed from the y-intercepts obtained as in (a) for all values of p. The values of T and N are fixed and equal in the two graphs.

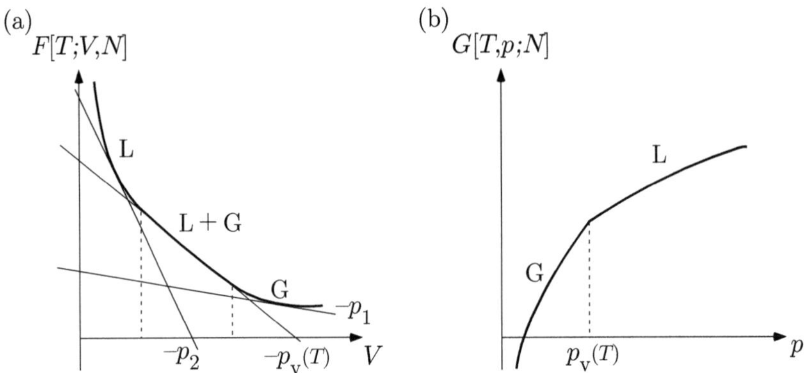

Figure 8.6 Construction of the Gibbs free energy from the Helmholtz free energy for a system exhibiting a phase transition. (a) The y-intercept of the line with slope $-p$ tangent to the graph of $F[T; V, N]$ is $G[T, p; N]$. (b) The graph of $G[T, p; N]$ constructed from the y-intercepts obtained as in (a) for all values of p. The slope of $G[T, p; N]$ changes discontinuously at the pressure corresponding to the phase transition, $p_{\mathrm{v}}(T)$. The values of T and N are fixed and equal in the two graphs.

In both the cases with and without a phase transition, $G[T, p; N]$ is a concave function of p (see Problem 8.5).[24] Also, note that the general forms of $G[T, p; N]$ in the two cases can be inferred directly from (8.17), recalling the behavior of V when considered as a function of p.

---

[24] A function $f(x)$ is said to be a *concave* function of x if $-f(x)$ is a convex function of x (see Appendix G).

8.4.2 Temperature dependence of the Gibbs free energy

As the final topic in this chapter, we study the behavior of $G[T, p; N]$ when we fix p (to a value below the critical pressure, p_c) and vary T. Here, we wish to study the temperature dependence of the Gibbs free energy in the neighborhood of the boiling point, $T_b(p)$, below which the fluid is a liquid, and above which it is a gas, as shown in Figure 7.7. Because we consider the case of fixed pressure, below we write $T_b(p)$ simply as T_b. Thus, in this discussion, $p_v(T_b)$ is identically the fixed value of the pressure itself.

As expressed in (8.14), we can determine the entropy $S(T, p; N)$ by taking the derivative of the Gibbs free energy with respect to the temperature. However, this derivative is undefined at T_b, and hence so too is $S(T, p; N)$. In order to properly describe the behavior here, we should therefore first re-express $S(T, p; N)$ in terms of $(T; V, N)$, because $S(T; V, N)$ is a continuous function at the boiling point.

First, with the pressure fixed, let us allow the temperature to approach T_b from the high-temperature side. In this case, the system is always a gas, and therefore we have[25]

$$\frac{\partial G[T_b, p; N]}{\partial T_+} := \lim_{T \searrow T_b} \frac{\partial G[T, p; N]}{\partial T} = - \lim_{T \searrow T_b} S(T, p; N)$$

$$= - \lim_{\substack{T \searrow T_b \\ V \searrow V_G(T_b; N)}} S(T; V, N) = -S(T_b; V_G(T_b; N), N) \,. \tag{8.32}$$

Similarly, approaching T_b from the low-temperature side, we have

$$\frac{\partial G[T_b, p; N]}{\partial T_-} := \lim_{T \nearrow T_b} \frac{\partial G[T, p; N]}{\partial T} = -S(T_b; V_L(T_b; N), N) \,. \tag{8.33}$$

Next, using the relation between the enthalpy of vaporization (heat of vaporization) and the entropy given in (7.55), we obtain

$$H_{\text{vap}}(p; N) = T \left\{ S(T_b; V_G(T_b; N), N) - S(T_b; V_L(T_b; N), N) \right\}$$

$$= T \left\{ \frac{\partial G[T_b, p; N]}{\partial T_-} - \frac{\partial G[T_b, p; N]}{\partial T_+} \right\} \,. \tag{8.34}$$

We have thus derived an expression for the enthalpy of vaporization in terms of the discontinuity exhibited by the derivative of the Gibbs free energy as T crosses

[25] The quantities $\partial G[T_b, p; N]/\partial T_+$ in (8.32) and $\partial G[T_b, p; N]/\partial T_-$ in (8.33) represent the *right derivative* and *left derivative* of $G[T_b, p; N]$, respectively. Although (8.32) and (8.33) are not the general definitions of these quantities, in the present case, these expressions are equivalent to the general definitions. (See Appendix G for further discussion.)

the boiling point.[26] Using (8.19), this can be rewritten as

$$h_{\text{vap}}(p) = T \left\{ \frac{\partial \mu(T_{\text{b}}, p)}{\partial T_-} - \frac{\partial \mu(T_{\text{b}}, p)}{\partial T_+} \right\} . \tag{8.35}$$

Equations (8.34) and (8.35) provide an important connection between the complete thermodynamic function $G[T_{\text{b}}, p; N]$ and experimentally measurable quantities.

Problems 8

8.1 (Sections 8.1 and 8.2 and Appendix F) Let us define the energy as a function of S, V and N through the inverse Legendre transformation as

$$U[S, V, N] = \max_T \{ F[T; V, N] + TS \} . \tag{8.36}$$

(This should be compared with (8.8).) As indicated by the square brackets, $U[S, V, N]$ is a complete thermodynamic function. Prove the extensivity of $U[S, V, N]$, expressed by $U[\lambda S, \lambda V, \lambda N] = \lambda U[S, V, N]$. Next, show that the equality $U(T; V, N) = U[S(T; V, N), V, N]$ holds wherever $S(T; V, N)$ is well defined (i.e., everywhere except at the triple point). Finally, evaluate the partial derivatives of $U[S, V, N]$ with respect to S, V and N.

8.2 (Section 8.1) Using (8.36) and the variational inequality for the Helmholtz free energy appearing in (7.30), demonstrate the variational inequality

$$U[S, V, N] \leq U[S_1, V_1, N_1] + U[S_2, V_2, N_2] \tag{8.37}$$

for arbitrary S_1, S_2, V_1, V_2, N_1 and N_2 satisfying the relations $S = S_1 + S_2$, $V = V_1 + V_2$, and $N = N_1 + N_2$. Show that from this result and the extensivity of $U[S, V, N]$, it follows that $U[S, V, N]$ is a convex function of S, V and N (see Section G.2).

8.3 (Section 8.1) Consider two systems with entropies, volumes and amounts of substance S_1, V_1 and N_1 and S_2, V_2 and N_2, respectively, and suppose that these systems are separated by a wall. Then, suppose that we carry out an adiabatic operation under which this wall is removed, thereby obtaining a system with entropy S, volume $V = V_1 + V_2$, and amount of substance $N = N_1 + N_2$. Using the law of energy conservation and the variational inequality (8.37), prove the inequality $S \geq S_1 + S_2$.

Hint: Use the fact that $U[S, V, N]$ is an increasing function of S (see Problem 8.1).

---

[26] When we defined the enthalpy in (7.55), we were considering the situation in which the temperature is fixed, and this is the reason that we used the notation $H_{\text{vap}}(T; N)$. In the present case, however, we are considering the case in which the pressure is fixed and the temperature is being varied. Thus, in this case, we express the same quantity as $H_{\text{vap}}(p; N)$. (To be entirely rigorous, we should use different notation, but this is probably unnecessary.) The enthalpy of vaporization is defined only at the liquid-gas coexistence point, and therefore if either T or p is known, the other is automatically determined. (As discussed in Section 7.5.4, because the independent variables here are T, p and N, the coexistence "region" consists of a single point.) Mathematically, this situation is expressed as follows: $H_{\text{vap}}(p_{\text{v}}(T); N) = H_{\text{vap}}(T; N)$; $H_{\text{vap}}(T_{\text{b}}(p); N) = H_{\text{vap}}(p; N)$.

8.4 (Section 8.1) The Helmholtz free energy as a function of T, p and N can be defined as

$$F(T,p;N) = F[T; V(T,p;N), N] \,. \tag{8.38}$$

Here, $V(T,p;N)$ is understood to be the volume as a function of T, p and N satisfying (8.11). In this problem, we demonstrate that $F(T,p;N)$ is not a complete thermodynamic function through investigation of a concrete example.

Consider the free energy

$$F[T; V, N] = -NRT \log \left\{ \left(\frac{T}{T^*} \right)^c \frac{V - aN}{v^* N} \right\} , \tag{8.39}$$

where a is a positive constant. We can regard this to be the free energy of an ideal gas (see (7.9)) augmented in a simple manner to account for the effect of the molecular volume. From this $F[T; V, N]$, derive $p(T; V, N)$ and $V(T,p;N)$ and then compute $F(T,p;N)$. The fact that a is not included in $F(T,p;N)$ implies that it is not a complete thermodynamic function. For comparison, derive $G[T,p;N]$ and show how the situation differs for it.

8.5 (Sections 8.1 and 10.3) Using the fact that $F[T; V, N]$ is a concave function of T (which follows from the fact that $S(T; V, N)$ is an increasing function of T), show that $G[T,p;N]$ is a concave function of both T and p. Specifically, prove the inequality

$$G[\lambda T_1 + (1 - \lambda)T_2, \lambda p_1 + (1 - \lambda)p_2; N]$$
$$\geq \lambda\, G[T_1, p_1; N] + (1 - \lambda)\, G[T_2, p_2; N] \tag{8.40}$$

for arbitrary T_1, T_2, p_1 and p_2 and arbitrary λ satisfying $0 \leq \lambda \leq 1$. What does this result imply for the Gibbs free energies of a multi-component fluid system, given in (9.17), and of a ferromagnetic system, given in (10.16)?

8.6 (Section 8.2) Here, we reconsider the Joule-Thomson experiment, previously treated in Problem 7.10. Recalling the notation used there, let us rewrite p_{H} as p, p_{L} as $p - \Delta p$, and T' as $T - \Delta T$. Then, we define the Joule-Thomson coefficient as $\mu_{\mathrm{JT}}(T,p) = \lim_{\Delta p \searrow 0}(\Delta T / \Delta p)$. The meaning of this definition is that if we decrease the pressure by a small amount Δp, the temperature decreases by an amount approximately equal to $\mu_{\mathrm{JT}}(T,p)\Delta p$.

Show that the enthalpy as a function of T, p and N can be written as $H(T,p;N) = G[T,p;N] + T\, S(T,p;N)$. Next, using this expression, demonstrate the relation

$$\mu_{\mathrm{JT}}(T,p) = \frac{1}{C_{\mathrm{p}}(T,p;N)} \left\{ T \frac{\partial V(T,p;N)}{\partial T} - V(T,p;N) \right\}$$
$$= \frac{1}{c_{\mathrm{p}}(T,p)} \left\{ T \frac{\partial v(T,p)}{\partial T} - v(T,p) \right\} , \tag{8.41}$$

where $c_{\mathrm{p}}(T,p) = C_{\mathrm{p}}(T,p;N)/N$ and $v(T,p) = V(T,p;N)/N$.

8.7 Referring to the previous problem, show that $\mu_{\mathrm{JT}}(T,p) = 0$ holds in the case of an ideal gas. Then, expanding the van der Waals equation of state

(3.37) to first order in bN/V and aN/RTV, derive the approximate equality $V(T, p; N) \simeq NRT/p - aN/RT + bN$. (This corresponds to a situation of much higher temperature and much lower pressure than at the phase transition.) Derive an expression for $\mu_{\mathrm{JT}}(T, p)$ valid in this case and discuss its meaning. (The function $c_{\mathrm{p}}(T, p)$ may be left as an unknown.)

8.8 (Section 8.2) As shown in Section 4.4.2, when an ideal gas makes a transition from a state $(T; V, N)$ to a state $(T'; V', N)$ under a quasi-static adiabatic operation, T, V, T' and V' are connected through the Poisson relation (4.45). Here we obtain the same result using the entropy written as a function of T, p and N.

First, consider a system consisting of an amount N of an ideal gas in equilibrium with pressure p under isothermal conditions at temperature T. Determine the entropy $S(T, p; N)$ of this system. Suppose that next the system is enclosed within adiabatic walls, and subsequently its pressure is slowly varied from p to p'. (This corresponds to a quasi-static adiabatic operation carried out by moving a piston in the case that the independent variables are T, V and N.) Using the fact that the entropy does not change under this operation, determine the final temperature, T'. Show that this result is consistent with (4.45).

In Problem 10.4, we consider a similar situation for a magnetic system. (As we will see, that case is more interesting.)

8.9 (Section 8.3) Let us write the heat capacity at constant volume as a function of T, p and N as

$$C_{\mathrm{v}}(T, p; N) = C_{\mathrm{v}}(T; V(T, p; N), N) . \tag{8.42}$$

In this problem, we derive the following relation between the heat capacity at constant volume and the heat capacity at constant pressure:

$$C_{\mathrm{p}}(T, p; N) - C_{\mathrm{v}}(T, p; N) = -T \left\{ \frac{\partial V(T, p; N)}{\partial T} \right\}^2 \left\{ \frac{\partial V(T, p; N)}{\partial p} \right\}^{-1} . \tag{8.43}$$

Recalling Result 7.1 (p. 146), from which it follows that the quantity

$$\left\{ \frac{\partial V(T, p; N)}{\partial p} \right\}^{-1} = \frac{\partial p(T; V, N)}{\partial V} \tag{8.44}$$

is no greater than 0, we know that the right-hand side of (8.43) is no less than 0. Hence, this relation implies the inequality

$$C_{\mathrm{p}}(T, p; N) \geq C_{\mathrm{v}}(T, p; N) . \tag{8.45}$$

To begin, using the first equality in (8.27) and the Maxwell relation (7.19), demonstrate the following:

$$C_{\mathrm{p}}(T, p; N) - C_{\mathrm{v}}(T, p; N) = T \left. \frac{\partial V(T, p; N)}{\partial T} \frac{\partial p(T; V, N)}{\partial T} \right|_{V = V(T, p; N)} . \tag{8.46}$$

Next, show that by applying the general relation demonstrated in Problem 8.10 to T, p and V here, the desired equation, (8.43), can be derived.

It is worth noting that the two quantities on the right-hand side of (8.43), $\partial V/\partial T$ (representing the temperature dependence of the volume at constant pressure) and $\partial V/\partial p$ (representing the pressure dependence of the volume at constant temperature), can both, at least in principle, be determined directly through experimental measurements.

8.10 Consider three arbitrary quantities x, y and z related in such a manner that if the values of any two of them are given, then the value of the third is uniquely determined. For such quantities, under certain conditions regarding differentiability, the following equality holds:[27]

$$\frac{\partial x(y,z)}{\partial y}\frac{\partial y(z,x)}{\partial z}\frac{\partial z(x,y)}{\partial x} = -1 \ . \tag{8.47}$$

Demonstrate this equality. What are the conditions necessary to derive it?

Hint: Considering the function $x(y,z)$ for given y and z, choose some Δy and Δz such that x is unchanged. Then examine the implications of the equation expressing this invariance.

8.11 (Sections 8.4 and 7.5.4) The Clausius-Clapeyron relation, (7.59), is usually derived using the Gibbs free energy. This derivation is indeed the shortest, but it requires treating $G[T,p;N]$ where it is non-differentiable. This should be done with extra care, as outlined below.

The continuity of $G[T,p;N]$ at $T_{\mathrm{b}}(p)$ implies

$$\lim_{\tau\searrow 0} G[T_{\mathrm{b}}(p)+\tau,p;N] = \lim_{\tau\nearrow 0} G[T_{\mathrm{b}}(p)+\tau,p;N] \ . \tag{8.48}$$

Here, the left-hand side corresponds to the gas region, and the right-hand side corresponds to the liquid region. Now, let us rewrite this relation for a slightly different value of p:

$$\lim_{\tau\searrow 0} G[T_{\mathrm{b}}(p+\Delta p)+\tau,p+\Delta p;N] = \lim_{\tau\nearrow 0} G[T_{\mathrm{b}}(p+\Delta p)+\tau,p+\Delta p;N] \ .$$
$$\tag{8.49}$$

Derive the Clausius-Clapeyron relation by taking the difference between these two expressions. In order to make this into a rigorous mathematical proof, what must be assumed about the behavior of $G[T,p;N]$?

---

[27] Perhaps the reader is familiar with the better-known case of two quantities x and y that can each be regarded as a function of the other. In this case, it is easily shown that the equality $\{dx(y)/dy\}\{dy(x)/dx\} = 1$ holds (under certain conditions). This expresses the general relation between the derivatives of a function and its inverse. (The simplest way to understand this relation is to consider some actual x and y and draw the graphs of $y(x)$ and $x(y)$.)

9

The Thermodynamics of
Multi-Component Systems

Our investigation of thermodynamics has to this point focused on single-component systems. In this chapter, we broaden our scope and investigate the thermodynamics of fluid systems consisting of mixtures of multiple chemical species.

We begin by considering multi-component systems in the absence of chemical reactions. First we derive the Helmholtz free energy and Gibbs free energy for such systems, with particular attention paid to the cases of ideal gases, dilute solutions and ideal solutions. In this treatment, we also derive the variational principle for the Gibbs free energy and the corresponding balance conditions. Then, as basic applications of the balance conditions, we treat osmotic pressure, the dissolution of gases in a dilute solution described by Henry's law, the partial pressures associated with an ideal solution described by Raoult's law, and boiling point elevation (ebullioscopy) and freezing point depression (cryoscopy) in dilute solutions. Our investigation of chemical reactions begins in Section 9.4. We first describe the operational construction of the free energies in this context. Next, employing the same operational approach, we derive the variational principle for the Gibbs free energy, which provides the chemical equilibrium condition. This is followed by discussion elucidating the proper physical interpretation of our operational treatment of chemical reactions. Next, we consider several applications. First, we give a brief treatment of Le Chatelier's principle. Next, we study ionization reactions in aqueous solution. Finally, as an introduction to equilibrium electrochemistry, we investigate a concentration cell, which generates electric current through a spontaneous redox reaction.

Before proceeding, let us first point out a somewhat delicate problem involved with the investigation of multi-component systems. In the treatment presented here, semipermeable walls play a fundamental role. However, at the present time, these are merely hypothetical structures.[1] In this book, we do not pursue this problem, but instead simply assume the existence of ideal semipermeable walls.[2] With this assumption, the thermodynamics of multi-component systems can be constructed through a straightforward extension of the theory developed to this point.

---

[1] Although there do exist both biological and synthetic semipermeable membranes of numerous types, as a generally utilizable class of structures, semipermeable walls must at this time be regarded as hypothetical.

[2] The problem of how to theoretically formalize the thermodynamics of such phenomena as mixing and chemical reactions without resorting to the use of semipermeable walls is investigated in Ref. [9].

Thermodynamics: A Modern Approach. Hal Tasaki and Glenn Paquette, Oxford University Press.
© Hal Tasaki and Glenn Paquette (2026). DOI: 10.1093/9780191878091.003.0009

9.1 Helmholtz free energy of a multi-component system

9.1.1 Determination of the free energy

Consider a fluid consisting of m chemical species contained in a vessel of volume V. We assign the indices $i = 1, 2, \ldots, m$ to these m substances and write the amount of the ith substance as N_i. For simplicity, we introduce the notation $\boldsymbol{N} = (N_1, N_2, \ldots, N_m)$ to collectively represent the amounts of all of the substances. The collective extensive variable for this system is $X = (V, \boldsymbol{N}) = (V, N_1, \ldots, N_m)$. We use the obvious analogy to vectors and introduce convenient expressions for mathematical operations involving $\boldsymbol{N}$, for example, defining the operations of addition and scalar multiplication as $\boldsymbol{N} + \boldsymbol{N}' = (N_1, N_2, \ldots, N_m) + (N_1', N_2', \ldots, N_m') = (N_1 + N_1', \ldots, N_m + N_m')$ and $\lambda \boldsymbol{N} = (\lambda N_1, \ldots, \lambda N_m)$.

Below, we obtain an expression for the Helmholtz free energy of this system in terms of the free energies of m pure systems, each consisting of one of the constituent substances, and the work necessary to quasi-statically mix all of these substances.

In the case that the amount of substance is 0 for all but one (say the ith) chemical species, we write the Helmholtz free energy as

$$F_i[T; V, N_i] = F[T; V, 0, \ldots, 0, N_i, 0, \ldots, 0] \,. \tag{9.1}$$

Because this is a single-component system, we can determine its free energy using the operational approach that we have employed in previous chapters. Thus, let us proceed by assuming that the free energy for each substance when considered in isolation, $F_i[T; V, N_i]$, is known.

The task now is to determine how the free energy for the system in which the m substances coexist, $F[T; V, \boldsymbol{N}]$, is obtained from the individual free energies $F_i[T; V, N_i]$. For simplicity, let us first consider the case $m = 2$. Suppose that the two species are initially placed into two separate vessels of volume V in thermal contact with the same environment at temperature T. We write the equilibrium state of this system as $(T; (V, N_1, 0), (V, 0, N_2))$. Then, suppose that there exists a quasi-static isothermal operation under which this system makes a transition to the equilibrium state in which the two species are both contained within a single vessel of volume V. Writing the work performed in this operation as

$$W_{\mathrm{mix}}(T; V, N_1, N_2) = W_{\mathrm{max}}(T; \{(V, N_1, 0), (V, 0, N_2)\} \to (V, N_1, N_2)) \,, \tag{9.2}$$

we can identify the quantity

$$F[T; V, N_1, N_2] = F_1[T; V, N_1] + F_2[T; V, N_2] - W_{\mathrm{mix}}(T; V, N_1, N_2) \tag{9.3}$$

as the Helmholtz free energy for the resulting system. In obtaining this result, we have used the relation between the maximum work and the Helmholtz free energy appearing in (3.27) and the additivity of the Helmholtz free energy, expressed by (3.25).

Although the above reasoning is theoretically appealing, experimentally realizing the mixing (or unmixing) of a two-component fluid through a quasi-static isothermal operation is not simple. Indeed, for many systems, it may be effectively impossible. However, if we introduce a class of walls with certain special properties, we are able to proceed—at least formally. These are hypothetical walls that are permeable to some substances and completely impermeable to all others. Although no actual wall can satisfy these conditions perfectly, we assume the existence of such walls as the idealization of actual walls that satisfy them only imperfectly. Walls of this kind are termed *semipermeable walls* or *semipermeable membranes*. With the concept of semipermeable walls, using, for example, the kind of experimental setup depicted in Figure 9.1, let us assume that it is possible to realize the mixing of a two-component fluid through a quasi-static isothermal operation and that the work performed in such an operation can be measured. Under these conditions, the Helmholtz free energy of the resulting two-component system, $F[T; V, N_1, N_2]$, can be operationally determined.

Next, similarly employing an array of m semipermeable walls, we can operationally determine the general form of the Helmholtz free energy for a system consisting of m chemical species, $F[T; V, \boldsymbol{N}] = F[T; V, N_1, \ldots, N_m]$:

$$F[T; V, N_1, \ldots, N_m] = F_1[T; V, N_1] + \cdots + F_m[T; V, N_m]$$
$$- W_{\text{mix}}(T; V, N_1, \ldots, N_m) \,. \tag{9.4}$$

The quantity $W_{\text{mix}}(T; V, N_1, \ldots, N_m)$ here is the work performed by the system in a quasi-static operation through which m pure systems in the states $(T; V, N_1), \ldots, (T; V, N_m)$ are mixed to realize the final state $(T; V, N_1, \ldots, N_m)$, as a generalization of the two-substance case described by Figure 9.1.

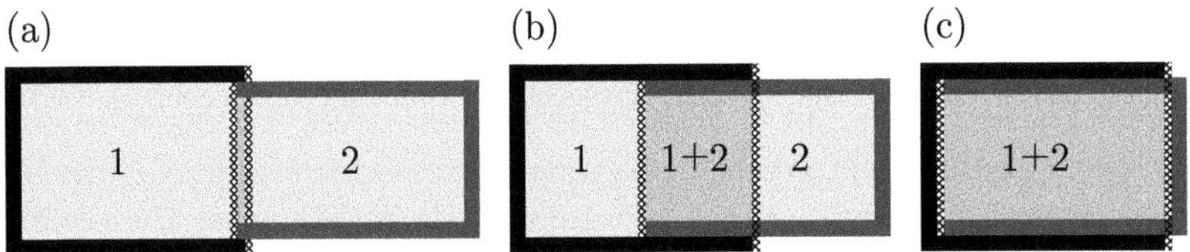

Figure 9.1 Idealized experimental apparatus employing semipermeable walls to carry out mixing and unmixing through quasi-static isothermal operations. This experimental setup implements the standard technique described in thermodynamics textbooks for realizing this class of operations. Initially, the chamber on the left contains only Species 1, and the chamber on the right contains only Species 2. The right-hand wall of the chamber on the left and the left-hand wall of the chamber on the right are both semipermeable, with the former allowing only Species 2 to pass through, and the latter allowing only Species 1 to pass through. (a) Initially, the two semipermeable walls form a single, two-layer wall, with no space between. The right-hand wall of the chamber on the left is positioned to the right of the left-hand wall of the chamber on the right. (b) With each semipermeable wall fixed in place on its chamber, the chamber on the right is pushed into the chamber on the left. (c) In the final configuration, the chamber that was initially on the right fits perfectly inside the other chamber, and together, they form a single vessel with no internal walls. This vessel contains a mixture of Species 1 and 2. If the experiment were subsequently carried out in reverse, beginning with the configuration in (c), the two substances would again become separated.

9.1.2 Properties of the Helmholtz free energy
for a multi-component system

As in the case of a single-component system, the Helmholtz free energy of a multi-component system possesses both extensivity and additivity with respect to the extensive variables V and $\boldsymbol{N}$. Explicitly, we have $F[T; \lambda V, \lambda \boldsymbol{N}] = \lambda F[T; V, \boldsymbol{N}]$ and $F[T; (V, \boldsymbol{N}), (V', \boldsymbol{N}')] = F[T; V, \boldsymbol{N}] + F[T; V', \boldsymbol{N}']$. Also, using an argument similar to that given in Section 7.3, it can be shown that the variational inequality

$$F[T; V, \boldsymbol{N}] \leq F[T; V_1, \boldsymbol{N}_1] + F[T; V_2, \boldsymbol{N}_2] \tag{9.5}$$

holds for arbitrary $V_1 + V_2 = V$ and $\boldsymbol{N}_1 + \boldsymbol{N}_2 = \boldsymbol{N}$. Furthermore, the situation regarding derivatives of the Helmholtz free energy for multi-component systems is almost identical to that for single-component systems. In this case, the relations among the various derivatives of $F[T; V, \boldsymbol{N}]$ can be determined from the following differential form, which corresponds to (7.12) in the single-component case:

$$dF = -S\,dT - p\,dV + \sum_{i=1}^{m} \mu_i dN_i \ . \tag{9.6}$$

The chemical potential $\mu_i(T; V, \boldsymbol{N}) := \partial F[T; V, \boldsymbol{N}]/\partial N_i$ can be understood as the "response" (resistance or assistance) that we would experience if we attempted to make a small change in the amount of Species i.

Also, with an approach analogous to that used in Section 7.4, we can derive the balance conditions for multi-component systems. Considering two systems containing the same m components that are in the states $(T; V, \widetilde{\boldsymbol{N}})$ and $(T; V', \widetilde{\boldsymbol{N}}')$, these conditions are as follows:

$$p(T; V, \widetilde{\boldsymbol{N}}) = p(T; V', \widetilde{\boldsymbol{N}}') \ , \tag{9.7}$$

$$\mu_i(T; V, \widetilde{\boldsymbol{N}}) = \mu_i(T; V', \widetilde{\boldsymbol{N}}') \quad \text{for every } i = 1, \ldots, m \ . \tag{9.8}$$

If we consider two such systems separated by a semipermeable wall, the conditions needed to realize equilibrium will depend on the nature of the wall, and will be some subset of the full set of balance conditions. For example, for two systems at temperature T placed into contact through a fixed wall that is permeable only to Species 1, the necessary and sufficient condition that these systems be in equilibrium is

$$\mu_1(T; V, \widetilde{\boldsymbol{N}}) = \mu_1(T; V', \widetilde{\boldsymbol{N}}') \ . \tag{9.9}$$

(This condition is derived in Problem 9.1.) With this single balance condition, employing an experimental setup like that depicted in Figure 9.2, we can (at least in principle) measure the chemical potential corresponding to one particular substance in a multi-component system. It is thus seen that even in the case of a multi-component system, the chemical potential is a state function that can be determined entirely operationally.

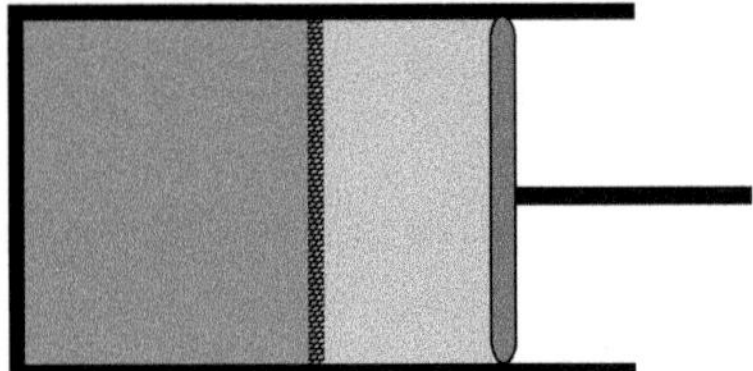

Figure 9.2 Hypothetical experiment for operationally measuring one component of the chemical potential, $\mu_i(T, V; \boldsymbol{N})$, for a multi-component system. The left and right compartments of the vessel are separated by a semipermeable wall that allows only Species i to pass through. The left compartment contains a mixture of all m substances in the state $(T, V; \boldsymbol{N})$. The right compartment contains only Species i, also at temperature T. Varying the position of the piston, the amount of Species i in the right compartment will also vary. In any case, in equilibrium, this amount will be such that the balance condition (9.9) is satisfied. Thus, because in the present situation, the chemical potential on one side of this equality is that for a single-component system, which we already know how to determine, this experiment provides a method for operationally determining a single component of the chemical potential in a multi-component system. Applying this method to each of the species independently, we can determine the total chemical potential for this system.

9.1.3 Helmholtz free energy of a multi-component ideal gas

Here, beginning from the general form of the Helmholtz free energy given in (9.4), we derive an explicit expression for the Helmholtz free energy of a multi-component ideal gas.

From empirical results, it is known that in the regime in which each component of a multi-component fluid can be regarded as an ideal gas, the work required for quasi-static mixing, W_{mix}, is negligible. Therefore, we infer that the Helmholtz free energy for an m-component ideal gas is given simply by the sum over the individual Helmholtz free energies of its constituent species as isolated ideal gases, obtained from (7.9):

$$
\begin{aligned}
F[T; V, N_1, \ldots, N_m] &= \sum_{i=1}^{m} F_i[T; V, N_i] \\
&= -RT \sum_{i=1}^{m} N_i \log \left\{ \left(\frac{T}{T_i^*} \right)^{c_i} \frac{V}{v_i^* N_i} \right\} + \sum_{i=1}^{m} N_i u_i .
\end{aligned} \tag{9.10}
$$

The quantity c_i here is a constant whose value is specific to the ith component gas. Note that this free energy retains the indeterminacy corresponding to arbitrary (and independent) variations of the additive energy constants u_i and the reference values of the molar volumes v_i^* and temperatures T_i^*.[3] Taking the derivative of (9.10) with respect to the volume, we obtain the following expression for the

---

[3] As pointed out previously, the quantities u_i are meaningful only in the context of chemical reactions.

pressure of an m-component ideal gas:

$$p(T; V, \boldsymbol{N}) = -\frac{\partial F[T, V; \boldsymbol{N}]}{\partial V} = \sum_{i=1}^{m} \frac{N_i RT}{V} \, . \tag{9.11}$$

This is known as *Dalton's law*.[4]

9.1.4 Helmholtz free energy of a dilute solution

We now derive a very useful expression for the Helmholtz free energy of a dilute solution.

A *solution* is a single-phase, homogeneous liquid or solid mixture of multiple substances.[5] Generally, a solution consists of a *solvent*, which is the substance in greatest abundance, and *solutes*, which are substances regarded as being dissolved into the solvent.[6] In a rough sense, a *dilute solution* is one for which the amount of the solvent species is much greater than the amounts of the solute species. This characterization is made precise in the following.

Because the work W_{mix} is an extensive quantity, the following equality holds for any solution:

$$W_{\mathrm{mix}}(T; V, N_1, \ldots, N_m) = N_1 \, W_{\mathrm{mix}}\!\left(T; \frac{V}{N_1}, 1, \frac{N_2}{N_1}, \ldots, \frac{N_m}{N_1}\right) . \tag{9.12}$$

Now, because W_{mix} is 0 in the case $N_2 = \cdots = N_m = 0$, for a dilute solution, we make the assumption that the function $W_{\mathrm{mix}}(T; V/N_1, 1, N_2/N_1, \ldots, N_m/N_1)$ can be expanded in the small quantities $N_2/N_1, \ldots, N_m/N_1$. In this case, we have the relation

$$W_{\mathrm{mix}}\!\left(T; \frac{V}{N_1}, 1, \frac{N_2}{N_1}, \ldots, \frac{N_m}{N_1}\right) \simeq -\sum_{i=2}^{m} \frac{N_i}{N_1} w_i\!\left(T; \frac{V}{N_1}\right) . \tag{9.13}$$

The quantity $-w_i(T; V/N_1)$ can be expressed as a derivative of W_{mix}, but the important point is that it is a function of T and V/N_1 alone. Combining (9.13) and (9.12), we obtain

$$W_{\mathrm{mix}}(T; V, N_1, \ldots, N_m) \simeq -\sum_{i=2}^{m} N_i \, w_i\!\left(T; \frac{V}{N_1}\right) . \tag{9.14}$$

Before the mixing operation under consideration here, there are m separate vessels each of volume V and each containing just one of the m substances. Because the amount of each of the solutes (i.e., Species 2 through m) is small, before the

---

[4] Because the quantity $N_i RT/V$ in the sum can be interpreted as the pressure exerted by the ith component gas, it is referred to as the *partial pressure* of this component. It should be noted, however, that partial pressure is a concept that is meaningful only to the extent that a gas can be regarded as ideal.

[5] Sometimes this term is defined broadly to include gas mixtures, but because solvation can be regarded as trivial in gas mixtures, we do not consider them to be solutions. This is also the convention of the International Union of Pure and Applied Chemists (IUPAC) [7].

[6] This distinction can become somewhat arbitrary when there is no clearly predominant substance, but generally no complication results in such situations.

operation, each can be treated as an ideal gas. Thus, using (9.3) and (9.14), we obtain the following generally valid approximate expression for the Helmholtz free energy of a dilute solution:

$$F[T; V, \boldsymbol{N}] \simeq F_1[T; V, N_1] + \sum_{i=2}^{m} F_i^{(\mathrm{id})}[T; V, N_i] + \sum_{i=2}^{m} N_i \, w_i\left(T; \frac{V}{N_1}\right). \qquad (9.15)$$

Here, the quantity

$$F_i^{(\mathrm{id})}[T; V, N_i] = -N_i RT \log\left\{\left(\frac{T}{T_i^*}\right)^{c_i} \frac{V}{v_i^* N_i}\right\} + N_i u_i \qquad (9.16)$$

is the Helmholtz free energy for an amount N_i of ideal gas composed of Species i. Because the expression in (9.15) depends on the undetermined functions w_i, it may seem that it is not particularly useful. However, as shown in the second half of this chapter, we are able to obtain numerous concrete results from the approximate expression for the Gibbs free energy of a dilute solution derived from (9.15) (see (9.40)).

It is important to understand here that (9.15) is valid only in the case that the expansion of the quantity $W_{\mathrm{mix}}(T; \frac{V}{N_1}, 1, \frac{N_2}{N_1}, \ldots, \frac{N_m}{N_1})$ can be carried out without encountering any singularity. Naively, it may seem that this should hold as long as the expansion parameters N_i/N_1 are all small, but there are situations in which the derivation given above is invalid even if this condition is satisfied. In particular, this is generally true in the case that the solvent undergoes phase separation due to the introduction of the solute[7] or in the case of an electrolyte solution.[8] However, ignoring such situations, the derivation given here can be regarded as generally valid for sufficiently small N_i/N.

Finally, we emphasize that the expression for the Helmholtz free energy of a dilute solution given in (9.15) is not simply an expansion of $F[T; V, N_1, \ldots, N_m]$ in N_i/N_1 ($i = 2, \ldots, m$). This is clear from the fact that the Helmholtz free energy of a multi-component ideal gas, given in (9.10), possesses terms containing $\log N_i$, which obviously cannot be expanded in N_i/N_1 about $N_i = 0$. With regard to this point, it is best simply to keep in mind that, in general, the Helmholtz free energy cannot be expanded in N_i/N_1.

---

[7] Although we have been considering the case in which the solvent consists of a single chemical species, as discussed in Footnote 5 on p. 27, a fluid consisting of multiple chemical species can be treated as a single-component fluid as long as it remains spatially homogeneous. The case of phase separation considered here is that in which a homogeneous solvent consisting of multiple chemical species becomes inhomogeneous upon the introduction of the solute. There are known to be cases in which such behavior occurs even with a very small amount of solute.

[8] An *electrolyte solution* is a solution that can conduct electricity through the motion of individually solvated ions. These ions in solution may result from an ionization reaction or from the dissociation of an ionic compound that occurs upon mixing of the solvent and solutes. For an electrolyte solution, if we properly treat the individual ion species as distinct substances, then in principle, in the sufficiently dilute case, the derivation given here will be valid. However, due to the strength of the Coulomb interaction, such a solution would generally be far too dilute to be of practical interest. (In order for (9.13) to hold, it must be the case that for each of the $m - 1$ solutes, a pure system with volume V consisting of this solute alone can be treated as an ideal gas. For an ionic solute, this situation will be realized only in the case of extreme dilution.) But even for solutions insufficiently dilute for this derivation to be valid, shielding effects exerted by the solvent on the solute can prevent interactions among the solute ions to such a degree that the dilute-solution form of the Gibbs free energy, given in (9.41) and (9.42), is valid. Indeed, empirically, it is known that the dilute-solution form of the Gibbs free energy is valid in many situations for electrolyte solutions even under conditions of practical interest. (See Section 9.5 for related discussion.)

9.2 Gibbs free energy of a multi-component system

In the investigation of thermodynamic behavior, actual experimental laboratories are most often under conditions of fixed temperature and pressure. For this reason, generally, the Gibbs free energy is more practically useful than the Helmholtz free energy. In view of this reality, the remainder of this chapter is presented from the point of view of the Gibbs free energy, rather than the Helmholtz free energy.

9.2.1 Definition and fundamental properties of the Gibbs free energy for a multi-component system

Repeating the arguments used to arrive at the definition of the Gibbs free energy in the case of single-component systems, (8.3), we immediately obtain the following as its generalization to the case of multi-component systems:

$$G[T, p; \boldsymbol{N}] := \min_{V}\{F[T; V, \boldsymbol{N}] + pV\} \ . \tag{9.17}$$

Similarly, the inverse transformation yielding the Helmholtz free energy from the Gibbs free energy in the multi-component case is obtained by simply replacing N with $\boldsymbol{N}$ in (8.8):

$$F[T; V, \boldsymbol{N}] = \max_{p}\{G[T, p; \boldsymbol{N}] - pV\} \ . \tag{9.18}$$

Also, in the case that there exists a unique solution $V(T, p; \boldsymbol{N})$ satisfying the equation $p = p(T; V, \boldsymbol{N})$, the multi-component version of (8.12),

$$G[T, p; \boldsymbol{N}] = F[T; V(T, p; \boldsymbol{N}), \boldsymbol{N}] + p\, V(T, p; \boldsymbol{N}) \ , \tag{9.19}$$

can be used as a shortcut definition in this case as well.

Clearly, the Gibbs free energy of a multi-component system possesses both extensivity and additivity with respect to the extensive variable $\boldsymbol{N}$. These properties are expressed by the relations $G[T, p; \lambda\boldsymbol{N}] = \lambda\, G[T, p; \boldsymbol{N}]$ and $G[T, p; \boldsymbol{N}, \boldsymbol{N}'] = G[T, p; \boldsymbol{N}] + G[T, p; \boldsymbol{N}']$, respectively.[9]

As the generalization of (8.22), we can express the various derivatives of the Gibbs free energy concisely using the differential form

$$dG = -S\, dT + V\, dp + \sum_{i=1}^{m} \mu_i\, dN_i \ . \tag{9.20}$$

However, it is important to note that, in contrast to the single-component case, for a multi-component system, the chemical potentials $\mu_i(T, p; \boldsymbol{N}) = \partial G[T, p; \boldsymbol{N}]/\partial N_i$ generally depend on $\boldsymbol{N}$, and thus are not functions of T and p alone.

---

[9] In the case of additivity, this relation means that if we consider two separate systems each under conditions of fixed temperature T and pressure p, then the sum of their Gibbs free energies is the same as the Gibbs free energy of the composite system obtained by placing these two systems in thermal contact.

Finally, carrying out a derivation similar to that applied in the single-component case, we obtain the multi-component version of the Euler equation for the Gibbs free energy:

$$G[T, p; \boldsymbol{N}] = \sum_{i=1}^{m} N_i \, \mu_i(T, p; \boldsymbol{N}) \; . \tag{9.21}$$

Within chemistry, it is often the case that theoretical investigations center on consideration of the chemical potentials, and for this reason, the above is a particularly important relation.

9.2.2 Variational principle and balance conditions

We now derive a variational principle for $G[T, p; \boldsymbol{N}]$.[10] From this variational principle, we obtain the balance conditions for multi-component systems with externally controlled temperature and pressure.

For arbitrary T, p and $\boldsymbol{N}$ and arbitrary $\boldsymbol{N}_1$ and $\boldsymbol{N}_2$ subject to the condition $\boldsymbol{N} = \boldsymbol{N}_1 + \boldsymbol{N}_2$, we have the inequality[11]

$$G[T, p; \boldsymbol{N}] \leq G[T, p; \boldsymbol{N}_1] + G[T, p; \boldsymbol{N}_2] \; . \tag{9.22}$$

This can be easily demonstrated by combining the definition of the Gibbs free energy (9.17) and the variational inequality for the Helmholtz free energy (9.5) as follows:

$$
\begin{aligned}
G[T, p; \boldsymbol{N}] &= \min_{V}\{F[T; V, \boldsymbol{N}] + pV\} \\
&= \min_{V_1, V_2}\{F[T; V_1 + V_2, \boldsymbol{N}_1 + \boldsymbol{N}_2] + p(V_1 + V_2)\} \\
&\leq \min_{V_1, V_2}\{F[T; V_1, \boldsymbol{N}_1] + F[T; V_2, \boldsymbol{N}_2] + p(V_1 + V_2)\} \\
&= \min_{V_1}\{F[T; V_1, \boldsymbol{N}_1] + pV_1\} + \min_{V_2}\{F[T; V_2, \boldsymbol{N}_2] + pV_2\} \\
&= G[T, p; \boldsymbol{N}_1] + G[T, p; \boldsymbol{N}_2] \; .
\end{aligned}
\tag{9.23}
$$

Now, let us consider a system consisting of an amount of fluid $\boldsymbol{N}$ in equilibrium at temperature T and pressure p. Suppose that we insert a diathermal partition into this system, creating two subsystems with amounts of substance $\widetilde{\boldsymbol{N}}_1$ and $\widetilde{\boldsymbol{N}}_2$. Because there is no change imparted to the system when the wall is inserted, we have $G[T, p; \boldsymbol{N}] = G[T, p; \widetilde{\boldsymbol{N}}_1, \widetilde{\boldsymbol{N}}_2]$. Applying the additivity property given above to this relation, we obtain $G[T, p; \boldsymbol{N}] = G[T, p; \widetilde{\boldsymbol{N}}_1] + G[T, p; \widetilde{\boldsymbol{N}}_2]$. Finally, using

---

[10] Obviously, we could carry out the same kind of investigation for a single-component system, but the result in that case is not useful. This is the reason that no such consideration was given in Chapter 8.

[11] From (9.22) and the extensivity of the Gibbs free energy, the inequality

$$G[T, p; \lambda \boldsymbol{N}_1 + (1 - \lambda)\boldsymbol{N}_2] \leq \lambda\, G[T, p; \boldsymbol{N}_1] + (1 - \lambda)G[T, p; \boldsymbol{N}_2]$$

can be derived for arbitrary $\boldsymbol{N}_1$ and $\boldsymbol{N}_2$ and any λ satisfying $0 \leq \lambda \leq 1$. We thus see that $G[T, p; \boldsymbol{N}]$ is a convex function of $\boldsymbol{N}$ (see Appendix G).

(9.22), we arrive at the variational principle

$$G[T,p;\widetilde{\boldsymbol{N}}_1] + G[T,p;\widetilde{\boldsymbol{N}}_2] = \min_{\substack{\boldsymbol{N}_1,\boldsymbol{N}_2 \\ (\boldsymbol{N}_1+\boldsymbol{N}_2=\widetilde{\boldsymbol{N}}_1+\widetilde{\boldsymbol{N}}_2)}} \left\{ G[T,p;\boldsymbol{N}_1] + G[T,p;\boldsymbol{N}_2] \right\}. \quad (9.24)$$

This relation is a variational principle with respect to the transport of substance, corresponding to (7.31) in the single-component case. It can be interpreted as expressing the idea that if a hole is made in the wall separating two systems (at equal temperature and pressure), substance will flow between them so as to minimize the Gibbs free energy of the combined system.

The variational principle (9.24) was derived through consideration of operations that partition and unpartition a system through the insertion and removal of an ordinary wall. This entire procedure can be generalized to the case of semipermeable walls of arbitrary type to obtain a generalized version of (9.24). The meaning of this generalized variational principle is as follows: For a system under conditions of fixed T and p that is partitioned into separate parts, in the case that the permeability properties defining the partitioning wall are changed, and as a result there emerge new extensive degrees of freedom, the system will spontaneously evolve toward a new equilibrium state that minimizes the Gibbs free energy of the total system under the new conditions.

Next, from (9.24), using arguments similar to those given in Section 7.4, we conclude that the necessary and sufficient condition for two systems separately in equilibrium with amounts of substance $\widetilde{\boldsymbol{N}}_1$ and $\widetilde{\boldsymbol{N}}_2$ at fixed temperature T and pressure p to be balanced is that the equality

$$\mu_i(T,p;\widetilde{\boldsymbol{N}}_1) = \mu_i(T,p;\widetilde{\boldsymbol{N}}_2) \quad (9.25)$$

hold for every $i = 1,\ldots,m$.[12] This set of balance conditions is of fundamental importance in the treatment of phase equilibrium. It provides a powerful tool often applied to problems in both physics and chemistry.

9.2.3 Gibbs free energy of a multi-component ideal gas

We now derive an explicit expression for the Gibbs free energy of a multi-component ideal gas. First, we substitute the formulas (9.10) and (9.11) for the Helmholtz free energy and pressure into the shortcut definition of the Gibbs free energy (9.19). Then, using $N = \sum_{i=1}^{m} N_i$ to represent the total amount of substance (total number of moles), replacing V with NRT/p, and simplifying, we obtain the following:

$$G[T,p;\boldsymbol{N}] = NRT - RT \sum_{i=1}^{m} N_i \log\left\{ \left(\frac{T}{T_i^*}\right)^{c_i} \frac{NRT}{pv_i^* N_i} \right\} + \sum_{i=1}^{m} N_i u_i$$

$$= NRT + RT \sum_{i=1}^{m} N_i \log\left\{ \left(\frac{T_i^*}{T}\right)^{c_i+1} \frac{p}{p_i^*} \frac{N_i}{N} \right\} + \sum_{i=1}^{m} N_i u_i. \quad (9.26)$$

---

[12] Here, as in the situation described by Result 7.2 (p. 148), the conditions of local balance and global balance are equivalent. This can be proved similarly to Result 7.2, or by applying Theorem G.10.

Here we have introduced the reference pressures $p_i^* = RT_i^*/v_i^*$. Although these pressures can be regarded as arbitrary and independent, in chemistry it is conventional (and convenient) to assign to each a single standard value, which is usually taken to be 1 atm ($= 1.01325 \times 10^5$ Pa) or 1 bar ($= 10^5$ Pa). We adopt this convention from this point, representing this value of the pressure by $p^{\ominus}$.[13]

In order to identify the contribution to (9.26) due solely to the mixing of distinct substances, we rewrite this Gibbs free energy as the sum of the corresponding single-component ideal gas Gibbs free energies, as given by (8.13), and the remaining part:

$$G[T, p; \boldsymbol{N}] = NRT + RT \sum_{i=1}^{m} N_i \log \left\{ \left(\frac{T_i^*}{T}\right)^{c_i+1} \frac{p}{p^{\ominus}} \right\}$$

$$+ \sum_{i=1}^{m} N_i u_i + RT \sum_{i=1}^{m} N_i \log x_i \, . \tag{9.27}$$

The quantity $x_i = N_i/N$ here is the *mole fraction* of Species i. The term $RT \sum_{i=1}^{m} N_i \log x_i$ can be regarded as the contribution from the mixing of the m substances, and it is sometimes called the *free energy of mixing* (of an ideal gas).[14]

Taking the derivative of (9.27) with respect to N_i and using (8.18), we can obtain the ith chemical potential. Carrying out this derivation, and using the identity

$$\frac{\partial}{\partial N_i} \left(\sum_{j=1}^{m} N_j \log \frac{N_j}{N} \right) = \frac{\partial}{\partial N_i} \left(\sum_{j=1}^{m} N_j \log N_j - N \log N \right)$$

$$= \left(\log N_i + 1\right) - \left(\log N + 1\right) = \log \frac{N_i}{N} \tag{9.28}$$

along the way, we find

$$\mu_i(T, p; \boldsymbol{N}) = RT + RT \log \left\{ \left(\frac{T_i^*}{T}\right)^{c_i+1} \frac{p}{p^{\ominus}} \right\} + u_i + RT \log \frac{N_i}{N}$$

$$= RT + RT \log \left\{ \left(\frac{T_i^*}{T}\right)^{c_i+1} \frac{p}{p^{\ominus}} \right\} + u_i + RT \log x_i \, . \tag{9.29}$$

---

[13] In chemistry, the superscript $\ominus$ is often used to indicate a reference value. In this chapter, it is used to indicate the reference values of several quantities.

[14] In the literature, the free energy of mixing (and the similarly defined entropy of mixing) is sometimes treated as a mysterious quantity, but in fact there is absolutely nothing mysterious about it. As seen in Section 9.1.3, in the situation that the volume is a controlled variable, the work required for mixing, W_{mix}, is 0 in the case of an ideal gas, and hence the effect of mixing is trivial. The fact that there is a (at first sight, perhaps mysterious) nonzero contribution of mixing to (9.27) can be attributed to the fact that in the situation to which this equation applies, it is the pressure that is controlled, and its value is fixed throughout the mixing process. For this reason, during this process, the volume available to the ith substance increases from $N_i RT/p$ to NRT/p. This contrasts with the mixing process considered in (9.2) through which $F[T, V, \boldsymbol{N}]$ is obtained from the individual free energies $F[T, V, N_i]$. Finally, we point out that in connection to the free energy of mixing (or entropy of mixing), there is a problem sometimes referred to as the *Gibbs paradox*. But despite its name, if this problem is treated properly, it is in no way paradoxical.

Comparing this expression with the corresponding expression for a single-component system, (8.20), we see that the term $RT \log x_i$ represents the effect of mixing.

We now introduce some convenient notation. In a number of treatments given in this chapter, it is useful (both practically and conceptually) to express the ith chemical potential $\mu_i(T, p; \mathbf{N})$ for a multi-component system in terms of the chemical potential of the corresponding pure system and the contribution due to its mixing. Of course, we can write the chemical potential of a pure system of Species i as $\mu_i(T, p; 0, \ldots, N_i, \ldots, 0)$, but in order to make explicit the fact that this quantity is independent of N_i,[15] we introduce the new function $\mu_i^{\ominus}(T, p) := \mu_i(T, p; 0, \ldots, N_i, \ldots, 0)$.

In terms of $\mu_i^{\ominus}(T, p)$, we can rewrite (9.29) as follows:

$$\mu_i(T, p; \mathbf{N}) = \mu_i^{\ominus}(T, p) + RT \log x_i \ . \tag{9.30}$$

Then, because for an ideal gas, we have $x_i = N_i/N = p_i/p$, where p_i is the partial pressure of the ith substance, this can be written as

$$\mu_i(T, p; \mathbf{N}) = \mu_i^{\ominus, (\mathrm{id})}(T) + RT \log \frac{p_i}{p^{\ominus}} \ , \tag{9.31}$$

where $\mu_i^{\ominus, (\mathrm{id})}(T) = RT + (c_i + 1)RT \log(T_i^*/T) + u_i$.[16] In the chemistry literature, this is the conventional form in which the chemical potentials of a multi-component ideal gas are generally expressed.

9.2.4 Gibbs free energy of a dilute solution

Next, we derive an explicit approximate expression for the Gibbs free energy of a dilute solution, whose Helmholtz free energy we derived in Section 9.1.4. The result we obtain here has numerous applications to diverse problems in chemistry and other fields. In the treatment presented below, we assume that the system is not near the phase transition point of the solvent. Then, using the expression for the Helmholtz free energy (9.15) and the shortcut definition of the Gibbs free energy (9.19), we derive the expression (9.40) for the Gibbs free energy of a dilute solution.

As above, $V(T, p; \mathbf{N})$ represents the unique solution to the equation $p = p(T; V, \mathbf{N})$, in other words, the volume realized by the system when T, p and $\mathbf{N}$ are given. We define $V_1(T, p; N_1) = V(T, p; N_1, 0, \ldots, 0)$ to be the volume of the system in the case that it contains N_1 moles of Species 1 (i.e., the solvent) and nothing else. Because $V_1(T, p; N_1)$ is extensive, $v_1(T, p) = V_1(T, p; N_1)/N_1$ is a function of only T and p. Next, we consider the situation in which $m - 1$ additional substances appear in amounts $N_2, N_3, \ldots, N_m$ each much smaller than N_1.

---

[15] See the discussion below (8.18) on p. 176.

[16] It may be thought that $\mu_i^{\ominus, (\mathrm{id})}(T)$ is simply $\mu_i^{\ominus}(T, p^{\ominus})$, i.e., the chemical potential of a pure system of Species i at $(T, p^{\ominus})$. Indeed, some chemistry textbooks make this assertion. However, this characterization is valid only if such a pure system can indeed be regarded as an ideal gas at $(T, p^{\ominus})$, which is not generally the case for systems described by the Gibbs free energy in (9.27). Thus, while for the purpose of obtaining an intuitive understanding of (9.31), this characterization is perhaps useful, it must be realized that it is not necessarily accurate.

We write the difference between the volumes of the system in the two cases as[17]

$$\Delta V = V(T, p; \mathbf{N}) - V_1(T, p; N_1) \,. \tag{9.32}$$

Evidently, $\Delta V / V(T, p; \mathbf{N})$ is a small quantity of the same order as (the largest among) the quantities N_i/N_1 $(i = 2, \ldots, m)$.[18]

Now, substituting the general form for the Helmholtz free energy of a dilute solution given in (9.15) into (9.19), and using (9.32), we obtain

$$G[T, p; \mathbf{N}] = F[T; V(T, p; \mathbf{N}), \mathbf{N}] + p\,V(T, p; \mathbf{N})$$

$$\simeq F_1[T; V_1(T, p; N_1) + \Delta V, N_1] + \sum_{i=2}^{m} F_i^{(\mathrm{id})}[T; V(T, p; \mathbf{N}), N_i]$$

$$+ \sum_{i=2}^{m} N_i\, w_i\!\left(T; \frac{V(T, p; \mathbf{N})}{N_1}\right) + p\,V(T, p; \mathbf{N}) \,. \tag{9.33}$$

Next, we expand the above expression to first order in ΔV. First, for the Helmholtz free energy of Species 1, using (3.31), we have

$$F_1[T; V_1(T, p; N_1) + \Delta V, N_1] \simeq F_1[T; V_1(T, p; N_1), N_1] - p_1(T; V_1(T, p; N_1), N_1)\, \Delta V$$

$$= F_1[T; V_1(T, p; N_1), N_1] - p\, \Delta V \,. \tag{9.34}$$

Then, recalling that the contribution from each ideal gas Helmholtz free energy for $i = 2, \ldots, m$ is already a small quantity of order N_i/N_1, in the presently considered expansion, we can approximate this contribution as follows:

$$F_i^{(\mathrm{id})}[T; V(T, p; \mathbf{N}), N_i] = -RTN_i \log\left\{ \left(\frac{T}{T_i^*}\right)^{c_i} \frac{V(T, p; \mathbf{N})}{v_i^* N_i} \right\} + N_i u_i$$

$$\simeq -RTN_i \log\left\{ \left(\frac{T}{T_i^*}\right)^{c_i} \frac{V_1(T, p; N_1)}{v_i^* N_i} \right\} + N_i u_i$$

$$= -RTN_i \log\left\{ \left(\frac{T}{T_i^*}\right)^{c_i} \frac{v_1(T, p) N_1}{v_i^* N_i} \right\} + N_i u_i \,. \tag{9.35}$$

Similarly, for each term containing a w_i, we can use the approximation

$$w_i\!\left(T; \frac{V(T, p; \mathbf{N})}{N_1}\right) \simeq w_i\!\left(T; \frac{V_1(T, p; N_1)}{N_1}\right) = w_i(T; v_1(T, p)) \,. \tag{9.36}$$

---

[17] Clearly, ΔV is a function of T, p and $\mathbf{N}$, but for conciseness, we omit its arguments.

[18] Given the validity of the dilute solution form of the Helmholtz free energy (see Section 9.1.4), the only assumption made in the present derivation is that ΔV is a small quantity. Ordinarily, of course, this is a valid assumption, but if the state $(T; V(T, p; \mathbf{N}), \mathbf{N})$ of the solution and the state $(T; V_1(T, p; N_1), N_1)$ of the pure solvent are on opposite sides of a phase transition point, or if one of these coincides with a phase transition point, then this will not be a valid assumption. (In the former case, ΔV will not be small, and in the latter case, it will not be well defined.)

Substituting (9.34), (9.35) and (9.36) into (9.33), and using (9.32), we find

$$
\begin{aligned}
G[T,p;\boldsymbol{N}] &\simeq F_1[T;V_1(T,p;N_1),N_1] + p\,V_1(T,p;N_1) \\
&\quad - RT \sum_{i=2}^{m} N_i \log\left\{ \left(\frac{T}{T_i^*}\right)^{c_i} \frac{v_1(T,p)N_1}{v_i^* N_i} \right\} \\
&\quad + \sum_{i=2}^{m} N_i\, w_i(T;v_1(T,p)) + \sum_{i=2}^{m} N_i u_i \\
&= G_1[T,p;N_1] + \sum_{i=2}^{m} N_i\, \mu_i^{\ominus,(\mathrm{dil})}(T,p) + RT \sum_{i=2}^{m} N_i \log \frac{N_i}{N_1} - RT \sum_{i=2}^{m} N_i \,.
\end{aligned}
$$

$$(9.37)$$

In deriving this expression, we have employed the shortcut definition of the Gibbs free energy for a single-component system, (8.12). The quantities $\mu_i^{\ominus,(\mathrm{dil})}(T,p)$ ($i = 2,\ldots,m$) appearing here are as follows:

$$
\mu_i^{\ominus,(\mathrm{dil})}(T,p) = RT - RT \log\left\{ \left(\frac{T}{T_i^*}\right)^{c_i} \frac{v_1(T,p)}{v_i^*} \right\} + w_i(T;v_1(T,p)) + u_i \,. \quad (9.38)
$$

It is noteworthy that these are functions of only T and p.[19]

The expression in (9.37) can be further simplified. For this purpose, we first rewrite the third term in accordance with the following relation (in which the total amount of substance $\sum_{i=1}^{m} N_i$ is represented by N):

$$
\begin{aligned}
\sum_{i=2}^{m} N_i \log \frac{N_i}{N_1} &= \sum_{i=2}^{m} N_i \log \frac{N_i}{N} - \sum_{i=2}^{m} N_i \log \frac{N_1}{N} \\
&= \sum_{i=1}^{m} N_i \log \frac{N_i}{N} - N \log \frac{N_1}{N} \\
&= \sum_{i=1}^{m} N_i \log \frac{N_i}{N} - N \log \left\{ 1 - \frac{1}{N} \sum_{i=2}^{m} N_i \right\} \\
&\simeq \sum_{i=1}^{m} N_i \log \frac{N_i}{N} + \sum_{i=2}^{m} N_i \,.
\end{aligned}
$$

$$(9.39)$$

To obtain the final form here, we have used the relation $N_i/N \ll 1$ for all $i > 1$ and expanded the logarithm in the second term of the second-to-last line in a Taylor series. Finally, substituting the above approximate expression into (9.37)

---

[19] Some chemistry textbooks attempt to give physical interpretations of the quantities $\mu_i^{\ominus,(\mathrm{dil})}(T,p)$, but those interpretations are based on fictional states, and for this reason they are not particularly illuminating. As far as we know, there is no physically realistic picture that provides a simple, intuitive way to understand these quantities. Perhaps $\mu_i^{\ominus,(\mathrm{dil})}(T,p)$ is best characterized as simply the part of the dilute-solution solute chemical potential, (9.42), that is independent of the amounts of substance.

and simplifying, we obtain

$$G[T, p; \boldsymbol{N}] \simeq G_1[T, p; N_1] + \sum_{i=2}^{m} N_i\, \mu_i^{\ominus, \text{(dil)}}(T, p) + RT \sum_{i=1}^{m} N_i \log x_i \,. \qquad (9.40)$$

As above, the quantity $x_i = N_i/N$ here is the mole fraction of the ith substance. This is the final form in our derivation of the approximate Gibbs free energy for a dilute solution. The third term in this expression represents the effect of mixing. Note that this term has the same form as the corresponding term in the Gibbs free energy for a multi-component ideal gas, given in (9.27).

Next, applying (9.20) and (8.18) to the above form of $G[T, p; \boldsymbol{N}]$, we can derive the chemical potentials of a dilute solution. For the solvent, we find

$$\mu_1(T, p; \boldsymbol{N}) \simeq \mu_1^{\ominus}(T, p) + RT \log x_1 \,, \qquad (9.41)$$

where $\mu_1^{\ominus}(T, p)$ is the chemical potential of a system consisting of the solvent alone, and for the solutes, we find

$$\mu_i(T, p; \boldsymbol{N}) \simeq \mu_i^{\ominus, \text{(dil)}}(T, p) + RT \log x_i \,. \qquad (9.42)$$

Here, in taking the derivatives of the log terms, we have used (9.28).

9.2.5 Gibbs free energy of an ideal solution

An *ideal solution* is a solution described by a Gibbs free energy $G[T, p; \mathbf{N}] = \sum_{i=1}^{m} N_i \mu_i(T, p; \mathbf{N})$ in which the individual chemical potentials take the simple form

$$\mu_i(T, p; \mathbf{N}) = \mu_i^{\ominus}(T, p) + RT \log x_i \qquad (9.43)$$

for each i, where $\mu_i^{\ominus}(T, p)$ is the chemical potential of a pure system of Species i at (T, p).

For a solution consisting of chemically similar substances, we can expect that inter-particle interactions will depend only weakly on the species involved. An ideal solution represents the idealization of this situation, in which the species dependence of these interactions vanishes. Thus, for an ideal solution, we have the equalities

$$V(T, p; N_1, \ldots, N_m) = \sum_{i=1}^{m} V_i(T, p; N_i) \qquad (9.44)$$

and

$$U(T, p; N_1, \ldots, N_m) = \sum_{i=1}^{m} U_i(T, p; N_i) \,, \qquad (9.45)$$

where $V_i(T, p; N_i)$ and $U_i(T, p; N_i)$ are the volume and energy of a pure system of Species i. In fact, although we define an ideal solution through (9.43), we could instead use these two equalities.[20]

---

[20] As the description of these equalities, it is said that for an ideal solution, the volume of mixing and energy of mixing vanish. The vanishing of both of these quantities implies the vanishing also of the enthalpy of mixing.

9.2.6 Definition of the activity

Above, we considered the chemical potentials of the substances composing multi-component systems in three idealized cases: ideal gases, dilute solutions and ideal solutions. These chemical potentials are generalized to non-idealized cases through the introduction of a quantity called the *activity*. There are several ways in which activities are defined, depending on the type of system under consideration. Conventionally, the defining equations are obtained from (9.31) in the case of a gas and from (9.43) or from (9.41) and (9.42) in the case of a solution by simply replacing the mole fraction, x_i, with the activity, a_i, and stipulating that the resulting equations hold exactly. Thus, for a gas, we use

$$\mu_i(T, p; \mathbf{N}) = \mu_i^{\ominus,(\mathrm{id})}(T) + RT \log a_i(T, p; \mathbf{N}) \quad (i = 1, \ldots, m) , \tag{9.46}$$

while for a solution, we use either

$$\mu_i(T, p; \mathbf{N}) = \mu_i^{\ominus}(T, p) + RT \log a_i(T, p; \mathbf{N}) \quad (i = 1, \ldots, m) , \tag{9.47}$$

or

$$\mu_1(T, p; \mathbf{N}) = \mu_1^{\ominus}(T, p) + RT \log a_1(T, p; \mathbf{N}) , \tag{9.48}$$

$$\mu_i(T, p; \mathbf{N}) = \mu_i^{\ominus,(\mathrm{dil})}(T, p) + RT \log a_i(T, p; \mathbf{N}) \quad (i = 2, \ldots, m) . \tag{9.49}$$

Generally, whether the first or second convention is more appropriate for a given solution depends on whether it is closer to the ideal limit or the dilute limit.

Often, the activity is expressed as $a_i = f_i/p^{\ominus}$ or $a_i = \varphi_i p_i/p^{\ominus}$ for a gas and $a_i = \gamma_i x_i$ for a solution. The quantities $f_i(T, p; \mathbf{N})$, $\varphi_i(T, p; \mathbf{N})$ and $\gamma_i(T, p; \mathbf{N})$ are referred to as the *fugacity, fugacity coefficient* and *activity coefficient*, respectively. They represent the deviation from the relatively simple behavior realized in the ideal and dilute limits. In the general case, all of these quantities, along with the activity, are complicated functions of T, p and $\mathbf{N}$.

9.3 Applications of balance conditions

In this section, we consider five applications of the balance conditions for multi-component systems.

9.3.1 Osmotic pressure

Here we investigate the phenomenon of *osmotic pressure*, which arises in solution systems with semipermeable walls. Also, as a technical point, we elucidate how clearly distinguishing between $(T; V, \boldsymbol{N})$ and $(T, p; \boldsymbol{N})$ as the independent variables of thermodynamic quantities and employing each in its proper role naturally results in clarity.[21]

---

[21] In Problem 9.3, we consider a derivation in which the "wrong" independent variables are used.

General case

Let us consider a solution consisting of two substances, with Species 1 acting as the solvent and Species 2 acting as the solute. As depicted in Figure 9.3, this system is contained in a closed vessel of fixed volume that is partitioned by a semipermeable wall into two compartments of fixed volumes V and V'. This wall is permeable to Species 1, but completely impermeable to Species 2. In Section 9.1, we introduced the concept of a semipermeable wall as a theoretical device to aid in the derivations presented there, but here, our interest is in elucidating the properties of physical systems that contain actual semipermeable walls. For example, in the case that Species 1 is water and Species 2 is some kind of polymer substance, there exist materials that can function as nearly ideal semipermeable walls.

Let N_1^{tot} and N_2 denote the total amounts of Species 1 and 2. For simplicity, we assume that Species 2 is contained entirely in the left-hand compartment (that of volume V), while Species 1 is able to move freely between the two compartments. We represent the amounts of Species 1 in the left-hand and right-hand compartments by N_1 and N_1', respectively.

Let us write the values of N_1 and N_1' in the equilibrium state at temperature T as $\widetilde{N}_1$ and $\widetilde{N}_1'$. For this substance, the balance condition (9.9) applies unaltered, and therefore the values of $\widetilde{N}_1$ and $\widetilde{N}_1'$ are uniquely determined by the relation

$$\mu_1(T; V, \widetilde{N}_1, N_2) = \mu_1(T; V', \widetilde{N}_1', 0) , \qquad (9.50)$$

along with $\widetilde{N}_1 + \widetilde{N}_1' = N_1^{\text{tot}}$.

Now, let us write the pressures in the left and right compartments in the state of balance as $p_{\text{L}} = p(T; V, \widetilde{N}_1, N_2)$ and $p_{\text{R}} = p(T; V', \widetilde{N}_1', 0)$. Because the wall separating these compartments is impermeable to Species 2, these two pressures will generally be unequal. The difference between the pressures on the two sides of the wall,

$$p_{\text{osmo}} = p_{\text{L}} - p_{\text{R}} , \qquad (9.51)$$

is referred to as the *osmotic pressure*.

Let us assume that the system we are considering is sufficiently far from the phase transition point of the solvent. In this case, the relation $p = p(T; V, \boldsymbol{N})$, regarded as an equation for V, possesses a unique solution $V(T, p; \boldsymbol{N})$. Therefore,

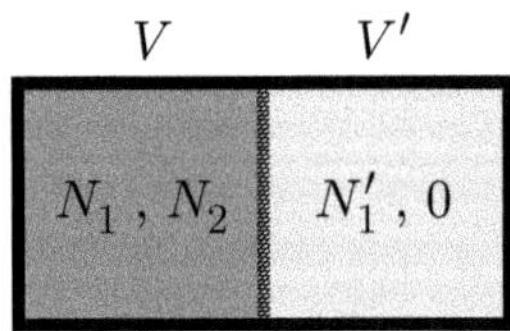

Figure 9.3 A situation in which osmotic pressure arises. Partitioning the vessel into two compartments with a fixed semipermeable wall that is impermeable to Species 2, the pressures on the two sides will differ in equilibrium. In the case of a dilute solution, this pressure difference has a universal form that is independent of the types of substances composing the system. The existence of a nonzero osmotic pressure means that the pressures on either side of the wall are unequal, in which case there is a net force exerted on the wall. As a result, if this wall possesses some degree of elasticity, it will bulge out to the right.

as discussed in Section 9.2, we can define a chemical potential as a function of T, p and $\boldsymbol{N}$ according to the prescription $\mu_i(T, p; \boldsymbol{N}) = \mu_i(T; V(T, p; \boldsymbol{N}), \boldsymbol{N})$. Then, using the pressures p_L and p_R given above for the left and right sides, the balance condition (9.50) can be rewritten as

$$\mu_1(T, p_\mathrm{L}; \widetilde{N}_1, N_2) = \mu_1^\ominus(T, p_\mathrm{R}) . \tag{9.52}$$

It is important to note here that although we are now regarding T, p and $\boldsymbol{N}$ as the independent variables of the chemical potential, this does *not* imply that the pressure is now a controllable parameter. We are still considering the original system, for which the volumes are controlled (in fact, fixed). What we have done here is merely to shift the point of view regarding functional dependence from one set of independent variables to another.

Next, if we substitute the relation defining the activity of the solvent, (9.48), into the balance condition (9.52), we obtain

$$\mu_1^\ominus(T, p_\mathrm{L}) + RT \log a_1(T, p_\mathrm{L}; \widetilde{N}_1, N_2) = \mu_1^\ominus(T, p_\mathrm{R}) . \tag{9.53}$$

From the relations (8.19) and (8.16) for a single-component system—the former expressing the Gibbs free energy in terms of the chemical potential, and the latter giving the derivative of the Gibbs free energy with respect to the pressure—we know that the quantity $\partial \mu_1^\ominus(T, p)/\partial p$ is equal to $v_1(T, p)$, the volume per unit amount of pure Species 1. Thus, we can rewrite (9.53) as

$$\int_{p_\mathrm{R}}^{p_\mathrm{L}} dp \, v_1(T, p) = -RT \log a_1(T, p_\mathrm{L}; \widetilde{N}_1, N_2) . \tag{9.54}$$

This is an *exact* equality relating the activity of the solvent and the pressures p_L and p_R. It is often used, for example, to measure the water activity in (not necessarily dilute) aqueous polymer solutions.

Dilute case

Next, we investigate the particular case in which the solution considered above is dilute. In this case, the activities are approximately equal to the mole fractions, and thus we have

$$\log a_1(T, p_\mathrm{L}; \widetilde{N}_1, N_2) \simeq \log \frac{\widetilde{N}_1}{\widetilde{N}_1 + N_2} \simeq \log(1 - \frac{N_2}{\widetilde{N}_1}) \simeq -\frac{N_2}{\widetilde{N}_1} . \tag{9.55}$$

Also, because the difference between the pressures p_L and p_R is small, the integral on the left-hand side of (9.54) can be evaluated as

$$\int_{p_\mathrm{R}}^{p_\mathrm{L}} dp \, v_1(T, p) \simeq (p_\mathrm{L} - p_\mathrm{R}) \, v_1(T, p_\mathrm{L}) \simeq p_\mathrm{osmo} \frac{V}{\widetilde{N}_1} . \tag{9.56}$$

The quantity $p_\mathrm{osmo} = p_\mathrm{L} - p_\mathrm{R}$ is the osmotic pressure that we seek. Substituting (9.55) and (9.56) into (9.54), we obtain

$$p_\mathrm{osmo} \simeq \frac{N_2 RT}{V} , \tag{9.57}$$

which is identical in form to the pressure of an ideal gas. We thus find that in the limit of a dilute solution, the osmotic pressure reduces to the pressure that would be realized in a system in which the solute alone exists as a dilute gas.

In his thermodynamics textbook [3], Enrico Fermi states that the result (9.57) is obvious from the point of view of molecular theory. However, this would appear to be mistaken.[22] Even in the case of a dilute solution, the presence of the solute introduces a non-trivial change in the concentration of the solvent (i.e., Species 1), which will generally affect the pressure. However, this change is somehow miraculously canceled by another change in the pressure introduced by the solute, which results from the alteration of the interactions within the system. As a result of this cancellation, we obtain the simple, universal result given above (see Problem 9.3). It is unclear (at least to us) how molecular theory could offer an intuitive understanding of this cancellation.

9.3.2 Henry's law

Consider a system consisting of a dilute solution and an ideal gas, with an arbitrary number of species contained in each. Choose an arbitrary substance, Species i, that is present in both the gas and the solution and is a solute in the latter. Here, as an application of the balance condition (9.25), we derive Henry's law, which relates the mole fraction of Species i in the solution with the partial pressure of Species i in the gas.

Following (9.25), we equate the expressions for the chemical potential of Species i in the gas, given by (9.31), and in the solution, given by (9.42):

$$\mu_i^{\ominus,(\mathrm{id})}(T) + RT \log \frac{p_i}{p^\ominus} = \mu_i^{\ominus,(\mathrm{dil})}(T,p) + RT \log x_i \ . \tag{9.58}$$

Here, it is important to keep in mind that the quantities on the left-hand side describe the gas, and the quantities on the right-hand side describe the solution. Rearranging, we obtain

$$p_i = \alpha(T,p)x_i \ , \tag{9.59}$$

where $\alpha(T,p) = p^\ominus \exp\left[\left(\mu_i^{\ominus,(\mathrm{dil})}(T,p) - \mu_i^{\ominus,(\mathrm{id})}(T)\right)/RT\right]$. This is Henry's law. Now, considering (9.38), we see that the p dependence of $\alpha(T,p)$ results entirely from that of $v_1(T,p)$, the molar volume of a system of pure solvent at (T,p). But because generally such a system can be regarded as nearly incompressible, under ordinary conditions we can, to good approximation, ignore this p dependence and regard α to be a function of T alone.

9.3.3 Raoult's law

Next, as another application of the balance condition (9.25), we derive a relation closely related to Henry's law. We again consider a system containing both a

---

[22] To be honest, we cannot make sense of Fermi's derivation of (9.57) given there.

solution and a gas, with an arbitrary number of species contained in each. However, in the present case, instead of a dilute solution, we consider an ideal solution.

As above, we consider an arbitrary substance, Species i, that appears in both the solution and the gas. Then, applying (9.25), we equate the expressions for its chemical potential in the gas, given by (9.31), and in the solution, given by (9.43):

$$\mu_i^{\ominus,(\mathrm{id})}(T) + RT \log \frac{p_i}{p^\ominus} = \mu_i^\ominus(T,p) + RT \log x_i \, . \tag{9.60}$$

As with (9.58), the left-hand side here describes the gas, and the right-hand side describes the solution.

Recall that $\mu_i^\ominus(T,p)$ above is the chemical potential for a pure sample of Species i at (T,p). We next simplify (9.60) by writing $\mu_i^\ominus(T,p)$ in a different form. For this purpose, we consider a second system, consisting of a liquid of pure Species i and a gas containing Species i at (T,p).[23] Applying (9.24) to this system, we have

$$\mu_i^\ominus(T,p) = \mu_i^{\ominus,(\mathrm{id})}(T) + RT \log \frac{p_i^*}{p^\ominus} \, , \tag{9.61}$$

where p_i^* is the partial pressure of the Species i vapor in this system. Combining (9.60) and (9.61), we have

$$\mu_i^{\ominus,(\mathrm{id})}(T) + RT \log \frac{p_i}{p^\ominus} = \mu_i^{\ominus,(\mathrm{id})}(T) + RT \log \frac{p_i^*}{p^\ominus} + RT \log x_i \, . \tag{9.62}$$

Simplifying, we obtain

$$p_i = x_i \, p_i^* \, . \tag{9.63}$$

This is Raoult's law.

Finally, note that the solvent in a dilute solution also obeys Raoult's law, because its chemical potential, given in (9.41), is identical in form to (9.43).

9.3.4 Boiling point elevation and freezing point depression in a dilute solution

As two further applications of the balance condition given in (9.25) and the two expressions for the chemical potentials of a dilute solution (9.41) and (9.42), here we investigate the closely related phenomena of boiling point elevation and freezing point depression.

Boiling point elevation

We consider a system consisting of two components, an amount N_1 of Species 1 and an amount N_2 of Species 2, under conditions of constant pressure p. The system is prepared at some temperature at which it consists entirely of a liquid solution, with Species 1 acting as the solvent and Species 2 acting as the solute. Let us suppose

---

[23] Of course, in general, this gas must contain at least one other substance. Thus, we are considering the situation in which any additional substances in the gas can be regarded as insoluble in the liquid of pure Species i.

that the amounts of substance satisfy the relation $N_1 \gg N_2$, and therefore that this can be treated as a dilute solution. With the system so prepared, we consider the situation in which its temperature is raised until it reaches its boiling point, and we seek to elucidate the effect of the solute on the observed behavior.

Let us write the boiling points (at p) of a single-component system consisting of only Species 1 as T_b and of the solution considered here as T_b'. Then, as discussed in Section 7.5, a pure system of Species 1 will be a liquid for $T < T_b$ and a gas for $T > T_b$, while at $T = T_b$, its liquid and gas phases will coexist (see Figure 7.7). On the basis of empirical observation, we postulate similar behavior for the solution considered here, and below we derive a relation between T_b and T_b'. To simplify the situation, let us assume that for the conditions under consideration, Species 2 can be regarded as completely nonvolatile and thus that the gas phase appearing for $T \geq T_b'$ will consist of only Species 1. Empirically, we know that there are systems for which this treatment provides a good approximation. For example, this is true in the case that Species 1 is water and Species 2 is sucrose at atmospheric pressure.

We are assuming that Species 2 exists only in the solution, and therefore, from (9.25), we obtain the following as the only balance condition governing the coexistence of the solution and gas at $T = T_b'$:[24]

$$\mu_1(T_b', p; N_1, N_2) = \mu_1^{\ominus}(T_b', p) \ . \tag{9.64}$$

The quantities on the left-hand and right-hand sides of this equation are the chemical potentials of Species 1 in the solution and in the single-component gas, respectively.

Because the liquid is a dilute solution with Species 2 acting as the solute, using the approximation for the chemical potential of the solvent appearing in (9.41), we have

$$\mu_1(T, p; N_1, N_2) \simeq \mu_1^{\ominus}(T, p) + RT \log(1 - x_2) \simeq \mu_1^{\ominus}(T, p) - RT x_2 \ , \tag{9.65}$$

where x_2 is the mole fraction of Species 2. As discussed in Footnote 18 on p. 200, for small x_2, the derivation of the dilute solution form of the Gibbs free energy is valid as long as systems consisting of the solution and of the solvent alone are on the same side of the phase transition for the temperature and pressure under consideration. Thus, assuming on the basis of empirical results that $T_b' > T_b$, it is seen that the above derivation is valid for $T < T_b$. Then, from the continuity of the two chemical potentials at T_b, we conclude that the relation (9.65) is also valid at $T = T_b$.

Figure 9.4 provides a clear understanding of the situation described above. The top curve represents the chemical potential of a pure sample of Species 1, $\mu_1^{\ominus}(T, p)$. As seen there, $\mu_1^{\ominus}(T, p)$ is a concave, continuous function of T, and its slope changes discontinuously at T_b.[25] The solid curve below $\mu_1^{\ominus}(T, p)$ in the region $T \leq T_b$ is

---

[24] The present treatment, in which we consider only one balance condition, should be understood as resulting from a more proper treatment in which, with the balance condition for μ_2 also included, we find that there is only a very small amount of Species 2 in the gas. However, because the result of such a more complicated, proper treatment is the same as that derived below, for the sake of conciseness, here we simply ignore the existence of Species 2 in the gas from the outset.

[25] In some textbooks, the treatment of boiling point elevation is carried out by introducing two independent chemical potentials describing a pure system of Species 1, one for the liquid phase, $\mu_1^{L}(T, p)$,

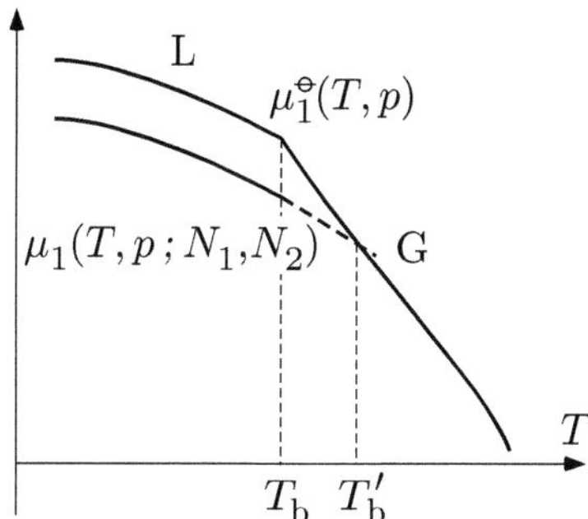

Figure 9.4 Boiling point elevation in a dilute solution. Here, $\mu_1^{\ominus}(T,p)$ is the chemical potential of a pure sample of Species 1. This quantity is plotted as a function of T, with p fixed. Also plotted is $\mu_1(T,p;N_1,N_2)$, the chemical potential of Species 1 in a solution in which a small amount of Species 2 is dissolved into Species 1. The low-temperature part plotted as a solid curve is the result obtained using the dilute solution approximation. The part plotted as a dashed curve is extrapolated from the low-temperature part in a "natural" way. The point at which $\mu_1^{\ominus}(T,p)$ and $\mu_1(T,p;N_1,N_2)$ intersect determines the boiling point of the solution, T_{b}'.

a plot of the approximate form of the chemical potential $\mu_1(T,p;N_1,N_2)$ given by (9.65). As shown in this figure, for values of T at which (9.65) is valid, the left-hand side of the balance condition (9.64), $\mu_1(T,p;N_1,N_2)$, is always less than the right-hand side, $\mu_1^{\ominus}(T,p)$. In other words, in equilibrium, the solution and gas cannot be balanced in the region $T \leq T_{\mathrm{b}}$. Thus, because we are assuming that the solution and gas do coexist at some temperature (at which (9.64) holds), this temperature must be greater than T_{b}. Between the boiling point of pure Species 1, T_{b}, and this new boiling point, T_{b}', the solution continues to exist as a liquid. However, we do not have any information about the behavior of $\mu_1(T,p;N_1,N_2)$

and one for the gas phase, $\mu_1^{\mathrm{G}}(T,p)$, which are defined on overlapping temperature intervals. Then, the desired result is obtained by considering the system with values of T between T_{b} and T_{b}' and writing the dilute solution form (9.65) in terms of $\mu_1^{\mathrm{L}}(T,p)$. In addition to the dubiousness of approximating $\mu_1(T,p;N_1,N_2)$ through an expansion about a fictional state (i.e., a pure system of Species 1 in the liquid phase with $T > T_{\mathrm{b}}$), this treatment is very unorthodox with regard to basic thermodynamics. Within ordinary thermodynamics, the chemical potential of a pure substance is never defined as multiple independent functions. It is defined as a single function, and that function alone describes solid, liquid and gas phases, and even possesses information describing phase transitions. Indeed, this can be regarded as one of the strengths of thermodynamic formalism. (As one possible way of formulating multiple chemical potential functions, it might be thought that we could, for example, construct a chemical potential $\mu_1^{\mathrm{L}}(T,p)$ in the liquid region and a chemical potential $\mu_1^{\mathrm{G}}(T,p)$ in the gas region and then extend the former into the gas region and the latter into the liquid region through analytic continuation. In this way, we could obtain independent, apparently physically meaningful liquid and gas chemical potentials, defined on overlapping domains. However, constructing such analytic continuations is generally quite difficult, and, in fact, it has been rigorously proved that their construction is mathematically impossible within certain simple standard models of statistical mechanics that exhibit phase transitions, including the Ising model. For detailed discussion, see Section 3.10.9 of S. Friedli and Y. Velenik, *Statistical Mechanics of Lattice Systems: A Concrete Mathematical Introduction*, Cambridge University Press, 2018.) Of course, for the purpose of investigating metastable states observed experimentally, for example in supercooled systems, we do seek meaningful multivalued chemical potentials. However, that is an entirely different problem, and much more important than the problem of simple calculational expediency discussed presently. In any case, as seen from the derivation given here, the single-valued chemical potential $\mu_1^{\ominus}(T,p)$ obtained from the ordinary thermodynamic formalism is sufficient to provide an understanding of boiling point elevation.

above T_b, and therefore we cannot continue our analysis beyond this point without the introduction of some additional assumptions.

If we assume that up to the temperature T_b', the properties of the solution do not undergo any kind of extreme change, then it should be possible to approximate the actual behavior of $\mu_1(T, p; N_1, N_2)$ in the region $T_b < T \leq T_b'$ by extrapolating its behavior for $T \leq T_b$ in a "natural" manner. The dashed curve in Figure 9.4 is meant to represent such an extrapolation of $\mu_1(T, p; N_1, N_2)$ from the $T \leq T_b$ region into the $T > T_b$ region. The temperature at which this dashed curve intersects the $\mu_1^{\ominus}(T, p)$ curve is T_b'. Below, we derive a natural extrapolation using the simplest considerations and thereby determine an expression for T_b'.

If the mole fraction of Species 2 in the solution is sufficiently small, the difference between T_b and T_b' should also be small. With this assumption, let us approximate $\mu_1(T, p; N_1, N_2)$ and $\mu_1^{\ominus}(T, p)$ between T_b and T_b' as linear functions. As the slopes of $\mu_1(T, p; N_1, N_2)$ and $\mu_1^{\ominus}(T, p)$ on this interval, it is reasonable to use the values[26]

$$\frac{\partial \mu_1^{\ominus}(T_b, p)}{\partial T_-} = \lim_{T \nearrow T_b} \frac{\partial \mu_1^{\ominus}(T, p)}{\partial T}, \quad \frac{\partial \mu_1^{\ominus}(T_b, p)}{\partial T_+} = \lim_{T \searrow T_b} \frac{\partial \mu_1^{\ominus}(T, p)}{\partial T}, \tag{9.66}$$

respectively. Then, with the relations

$$\mu_1(T_b', p; N_1, N_2) \simeq \mu_1^{\ominus}(T_b, p) - RT_b x_2 + \frac{\partial \mu_1^{\ominus}(T_b, p)}{\partial T_-}(T_b' - T_b), \tag{9.67}$$

$$\mu_1^{\ominus}(T_b', p) \simeq \mu_1^{\ominus}(T_b, p) + \frac{\partial \mu_1^{\ominus}(T_b, p)}{\partial T_+}(T_b' - T_b), \tag{9.68}$$

the condition (9.64) gives

$$\left(\frac{\partial \mu_1^{\ominus}(T_b, p)}{\partial T_-} - \frac{\partial \mu_1^{\ominus}(T_b, p)}{\partial T_+}\right)(T_b' - T_b) \simeq RT_b x_2. \tag{9.69}$$

(This should be compared with the behavior displayed in Figure 9.4.) Next, using (8.35), which relates the discontinuity exhibited by the derivative of the chemical potential and the enthalpy of vaporization, after simplification, we find

$$T_b' - T_b \simeq \frac{RT_b^2}{h_{vap}(p)} x_2. \tag{9.70}$$

We have thus obtained an expression for the boiling point elevation of a dilute solution. Because $h_{vap}(p)$ is generally positive, we conclude that the boiling point of a dilute solution increases in proportion to the mole fraction of the solute, x_2. In addition, we find that the proportionality coefficient contains only T_b and $h_{vap}(p)$ (along with the universal constant R), which reflect properties of the solvent alone.

---

[26] In order to express the first equation here in terms of $\mu_1^{\ominus}(T, p)$ rather than $\mu_1(T, p; N_1, N_2)$, we have used (9.65) and the fact that x_2 is small.

Finally, let us consider a concrete example. For this purpose, we put (9.70) into the following more standard form:[27]

$$T'_{\mathrm{b}} - T_{\mathrm{b}} \simeq \frac{MRT_{\mathrm{b}}{}^2}{h_{\mathrm{vap}}(p)}\, m = E_b\, m \,. \tag{9.71}$$

Here, M is the molar mass of the solvent and m is the molality (moles of solute per kilogram of solvent). The quantity E_b is referred to as the *ebullioscopic constant* or sometimes the *boiling point elevation coefficient*. For water at 1 atm, we have $E_b \approx 0.513\,\mathrm{K\,kg\,mol}^{-1}$. Thus, for example, if 1 mol of solute is dissolved in 1 kg of water, the boiling point will increase by approximately 0.5 K.

Freezing point depression

Note that the above treatment is based on three main assumptions: that the solute is present only in the liquid phase, that the dilute solution form of the chemical potential is valid for $T \le T_{\mathrm{b}}$, and that the presence of the solute increases the boiling point. Now, if instead of the liquid-gas phase transition, we consider the liquid-solid phase transition, applying analogous assumptions and repeating the above argument, we can derive a result analogous to (9.71) that predicts the amount by which the solute *decreases* the freezing point. In this case, we assume that the solute exists only in the liquid phase,[28] that the dilute solution form of the chemical potential is valid for $T \ge T_{\mathrm{f}}$ (where T_{f} is the freezing point of pure solvent), and that the presence of the solute decreases the freezing point. Then, proceeding as above, we obtain

$$T'_{\mathrm{f}} - T_{\mathrm{f}} \simeq -\frac{MRT_{\mathrm{f}}^2}{h_{\mathrm{fus}}(p)}\, m = -E_{\mathrm{f}}\, m \,, \tag{9.72}$$

where T'_{f} is the freezing point of the solution, $h_{\mathrm{fus}}(p)$ is the enthalpy of fusion, and E_{f} is the *cryoscopic constant*.

9.4 Equilibrium in chemical reactions

For the remainder of this chapter, we study the thermodynamics of chemical reactions. In this section, we first present the formalism through which we extend the theory of thermodynamics constructed to this point to allow the treatment of systems exhibiting chemical reactions. We then derive the *equilibrium condition* for such systems. This condition determines the composition of a system that has realized a state of *chemical equilibrium*[29] under conditions of constant temperature and pressure. Next, utilizing some of the results obtained in the previous sections of this chapter, we present several practically useful forms of the equilibrium condition. As an application of the equilibrium condition, we then study Le Chatelier's principle. Finally, we consider the equilibrium condition for systems exhibiting multiple chemical reactions.

---

[27] To obtain this form, we first approximate x_2 as N_2/N_1.

[28] Because the conditions required for solid solvation are relatively stringent, this situation is the norm.

[29] As pointed out in Footnote 28 on p. 150, the general state of equilibrium in physical systems can, in a rough sense, be regarded as possessing two aspects, phase equilibrium and chemical equilibrium.

Before moving on, let us make an important comment concerning notation. In this and the following section, we use tildes to indicate the values taken by quantities in equilibrium. For example, the extent of reaction, ξ, is a variable that generally can take a range of values. However, for a given system, in equilibrium, ξ takes a single, well-defined value, which we write $\tilde{\xi}$. In the present section and in Section 9.5, every quantity with a tilde should be interpreted as the equilibrium value.

9.4.1 The description and treatment of chemical reactions

Description of chemical reactions

Let us consider a system consisting of m chemical species. Representing the chemical formula (for example, CO_2, HCl, etc.) of Species i by A_i, we assume that the system exhibits a chemical reaction expressed as

$$n_1 A_1 + n_2 A_2 + \cdots + n_k A_k \rightleftarrows n_{k+1} A_{k+1} + \cdots + n_\ell A_\ell , \tag{9.73}$$

where $\ell \leq m$. Here, the coefficients $n_1, \ldots, n_\ell$ (the *stoichiometric coefficients*) are all positive integers. Species $1, \ldots, k$ are the reactants, and Species $k+1, \ldots, \ell$ are the products. If $\ell < m$, then Species $\ell+1, \ldots, m$ are spectators, whose amounts do not change under the reaction. For convenience, we define the quantities ν_i as follows:

$$\nu_i = \begin{cases} -n_i & \text{for } i = 1, \ldots, k \text{ (reactants)} , \\ n_i & \text{for } i = k+1, \ldots, \ell \text{ (products)} , \\ 0 & \text{for } i = \ell+1, \ldots, m \text{ (spectators)} . \end{cases} \tag{9.74}$$

Then, along with the abbreviated notation $\boldsymbol{N} = (N_1, \ldots, N_m)$, we also introduce $\boldsymbol{\nu} = (\nu_1, \ldots, \nu_m)$.

On the operational treatment of chemical reactions

The operational treatment of chemical reactions that we employ requires that we can clearly separate the mixing and reaction processes undergone by the chemical species in a system. In Sections 9.4.2 and 9.4.3, we present an analysis in which it is assumed that this separation can be made, and we show how this analysis allows for the treatment of systems exhibiting chemical reactions. There, in order to explicitly demonstrate this analysis, we consider the simple situation in which the separation of the mixing and reaction processes can be realized through the addition and removal of a catalyst. Then, in Section 9.4.4, we reconsider this analysis from a general point of view and evaluate its meaning as a treatment of chemical reactions in both abstract and concrete terms. There, we first discuss the physical meaning of this analysis as a general method for applying the framework of thermodynamics to systems exhibiting chemical reactions. Next, we consider the practical use of this analysis as a method for deriving the equilibrium condition, (9.82), which is its main purpose.

9.4.2 Construction of free energies

As discussed above, in this and the following subsection, we carry out the analysis of systems exhibiting chemical reactions by considering the case in which the separation of the mixing and reaction processes can be realized through the use of a catalyst.

Consider a system consisting of m chemical species that can undergo a chemical reaction. First, we assume that the rate of the reaction represented by (9.73) is extremely slow when there is no catalyst present. In this situation, preparing a system with an arbitrary composition $\boldsymbol{N}^{(0)}$, fixing the pressure to some value p, and placing it in an environment of temperature T, we can make the approximation that the mixing of the substances is completed on a time scale over which this composition is unchanged, and thus that the system reaches a phase equilibrium state $(T; V, \boldsymbol{N}^{(0)})$. This initial state is essentially the same as the equilibrium states of the multi-component systems considered to this point. Next, we assume that the addition of a catalyst allows the chemical reaction (9.73) to proceed spontaneously, with a greatly increased rate, and that the system eventually reaches a state of both phase and chemical equilibrium, in which the forward and backward reactions expressed in (9.74) occur at the same rate, and hence the amounts of substance are fixed in time.

In the situation described above, let us write the amount of the ith substance at an arbitrary time after the introduction of the catalyst as $N_i^{(0)} + \nu_i \xi$. This implies that, at this time, the amounts for all m species are collectively given by $\boldsymbol{N}^{(0)} + \boldsymbol{\nu}\xi$. The variable ξ characterizing the progression of the reaction is referred to as the *extent of reaction*. This is an extensive quantity that can take both positive and negative values.

It is important to understand that in an experiment of the type we are considering here, we have no control over the time evolution of ξ: Once the catalyst is introduced and the reaction is activated, ξ evolves autonomously, eventually reaching its equilibrium value. Now, let us assume that, in addition to this kind of autonomous evolution, the variation of ξ can also be carried out in a controlled manner through external operations of some type. This assumption is essential in the theoretical treatment given here. Just as it was useful in developing the operational treatment of mixing to consider the hypothetical experimental apparatus depicted in Figure 9.1, it is useful in developing the operational treatment of chemical reactions to consider a kind of apparatus through which we are able to control the chemical reaction (9.73), allowing us to slowly advance this reaction in the desired direction by the desired amount. The construction of such an apparatus is probably not generally feasible, but we can imagine the kind of hypothetical experimental setup employing semipermeable walls depicted in Figure 9.5.[30] Assuming the existence of an apparatus as described there and the ideal situation with regard to our ability to control the rate of reaction, the Helmholtz and Gibbs free energies of a system undergoing a chemical reaction can be constructed in the manner outlined below.

---

[30] There are simpler possibilities as well. For example, as discussed in Section 9.6, in the case of a chemical battery, the extent of reaction can in some sense be controlled by controlling the flow of current induced by the battery.

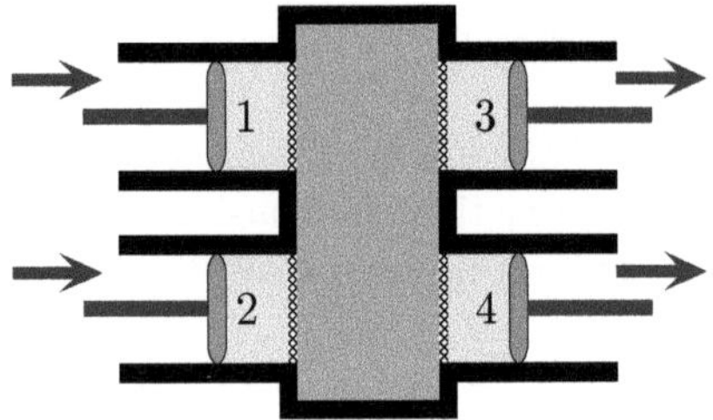

Figure 9.5 Hypothetical experimental apparatus for operationally controlling a chemical reaction referred to as a *van 't Hoff reaction box*. We consider a system exhibiting the reaction $n_1 A_1 + n_2 A_2 \rightleftarrows n_3 A_3 + n_4 A_4$. As shown in the figure, the reaction box consists of a central compartment and four reservoirs with pistons. The ith reservoir contains a pure sample of the ith substance, and it is separated from the central compartment by a semipermeable wall that allows only the ith substance to pass through. The central compartment contains a mixture of all four substances plus catalyst. The entire system begins in a state of (chemical and phase) equilibrium. Then, the pistons are slowly moved in such a manner to cause amounts $n_1\xi$ and $n_2\xi$ of Species 1 and 2, respectively, to move from the corresponding reservoirs to the central compartment and amounts $n_3\xi$ and $n_4\xi$ of Species 3 and 4, respectively, to move from the central compartment to the corresponding reservoirs. Accordingly, through the chemical reaction, the substances that were forced into the central compartment exactly replace the substances that were forced out, and as a result, the composition in the central compartment is unchanged. Thus, the net effect imparted by this procedure is limited to the reservoirs, whose amounts of substance and volumes have both been altered. In this way, a chemical reaction through which some quantities of pure Species 1 and pure Species 2 are converted into some quantities of pure Species 3 and pure Species 4 is realized as a quasi-static isothermal operation. Also, note that there is a well-defined amount of work associated with this operation—simply the sum of the amounts of work performed by the reservoirs on the four pistons.

As above, we consider a system consisting of m substances, k of which act as reactants, $\ell - k$ of which act as products, and $m - \ell$ of which act as spectators in a chemical reaction. The entire procedure considered here is understood as taking place under isothermal conditions at temperature T. First, suppose that the system contains no catalyst and is in an initial (phase equilibrium) state $(T; V^{(0)}, \boldsymbol{N}^{(0)})$. The content of the system is then completely separated into its constituent substances through a quasi-static operation using a mixing-unmixing apparatus, and we thereby obtain a set of m individual, single-component systems. We write the amount of work performed in this operation as W_1 and the amount of substance in the ith pure system as $N_i^{(0)}$. The ℓ pure systems containing the product and reactant species are to be used as the reservoirs for a reaction box, like that described in Figure 9.5, whose central compartment contains the same ℓ substances in a state of both phase and chemical equilibrium also at temperature T. Initially, each pure system is separated from the central compartment by an ordinary wall. With this configuration, the volume of each reservoir is quasi-statically adjusted until the chemical potential of the corresponding substance is the same on each side of this wall. Let us call the total amount of work performed by all of the reservoirs in these operations W_2. After these operations have been completed, the ordinary walls separating the reservoirs from the central compartment are replaced with semipermeable walls, each of which is permeable only to the substance in the corresponding reservoir. Then the volumes of the reservoirs are varied quasi-statically in some arbitrary manner, and as a result, the amounts of substance change from their initial values, $N_i^{(0)}$, to some values $N_i^{(0)} + \Delta N_i$. (Obviously, $\Delta N_i = 0$ for each spectator species.) While under such an operation, it is possible to realize a wide range of values of $\Delta \boldsymbol{N} = (\Delta N_1, \ldots, \Delta N_m)$, here we are interested only in the case in which $\Delta \boldsymbol{N}$ is given by $\boldsymbol{\nu}\xi$, for some (positive or negative) value ξ. In this case, the state of the central compartment is unchanged under this operation. We call the total work performed by all reservoirs in this operation W_3. Next, the reservoirs are again separated from the central compartment by ordinary walls and then removed from the reaction box. The pure systems are then reintroduced to the mixing-unmixing apparatus and mixed through a quasi-static operation (again without catalyst), through which the system performs an amount of work W_4. Finally, the volume of the system is adjusted quasi-statically to some desired final value, $V^{(0)} + \Delta V$, with the system performing an amount of work W_5. In this way, we have realized a transition from the initial phase equilibrium state $(T; V^{(0)}, \boldsymbol{N}^{(0)})$ to a final phase equilibrium state $(T; V^{(0)} + \Delta V, \boldsymbol{N}^{(0)} + \boldsymbol{\nu}\xi)$. Then because all of the operations performed in this procedure are quasi-static, while the state of the central compartment in the reaction box is unchanged, we have

$$F[T; V^{(0)} + \Delta V, \boldsymbol{N}^{(0)} + \boldsymbol{\nu}\xi] - F[T; V^{(0)}, \boldsymbol{N}^{(0)}] = -\sum_{j=1}^{5} W_j \, . \qquad (9.75)$$

Repeating this procedure, we can construct $F[T; V, \boldsymbol{N}^{(0)} + \boldsymbol{\nu}\xi]$ as a function of V and ξ, and with this, we can determine $G[T, p; \boldsymbol{N}^{(0)} + \boldsymbol{\nu}\xi]$ in the ordinary manner.

9.4.3 Variational principle and the equilibrium condition

Having assumed that we can operationally control the extent of reaction, we now derive the corresponding variational principle. For this purpose, we could proceed just as in Section 9.2.2, where the variational principle for the Gibbs free energy, (9.22), was derived from the corresponding variational principle for the Helmholtz free energy, (9.5), with the aid of the Legendre transformation. However, here we carry out a more direct derivation, based on the maximum work principle (9.76), which relates the change in the Gibbs free energy to the maximum work done by a thermodynamic system under conditions of fixed temperature and pressure.

Let us consider a system composed of multiple chemical species that is in contact with the atmosphere at constant temperature T and constant pressure p. In experimental studies, conditions of this type are very common. We write the composition of the system in the form $\boldsymbol{N}^{(0)} + \boldsymbol{\nu}\xi$, where now $\boldsymbol{N}^{(0)}$ is regarded as a fixed reference composition. (We could take $\boldsymbol{N}^{(0)}$ to be the initial composition, as in the previous subsection, but it is more convenient to regard it as a reference value, which need not be the initial composition.) Suppose that initially the system contains no catalyst and is in the phase equilibrium state $(T; V, \boldsymbol{N}^{(0)} + \boldsymbol{\nu}\xi_1)$ with pressure p. Then, suppose that an external agent carries out a mechanical operation through which the extent of reaction is changed from the initial value ξ_1 to some final value ξ_2.

We express the maximum possible amount of work performed by the system on the external agent in any such operation as $W_{\max}(T, p; \boldsymbol{N}^{(0)} + \boldsymbol{\nu}\xi_1 \to \boldsymbol{N}^{(0)} + \boldsymbol{\nu}\xi_2)$ and call this quantity the *maximum work under constant pressure.*

As discussed in Sections 2.1.1 and 8.1.1, in general, we model physical systems of the type considered here by regarding the pressure to be fixed by a mechanical device, such as the weight in Figure 8.1. Now, let W_{tot} be the total work performed by the system in an operation causing the transition $\boldsymbol{N}^{(0)} + \boldsymbol{\nu}\xi_1 \to \boldsymbol{N}^{(0)} + \boldsymbol{\nu}\xi_2$, with fixed T and p. With V_1 and V_2 representing the initial and final volumes of the system, an amount $p(V_2 - V_1)$ of this total work is performed on the device maintaining the constant pressure. Therefore, the amount of work performed on the external agent is $W = W_{\text{tot}} - p(V_2 - V_1)$. We seek the maximum value of this W over all possible operations. Below, we determine a general expression for this value by employing the reasoning used in our derivation of the Gibbs free energy given in Section 8.1.1.

As we explained when we introduced the Gibbs free energy through (8.3), the mechanical device maintaining the constant pressure can also be regarded as a thermodynamic system, and we can regard the operation as being performed on the composite thermodynamic system consisting of the fluid system and this mechanical device. The work done by the composite system in this operation is identically the work W done on the external agent that controls the chemical reaction, which is the quantity of interest. Therefore, the desired maximum work, $W_{\max}(T, p; \boldsymbol{N}^{(0)} + \boldsymbol{\nu}\xi_1 \to \boldsymbol{N}^{(0)} + \boldsymbol{\nu}\xi_2)$, is simply the maximum work (in the original sense, defined in Section 3.5) performed by the composite system. As expressed by the form (3.27) of the original principle of maximum work, this quantity is given by the difference between the values of the Helmholtz free energy of the composite system before and after the operation. Then, because (as discussed in Section 8.1.1) the Helmholtz free energy of the composite system is simply the Gibbs free energy of the fluid system alone, whose construction is described in Section 9.4.2, we find

$$W_{\max}(T, p; \boldsymbol{N}^{(0)} + \boldsymbol{\nu}\xi_1 \to \boldsymbol{N}^{(0)} + \boldsymbol{\nu}\xi_2) = G[T, p; \boldsymbol{N}^{(0)} + \boldsymbol{\nu}\xi_1] - G[T, p; \boldsymbol{N}^{(0)} + \boldsymbol{\nu}\xi_2] .$$
$$(9.76)$$

This is the desired principle of maximum work under constant pressure. Note that there is a perfect correspondence between this and the original principle of maximum work, as expressed by (3.27).

With the above preparation, we are now ready to give a complete treatment of the state of chemical equilibrium.

We consider a system whose pressure and temperature are fixed at some values p and T. Initially, the system contains no catalyst and is in a state of phase equilibrium but not chemical equilibrium, with composition $\boldsymbol{N}^{(0)} + \boldsymbol{\nu}\xi$. We write this state as $(T; V, \boldsymbol{N}^{(0)} + \boldsymbol{\nu}\xi)$. Then, a catalyst is added to the system, and it is left until it reaches the state of chemical (and phase) equilibrium. Finally, the catalyst is removed. This final state is denoted by $(T; \tilde{V}, \boldsymbol{N}^{(0)} + \boldsymbol{\nu}\tilde{\xi})$. The composition $\boldsymbol{N}^{(0)} + \boldsymbol{\nu}\tilde{\xi}$ represents the state of chemical equilibrium at temperature T and pressure p, and $\tilde{\xi}$ is the extent of reaction in this equilibrium state. Because the transition from the initial composition, $\boldsymbol{N}^{(0)} + \boldsymbol{\nu}\xi$, to the final composition, $\boldsymbol{N}^{(0)} + \boldsymbol{\nu}\tilde{\xi}$, is accomplished solely through the introduction and removal of the catalyst, the system performs no work on the mechanical world during this process.[31] With this observation, from the definition of the maximum work given above, we obtain

$$W_{\max}(T, p; \boldsymbol{N}^{(0)} + \boldsymbol{\nu}\xi \to \boldsymbol{N}^{(0)} + \boldsymbol{\nu}\tilde{\xi}) \geq 0 \,. \qquad (9.77)$$

Combining this with (9.76), we find that the relation

$$G[T, p; \boldsymbol{N}^{(0)} + \boldsymbol{\nu}\tilde{\xi}] \leq G[T, p; \boldsymbol{N}^{(0)} + \boldsymbol{\nu}\xi] \qquad (9.78)$$

holds for arbitrary ξ. This is the variational inequality for chemical equilibrium. Rewriting this in the form of a variational principle, we have

$$G[T, p; \boldsymbol{N}^{(0)} + \boldsymbol{\nu}\tilde{\xi}] = \min_{\xi} G[T, p; \boldsymbol{N}^{(0)} + \boldsymbol{\nu}\xi] \,. \qquad (9.79)$$

This equilibrium condition is of fundamental importance in the study of chemical reactions, as it determines the extent of reaction in the chemical equilibrium state, $\tilde{\xi}$. Because the derivative of $G[T, p; \boldsymbol{N}^{(0)} + \boldsymbol{\nu}\xi]$ with respect to ξ vanishes at the minimum, the equation [32]

$$\left. \frac{\partial G[T, p; \boldsymbol{N}^{(0)} + \boldsymbol{\nu}\xi]}{\partial \xi} \right|_{\xi = \tilde{\xi}} = 0 \qquad (9.80)$$

constitutes the local condition for chemical equilibrium. Using the differential form for the Gibbs free energy given in (9.20), we find

$$\frac{\partial G[T, p; \boldsymbol{N}^{(0)} + \boldsymbol{\nu}\xi]}{\partial \xi} = \sum_{i=1}^{m} \nu_i \, \mu_i(T, p; \boldsymbol{N}^{(0)} + \boldsymbol{\nu}\xi) \,, \qquad (9.81)$$

and thus we can rewrite (9.80) as

$$\sum_{i=1}^{m} \nu_i \, \mu_i(T, p; \boldsymbol{N}^{(0)} + \boldsymbol{\nu}\tilde{\xi}) = 0 \,. \qquad (9.82)$$

This relation plays a central role in chemistry, and it is the main result of the present section.[33] If we have sufficient information regarding the functional forms of the chemical potentials $\mu_i(T, p; \boldsymbol{N})$, then for any given initial composition $\boldsymbol{N}^{(0)} + \boldsymbol{\nu}\xi$, the equilibrium composition $\boldsymbol{N}^{(0)} + \boldsymbol{\nu}\tilde{\xi}$ can be uniquely determined by solving (9.82).

---

[31] In principle, the amount of work required to add and later extract the catalyst can be made arbitrarily small.

[32] Because $G[T, p; \boldsymbol{N}^{(0)} + \boldsymbol{\nu}\xi]$ is a convex function of ξ (see Theorem G.8), from Theorem G.6 it is concluded that the local equilibrium condition expressed by (9.80) is equivalent to the global condition given in (9.79) (see Result 7.2 (p. 148)).

[33] Use of the quantity $A(T, p; \boldsymbol{N}) = -\sum_{i=1}^{m} \nu_i \, \mu_i(T, p; \boldsymbol{N})$, termed the *affinity*, is quite common in chemistry, and the condition (9.82) is often expressed as $A(T, p; \boldsymbol{N}^{(0)} + \boldsymbol{\nu}\tilde{\xi}) = 0$.

9.4.4 Evaluation of the operational treatment of chemical reactions

In Sections 9.4.2 and 9.4.3, we presented an analysis through which it was demonstrated how the entire framework of thermodynamics could, in principle, be applied to systems exhibiting chemical reactions. There, for conciseness, we considered only the case of reactions whose spontaneous progression is very slow, and hence for which the conditions necessary for this purpose can be realized through the addition and removal of a catalyst. For systems of that type, this approach is indeed sufficient, and it is not difficult to imagine how it could be applied to some systems even with present-day technology. However, because many chemical reactions proceed too rapidly for this approach to be meaningful, reformulating the present analysis as a general treatment for chemical systems requires the possibility of artificially slowing rates of reaction. The framework of thermodynamics could then be applied quite generally to the description of systems exhibiting chemical reactions by employing the assumption that it is possible to effectively turn reactions off and on as desired. This may seem unrealistic, and indeed as a practical matter, at this time it is. However, it is important to realize that for analysis of the type presented above to be physically meaningful in the sense that it implies the existence of the free energies that it provides, it is not necessary that its experimental scenario be realizable in a practical sense. It is only necessary that it be realizable in principle. In other words, we do not propose to carry out actual experiments in which the chemical reaction is turned off and on and concrete free energies $F[T, V; \mathbf{N}]$ and $G[T, p; \mathbf{N}]$ are constructed. We simply hypothesize that the realization of such experiments does not violate any physical laws and hence that it is possible in principle to develop a technology that would allow for this realization. With this interpretation, it is seen how the analysis presented in Sections 9.4.2 and 9.4.3 (properly reformulated using a generalized on/off technology) provides a physically meaningful picture through which the framework of ordinary thermodynamics can be generalized and applied conceptually to systems exhibiting chemical reactions.

On a practical level, the purpose of this analysis is limited to the concrete task of deriving the equilibrium condition (9.82). Now, note that for this purpose, it is not necessary that we be able to construct $G[T, p; \mathbf{N}]$ for arbitrary composition. It is only necessary that this can be done in the neighborhood of equilibrium. This greatly simplifies the experimental situation because, as a general principle of chemical kinetics, it is known that the rate of a chemical reaction goes to zero as the equilibrium state is approached. Thus, because the mixing rate should remain essentially unchanged in the same limit, it follows that for a system in the neighborhood of equilibrium, the type of experimental procedure described in Sections 9.4.2 and 9.4.3 could be carried out without the need to artificially slow reactions (provided that we possess suitable semipermeable membranes). In the actual application of this procedure, we would fix T and p and then directly construct $G[T, p; \mathbf{N}]$ for values of $\mathbf{N}$ in the neighborhood of equilibrium by carrying out experiments in which the initial and final volumes, V_0 and $V_0 + \Delta V$, are chosen such that the pressure is unchanged. Then, considering (9.75), we properly identify $G[T, p; \mathbf{N}^{(0)} + \boldsymbol{\nu}\xi] - G[T, p; \mathbf{N}^{(0)}]$ as $-\sum_{j=1}^{5} W_j + p\Delta V$. (In order for this procedure to be valid, it is only necessary that at all times during the mixing and unmixing operations, the system remain sufficiently close to equilibrium that the extent of reaction can be regarded as fixed.) In this way, we can directly construct $G[T, p; \mathbf{N}]$ in the neighborhood of the equilibrium state and with this empirically derive the equilibrium condition, (9.82), in a concrete manner.[34]

9.4.5 Equilibrium condition for a chemical reaction in an ideal gas

If each of the substances in a system can be treated as an ideal gas, then the chemical potential takes the form appearing in (9.29), and the condition for chemical equilibrium given in (9.82) becomes

$$\sum_{i=1}^{m} \nu_i \left(1 + \log \left\{ \left(\frac{T_i^*}{T} \right)^{c_i+1} \frac{p}{p^\ominus} \right\} + \frac{u_i}{RT} + \log \tilde{x}_i \right) = 0 \ . \tag{9.83}$$

The quantity $\tilde{x}_i = (N_i^{(0)} + \nu_i \tilde{\xi}) / \left(\sum_{j=1}^{m} (N_j^{(0)} + \nu_j \tilde{\xi}) \right)$ here represents the mole fraction of Species i in chemical equilibrium.

In practical terms, (9.83) can be used as follows. For a given initial composition $\mathbf{N}^{(0)}$, fixing p and carrying out experiments using various values of T (such that the system remains in the ideal gas regime), from (9.83) we can obtain one condition relating the m constants $T_1^*, \ldots, T_m^*$ and one condition relating the m constants $u_1, \ldots, u_m$. Using these conditions, then, we can predict $\tilde{\xi}$ for any given values of $\mathbf{N}^{(0)}$, T and p in the ideal gas regime.[35]

9.4.6 Conventional form of the equilibrium condition in the general case

Here we present the general equilibrium condition (9.82) in its conventional form. For this purpose, we first introduce some standard notation and define some useful quantities.

In the chemistry literature, the quantity $\partial G[T, p; \mathbf{N}^{(0)} + \boldsymbol{\nu}\xi] / \partial \xi$ is generally written $\Delta_r G$. We refer to $\Delta_r G$ as the *reaction Gibbs energy*.[36] As discussed below, it is useful to write $\Delta_r G$ as

$$\Delta_r G = \Delta_r G^\ominus + RT \log Q \ , \tag{9.84}$$

where $Q = \prod_{i=1}^{m} a_i^{\nu_i}$ is the *reaction quotient*, and $\Delta_r G^\ominus$ (which is defined through this equation) is the *standard reaction Gibbs energy*. Comparing (9.81) and (9.84) with the conventional definitions of the activity given in (9.46)–(9.49), it is found

---

[34] Let us note that within our treatment of chemical reactions, the procedure described here serves only as a practical reinterpretation of that described in Sections 9.4.2 and 9.4.3. This procedure is not applied beyond this point.

[35] **Advanced note**: The generality of this statement implicitly relies on the validity of the third law (the Nernst-Planck postulate). Without the third law, we can only make this claim for initial compositions that can be obtained from $\mathbf{N}^{(0)}$ by varying ξ, i.e., initial compositions of the form $\mathbf{N}^{(0)} + \boldsymbol{\nu}\xi$ for arbitrary ξ.

[36] This quantity goes by various names in the literature.

that for a gas we have

$$\Delta_r G^{\ominus}(T) = \sum_{i=1}^{m} \nu_i \mu_i^{\ominus,(\mathrm{id})}(T) \,, \tag{9.85}$$

where $\mu_i^{\ominus,(\mathrm{id})}(T) = RT + (c_i + 1)RT \log(T_i^*/T) + u_i$, and for a solution we have either

$$\Delta_r G^{\ominus}(T,p) = \sum_{i=1}^{m} \nu_i \mu_i^{\ominus}(T,p) \,, \tag{9.86}$$

where $\mu_i^{\ominus}(T,p)$ is the chemical potential of a pure system of Species i at (T,p), or

$$\Delta_r G^{\ominus}(T,p) = \nu_1 \mu_1^{\ominus}(T,p) + \sum_{i=2}^{m} \nu_i \mu_i^{\ominus,(\mathrm{dil})}(T,p) \,, \tag{9.87}$$

where $\mu_1^{\ominus}(T,p)$ is the chemical potential of a pure system of the solvent at (T,p), and $\mu_i^{\ominus,(\mathrm{dil})}(T,p)$ is the quantity given in (9.38) for the ith solute.

Now, with the quantities introduced above, we can write the general equilibrium condition, (9.82), as follows:

$$\Delta_r G^{\ominus} = -RT \log \widetilde{Q} \,. \tag{9.88}$$

The equilibrium value of the reaction coefficient, $\widetilde{Q} = \prod_{i=1}^{m} \tilde{a}_i^{\nu_i}$, is generally written K, and referred to as the *equilibrium constant*.[37]

Equation (9.88) is the conventional form of the equilibrium condition. This form is useful not merely because of its conciseness, but because $\Delta_r G^{\ominus}$ and K are independently measurable quantities that depend only on T and p. Of course, because in the general case we do not have explicit forms for $\Delta_r G^{\ominus}$ or K, the practical application of (9.88) is far less straightforward than that of (9.83), describing an ideal gas. But because there exist various experimental and theoretical methods that can be used to independently determine all of the relevant quantities here— not only $\Delta_r G^{\ominus}$ and $\tilde{a}_i$, but also $\tilde{x}_i$ and $\tilde{p}_i$—(9.88) can indeed provide a great deal of useful information. (For example, see Ref. [8] for detailed discussion.)

9.4.7 Le Chatelier's principle

Let us consider a system composed of m chemical species undergoing a chemical reaction under general conditions. The extent of reaction in chemical equilibrium, determined by (9.82), obviously depends on the imposed temperature and pressure. In the following, we write this quantity as $\tilde{\xi}(T,p)$ and investigate its dependence on T and p (with $N^{(0)}$ fixed).

---

[37] It is worth recalling the limiting forms of the activities in the cases considered here: In the ideal limit of a gas, a_i reduces to $p_i/p^{\ominus}$, while in either the ideal or dilute limit of a solution, a_i reduces to x_i.

First, we take the derivative with respect to T of the condition for chemical equilibrium given in (9.80):

$$\frac{\partial}{\partial T}\left\{\left.\frac{\partial G[T,p;\boldsymbol{N}^{(0)}+\boldsymbol{\nu}\xi]}{\partial \xi}\right|_{\xi=\tilde{\xi}(T,p)}\right\}=0\,. \tag{9.89}$$

Assuming that $G[T,p;\boldsymbol{N}]$ is twice differentiable and using the differential form for the Gibbs free energy appearing in (9.20), we can rewrite this as follows:

$$-\left.\frac{\partial S(T,p;\boldsymbol{N}^{(0)}+\boldsymbol{\nu}\xi)}{\partial \xi}\right|_{\xi=\tilde{\xi}(T,p)}+\frac{\partial \tilde{\xi}(T,p)}{\partial T}\left.\frac{\partial^2 G[T,p;\boldsymbol{N}^{(0)}+\boldsymbol{\nu}\xi]}{\partial \xi^2}\right|_{\xi=\tilde{\xi}(T,p)}=0\,. \tag{9.90}$$

Next, using (6.7), which relates the maximum heat and the entropy, we find that the amount of energy absorbed by the system from the environment in the form of heat when the extent of reaction changes quasi-statically by a small amount[38] $\Delta\xi$ from $\tilde{\xi}(T,p)$ is given by[39]

$$\Delta Q = T\left\{S(T,p;\boldsymbol{N}^{(0)}+\boldsymbol{\nu}(\tilde{\xi}(T,p)+\Delta\xi))-S(T,p;\boldsymbol{N}^{(0)}+\boldsymbol{\nu}\tilde{\xi}(T,p))\right\}$$

$$= T\Delta\xi\left.\frac{\partial S(T,p;\boldsymbol{N}^{(0)}+\boldsymbol{\nu}\xi)}{\partial \xi}\right|_{\xi=\tilde{\xi}(T,p)}+O((\Delta\xi)^2)\,. \tag{9.91}$$

Using this relation, we can rewrite (9.90) as

$$\frac{\partial \tilde{\xi}(T,p)}{\partial T}=\frac{1}{T}\frac{\Delta Q}{\Delta\xi}\left(\left.\frac{\partial^2 G[T,p;\boldsymbol{N}^{(0)}+\boldsymbol{\nu}\xi]}{\partial \xi^2}\right|_{\xi=\tilde{\xi}(T,p)}\right)^{-1}+O(\Delta\xi)\,. \tag{9.92}$$

Then, taking the $\Delta\xi\to 0$ limit, this becomes

$$\frac{\partial \tilde{\xi}(T,p)}{\partial T}=\frac{Q'}{T}\left(\left.\frac{\partial^2 G[T,p;\boldsymbol{N}^{(0)}+\boldsymbol{\nu}\xi]}{\partial \xi^2}\right|_{\xi=\tilde{\xi}(T,p)}\right)^{-1}\,, \tag{9.93}$$

where Q' represents $\left.\frac{dQ(\xi)}{d\xi}\right|_{\xi=\tilde{\xi}(T,p)}$, the total derivative of Q with respect to ξ.[40] Now, noting that because $G[T,p;\boldsymbol{N}^{(0)}+\boldsymbol{\nu}\xi]$ is a convex function of ξ,

---

[38] From the condition for chemical equilibrium given in (9.80), we know that the work done by the system on the external world (i.e., the change in the Gibbs free energy) when ξ changes by the small amount $\Delta\xi$ is $O((\Delta\xi)^2)$. Therefore, we can assume that the only change in energy experienced by the system is due to the absorption of heat.

[39] It is standard to use ΔH, the change in the enthalpy, in place of ΔQ in (9.92) and (9.94). The equality of these two quantities in the present context can be seen as follows. Defining the enthalpy as a function of T, p and $\boldsymbol{N}$ through the relation $H(T,p;\boldsymbol{N})=G[T,p;\boldsymbol{N}]+T\,S(T,p;\boldsymbol{N})$, for any transition under conditions of fixed T and p, we have $\Delta H=\Delta G+T\Delta S=(\Delta F+p\Delta V)+(\Delta U-\Delta F)=p\Delta V+\Delta U=\Delta Q$.

[40] Although, in general, Q is not a state function, because here we consider the quasi-static variation of ξ, in this context Q can be regarded as a function of ξ.

from Theorem G.1 we know that $\partial^2 G[T, p; \mathbf{N}^{(0)} + \boldsymbol{\nu}\xi]/\partial\xi^2$ is everywhere non-negative. Thus, assuming that it is also nonzero at $\tilde{\xi}(T, p)$, we obtain the following inequalities:

$$\frac{\partial \tilde{\xi}(T, p)}{\partial T} \begin{cases} > 0 & \text{for } Q' > 0 \ , \\ < 0 & \text{for } Q' < 0 \ . \end{cases} \tag{9.94}$$

Let us first consider the case $Q' > 0$. Then, if we increase the temperature, ξ increases, and thus both $\Delta\xi$ and ΔQ are positive. Hence, increasing the temperature causes the system to move in the endothermic direction. Next, if we decrease the temperature, ξ decreases, and thus both $\Delta\xi$ and ΔQ are negative. Hence, decreasing the temperature causes the system to move in the exothermic direction. Applying similar considerations to the case $Q' < 0$, we again find that increasing T causes the system to move in the endothermic direction, and decreasing T causes the system to move in the exothermic direction. It is thus seen that the tendency of the system is such that the chemical reaction moves in whichever direction opposes the change in temperature of the environment. This tendency is one example of a general class of behavior exhibited by thermodynamic systems. The characterization of this class of behavior is provided by the following assertion, known as Le Chatelier's principle: The spontaneous evolution exhibited by a thermodynamic system in response to a change in the ambient conditions is always in the direction that partially counteracts this change. (Le Chatelier's principle is treated in Problem 7.7.)

The situation is similar when the externally controlled pressure is varied, as can be seen from the following relation, derived analogously to (9.93):

$$\frac{\partial \tilde{\xi}(T, p)}{\partial p} = - \left. \frac{\partial V(T, p; \mathbf{N}^{(0)} + \boldsymbol{\nu}\xi)}{\partial\xi} \right|_{\xi=\tilde{\xi}(T,p)} \left(\left. \frac{\partial^2 G[T, p; \mathbf{N}^{(0)} + \boldsymbol{\nu}\xi]}{\partial\xi^2} \right|_{\xi=\tilde{\xi}(T,p)} \right)^{-1} . \tag{9.95}$$

From this equation, we find that if the pressure is increased, the reaction progresses in the direction of decreasing volume, while if the pressure is decreased, the reaction progresses in the direction of increasing volume. Hence, in each case, after the externally controlled pressure has been changed, the system evolves in such a manner as to partially counteract this change, as predicted by Le Chatelier's principle.

9.4.8 Treatment of systems exhibiting multiple chemical reactions

Before ending this section, we briefly consider systems exhibiting multiple simultaneously occurring chemical reactions. Here, we limit our treatment to the case of two reactions, but the general case of an arbitrary number of reactions can be understood as a straightforward generalization.

We represent the two reactions by the coefficients $\nu_1, \ldots, \nu_m$ defined in (9.74) and the analogously defined $\kappa_1, \ldots, \kappa_m$. As above, we write these coefficients collectively as $\boldsymbol{\nu} = (\nu_1, \ldots, \nu_m)$ and $\boldsymbol{\kappa} = (\kappa_1, \ldots, \kappa_m)$. The composition of the system is expressed as $\mathbf{N}^{(0)} + \boldsymbol{\nu}\xi + \boldsymbol{\kappa}\eta$, where $\mathbf{N}^{(0)}$ is the reference composition, and ξ and η are the extents of reaction for the two reactions.

Assuming that ξ and η can be controlled independently, with considerations similar to those applied to the case of a single reaction, it is found that the values realized by ξ and η in chemical equilibrium, $\tilde{\xi}$ and $\tilde{\eta}$, are those that minimize the Gibbs free energy $G[T, p; \boldsymbol{N}^{(0)} + \boldsymbol{\nu}\xi + \boldsymbol{\kappa}\eta]$, with T and p fixed. In this way, corresponding to (9.80), we obtain the following two conditions characterizing equilibrium:

$$
\left.\frac{\partial G[T, p; \boldsymbol{N}^{(0)} + \boldsymbol{\nu}\xi + \boldsymbol{\kappa}\eta]}{\partial \xi}\right|_{\xi=\tilde{\xi},\,\eta=\tilde{\eta}} = 0 \,, \qquad
\left.\frac{\partial G[T, p; \boldsymbol{N}^{(0)} + \boldsymbol{\nu}\xi + \boldsymbol{\kappa}\eta]}{\partial \eta}\right|_{\xi=\tilde{\xi},\,\eta=\tilde{\eta}} = 0 \,,
$$
$$
(9.96)
$$

or, equivalently,

$$
\sum_{i=1}^{m} \nu_i\, \mu_i(T, p; \boldsymbol{N}^{(0)} + \boldsymbol{\nu}\tilde{\xi} + \boldsymbol{\kappa}\tilde{\eta}) = 0 \,, \qquad
\sum_{i=1}^{m} \kappa_i\, \mu_i(T, p; \boldsymbol{N}^{(0)} + \boldsymbol{\nu}\tilde{\xi} + \boldsymbol{\kappa}\tilde{\eta}) = 0 \,.
$$
$$
(9.97)
$$

Because we have two conditions and two unknowns, $\tilde{\xi}$ and $\tilde{\eta}$, the chemical equilibrium state is uniquely determined.

9.5 Chemical reactions in aqueous solutions

One of the most important applications of the thermodynamic theory of chemical reactions is to the innumerably many types of reactions that take place in aqueous solutions, i.e., solutions in which the solvent is water. In this section, we treat aqueous solutions of acids, but the approach used here is of general validity and can be applied to the analysis of more complicated reactions as well. In this and the following section, we also present an overview of equilibrium electrochemistry.

9.5.1 On the treatment of ionic species as thermodynamic substances

In the thermodynamic treatment of electrolyte solutions, we encounter a subtle problem that we have not seen to this point. Here we discuss this problem.

Let us consider a system in which some arbitrary metallic salt, denoted here by MA, is dissolved in water. Suppose that the entire amount of MA then undergoes complete dissociation as $\mathrm{MA} \to \mathrm{M}^+ + \mathrm{A}^-$, and the system thus comes to form an electrolyte solution. If we then introduce two electrodes into the solution and externally apply a voltage across them, the solution will undergo electrolysis, with the anionic species and cationic species migrating to the positive electrode (anode) and negative electrode (cathode), respectively. On the basis of this macroscopic behavior, it is inferred that M^+ and A^- behave as separate substances in water. For this reason, the system considered here should be regarded as consisting of three substances, with $\mathrm{H_2O}$ acting as the solvent, and M^+ and A^- acting as the solutes. Hence, it is also natural to consider three chemical potentials, $\mu_{\mathrm{H_2O}}$, μ_{M^+} and μ_{A^-}, and three activities, $a_{\mathrm{H_2O}}$, a_{M^+} and a_{A^-}. It is seen below how extending the concept of a substance in this manner allows us to analyze numerous novel problems.

The treatment of ionic species, however, involves the distinctive complication that cations and anions strongly attract each other through the Coulomb force. Because of this strong interaction, it is unfeasible to physically separate the individual ionic species in a thermodynamic system into isolated cation and anion subsystems, and, for example, using an apparatus like that depicted in Figure 9.2 to measure the chemical potentials of individual ionic species would be extremely difficult.[41] For this reason, in most contexts, the chemical potentials and activities of individual ions play only formal, theoretical roles, while it is the sums of individual chemical potentials and the products of individual activities corresponding to electrically neutral pairs of ions that are the physically relevant, measurable quantities. Indeed, in the conditions for chemical equilibrium derived in this section and in the expression for the electromotive force (emf) of a concentration battery given in (9.129), the activities of cationic species and anionic species appear only in products of pairs.

In spite of the challenges inherent in their measurement, however, the activities of individual ionic species sometimes do appear in unpaired forms even within the context of practical applications. For example, in the definition of the pH, given in (9.104), the activity of H_3O^+ appears by itself.[42] The questions of how we should understand this definition and how we could measure the activity of an individual ionic species are quite delicate, and at this time, there seem to be no definitive answers. While these are important questions, we do not study them further here.

Below, we investigate two types of electrolyte solutions employing the dilute solution form of the equilibrium condition. As discussed in Footnote 8 on p. 194, in practical situations, the validity of this form for electrolyte solutions does not follow from the validity of the derivation given in Section 9.1.4, because the expansion of W_{mix} will break down due to the strength of the Coulomb interaction. Instead, this should be regarded as an experimental result that must be confirmed on a system-by-system basis. Empirically, it is known that the dilute solution approximation is generally valid for electrolyte solutions with molarities (moles of solute per liter of solution) of approximately 10^{-3} mol/L or less.[43]

9.5.2 Chemical equilibrium in the self-ionization of water

We begin by considering the simplest example, a system consisting of water alone. Even this very simple system exhibits an ionization reaction, which is expressed as[44]

---

[41] This is not to say that this kind of measurement would be impossible, however. If we could somehow overcome the strong Coulomb force and macroscopically isolate the individual ionic species in such a solution, then a method like that described by Figure 9.2 could, in principle, be used to measure the chemical potentials of these individual species. However, in that case, it would be necessary to consider the electrochemical potential (discussed near the end of this section), in order to account for the electric potential energy.

[42] The standard primary method for experimentally evaluating the pH is briefly explained in Footnote 48 on p. 226.

[43] For larger concentrations, some other approximation scheme, for example Debye-Hückel theory, must be used. (For the interested reader, we recommend the detailed treatments of electrochemistry given in Refs. [5; 10].)

[44] In an aqueous solution, the ion H^+ generally does not exist by itself, but instead combines with a water molecule to form the H_3O^+ ion. In this book, we assume that H^+ in aqueous solution appears *exclusively* as a part of H_3O^+. Even with such an understanding, however, the chemical reaction represented by (9.98) is sometimes written simply as $H_2O \rightleftharpoons H^+ + OH^-$. In this case, the mole fraction expressed as x_{H^+} and the activity expressed as a_{H^+} actually represent those of H_3O^+.

$$H_2O + H_2O \rightleftarrows H_3O^+ + OH^- \ . \tag{9.98}$$

We write the amounts of water (H_2O), hydronium (H_3O^+) and hydroxide (OH^-) as N_{H_2O}, $N_{H_3O^+}$ and N_{OH^-}, and express them collectively as $\boldsymbol{N} = (N_{H_2O}, N_{H_3O^+}, N_{OH^-})$. Of course, here H_2O is regarded as the solvent, and the two ionic species are regarded as the solutes. In equilibrium, the activities of the three substances, $a_{H_2O}(T, p; \boldsymbol{N})$, $a_{H_3O^+}(T, p; \boldsymbol{N})$ and $a_{OH^-}(T, p; \boldsymbol{N})$, satisfy the condition[45]

$$\frac{a_{H_3O^+}(T, p; \boldsymbol{N})\, a_{OH^-}(T, p; \boldsymbol{N})}{\{a_{H_2O}(T, p; \boldsymbol{N})\}^2} = K(T, p) \ , \tag{9.99}$$

where $K(T, p)$ is the equilibrium constant.

In Section 9.2, we defined the activities a_i of the species in a solution such that in the dilute limit, each reduces to the corresponding mole fraction, x_i. In this situation, it is said that we are using the *mole-fraction basis* for the activities. However, in the treatment of aqueous solutions, it is more common for the solutes (but *not* the solvent) to use the *molality basis* or *molarity basis*, in which the activity of a solute is defined such that it reduces to the dimensionless molality (molality divided by the standard-state molality, $1\,\mathrm{mol/kg}$) and dimensionless molarity (molarity divided by the standard-state molarity, $1\,\mathrm{mol/L}$) in the dilute limit. Because the molality basis provides several practical advantages (due primarily to the fact that the molality has a weaker temperature dependence than the molarity[46]), standard definitions and empirical reference values of relevant quantities are generally given in terms of the molality, and for this reason, we use the molality basis here. For a given solute, we write the molality as m_i and the activity and activity coefficient on the molality basis as a_i^{m} and γ_i^{m}.

The relation between a_i and a_i^{m} is easily determined. First, let us reconsider (9.49) in the general case of a (not necessarily dilute, not necessarily electrolyte) solution, rewriting $a_i(T, p; \mathbf{N})$ explicitly in terms of $\gamma_i(T, p; \mathbf{N})$:

$$\mu_i(T, p; \mathbf{N}) = \mu_i^{\ominus,(\mathrm{dil})}(T, p) + RT \log[\gamma_i(T, p; \mathbf{N})x_i] \ . \tag{9.100}$$

Then, substituting $m_i x_1 M_1$ for x_i, where M_1 is the molar mass of the solvent, and applying some straightforward manipulations, we obtain

$$\mu_i(T, p; \mathbf{N}) = \mu_i^{\ominus,(\mathrm{dil})}(T, p) + RT \log[m^{\ominus} M_1] + RT \log[\gamma_i(T, p; \mathbf{N})x_1(m_i/m^{\ominus})] \ , \tag{9.101}$$

where $m^{\ominus} = 1\,\mathrm{mol/kg}$ is the standard-state molality. From this expression, we can identify the activity coefficient and activity on the molality basis as $\gamma_i^{\mathrm{m}}(T, p; \mathbf{N}) = \gamma_i(T, p; \mathbf{N})x_1$ and $a_i^{\mathrm{m}}(T, p; \mathbf{N}) = a_i(T, p; \mathbf{N})/(m^{\ominus} M_1)$. Note that these choices are unique (up to the definition of $m^{\ominus}$) given the stipulations that $a_i^{\mathrm{m}}(T, p; \mathbf{N})$ reduce

---

[45] Because the left-hand side here is the reaction quotient, Q, and the right-hand side is the equilibrium constant, which is defined simply as the equilibrium value of Q, this equation may appear to be trivial. However, here we are using $K(T, p)$ to represent $\exp[-\Delta_r G^{\ominus}/RT]$, through application of the equilibrium condition (9.88). This is common practice in the chemistry literature.

[46] In some chemistry textbooks, it is claimed that the molality is independent of temperature, but this is not generally true, as it has a temperature dependence through $\xi(T, p)$.

to $m_i/m^{\ominus}$ in the dilute limit and that all of the $\mathbf{N}$ dependence of $\mu_i(T,p;\mathbf{N})$ be contained in $a_i^{\mathrm{m}}(T,p;\mathbf{N})$.

With the newly defined activities a_i^{m}, the condition (9.99) becomes

$$\frac{a_{\mathrm{H_3O^+}}^{\mathrm{m}}(T,p;\mathbf{N})\, a_{\mathrm{OH^-}}^{\mathrm{m}}(T,p;\mathbf{N})}{\{a_{\mathrm{H_2O}}(T,p;\mathbf{N})\}^2} = K(T,p)/(m^{\ominus} M_{\mathrm{H_2O}})^2 = K_{\mathrm{W}}(T,p)\,. \qquad (9.102)$$

Although here we are considering a system consisting of water alone, the reaction (9.98) is exhibited by all aqueous solutions, and for this reason, the condition (9.102) applies to all aqueous solutions. The universal quantity $K_{\mathrm{W}}(T,p)$ is called the *ionic product of water*.[47] At 1 atm and 25°C, it is given by $K_{\mathrm{W}}(T,p) \approx 1.012 \times 10^{-14}$.

Pure water can be regarded as an extremely dilute solution, with $\tilde{a}_{\mathrm{H_3O^+}}^{\mathrm{m}} \simeq \tilde{m}_{\mathrm{H_3O^+}}/m^{\ominus}$, $\tilde{a}_{\mathrm{OH^-}}^{\mathrm{m}} \simeq \tilde{m}_{\mathrm{OH^-}}/m^{\ominus}$ and $\tilde{a}_{\mathrm{H_2O}} \simeq 1$. Then, because under the reaction (9.98), the relation $N_{\mathrm{H_3O^+}} = N_{\mathrm{OH^-}}$ is always satisfied, at 1 atm and 25°C, we have

$$\tilde{m}_{\mathrm{H_3O^+}}/m^{\ominus} \simeq \sqrt{K_{\mathrm{W}}(T,p)} \simeq 1.006 \times 10^{-7}\,. \qquad (9.103)$$

As a measure of the acidity of a solution, it is conventional to use a quantity called the pH, defined as[48]

$$\mathrm{pH}(T,p;\mathbf{N}) := -\log_{10} a_{\mathrm{H_3O^+}}^{\mathrm{m}}\,. \qquad (9.104)$$

With the above value of $\tilde{m}_{\mathrm{H_3O^+}}/m^{\ominus}$, we find that the pH of water in equilibrium at 1 atm and 25°C is approximately 6.997. For this reason, it is conventional to regard 7 as the neutral value of the pH under standard conditions.[49]

---

[47] In the chemistry literature, it is often stated that the ionic product of water is the quantity $[\mathrm{H_3O^+}][\mathrm{OH^-}]$ (i.e., the product of the molarities—sometimes regarded as the molarities themselves and sometimes as the dimensionless molarities). It is important to understand that the ionic product of water is defined as $K_{\mathrm{W}}(T,p) := \tilde{a}_{\mathrm{H_3O^+}}^{\mathrm{m}}(T,p;\mathbf{N})\tilde{a}_{\mathrm{OH^-}}^{\mathrm{m}}(T,p;\mathbf{N})/\{\tilde{a}_{\mathrm{H_2O}}(T,p;\mathbf{N})\}^2$, with the approximation $K_{\mathrm{W}}(T,p) \simeq [\mathrm{H_3O^+}][\mathrm{OH^-}]$ holding only in the case of a dilute solution.

[48] There is no unique definition of the pH. We follow the IUPAC [7]. Other sources replace the activity on the molality basis with the activity on the molarity basis or the dimensionless molarity itself. For dilute solutions, the differences among the actual values of the pH resulting from these various definitions are negligible. However, for non-dilute solutions, these differences may be significant. Also, while the IUPAC defines the pH only for aqueous solutions, because this concept can be extended to other systems, some sources define it more generally. Today, it is standard to evaluate the pH using methods of electrochemistry. These methods take advantage of the fact that when an electrode is placed in an electrolyte solution, an electric potential will develop at the electrode. This potential depends on the compositions of both the electrode and the solution surrounding it. Therefore, connecting two electrodes for which one or both of these compositions differ and measuring the potential difference between them, we can obtain information about the solution(s) in which they are placed. Various methods exploiting this behavior have been developed to estimate the pH as defined above. The primary method (employing the Harned cell) consists of a rather involved process that combines potential difference measurements taken in multiple solutions and approximations provided by theoretical models (most importantly the Debye-Hückel theory) to derive a value for $a_{\mathrm{H_3O^+}}^{\mathrm{m}}$ in the solution of interest.

[49] It is a common misconception that the value 7 itself defines the neutral pH. The actual situation can be understood from the above considerations. With a neutral aqueous solution defined by the relation $a_{\mathrm{H_3O^+}}^{\mathrm{m}} = a_{\mathrm{OH^-}}^{\mathrm{m}}$, from (9.102) we find the general expression $\mathrm{pH} = -\log \tilde{a}_{\mathrm{H_2O}}(T,p;\mathbf{N}) - \frac{1}{2}\log K_{\mathrm{W}}(T,p)$ for the neutral pH in equilibrium. It is thus seen that the neutral pH depends on the concentrations of all species, as well as temperature and pressure. If we consider the case of a dilute solution, however, the first term can be ignored, and we therefore have $\mathrm{pH} \simeq -\frac{1}{2}\log K_{\mathrm{W}}(T,p)$. Between 0°C and 100°C (and under ordinary pressures), this varies from approximately 7.5 to approximately 6.1.

9.5.3 Chemical equilibrium in an aqueous solution of acid

Next, we consider systems in which a small amount of acid, for example, HCl, HF or HCN, is dissolved in water. For the sake of generalness, we consider an arbitrary acid, represented by HA.

In water, an acid undergoes the ionization reaction

$$HA + H_2O \rightleftarrows A^- + H_3O^+ , \tag{9.105}$$

which will be accompanied by the self-ionization reaction of water, (9.98). We write the amounts of the acid HA and the base A^- as N_{HA} and N_{A^-}, and their activities as $a_{HA}(T, p; \boldsymbol{N})$ and $a_{A^-}(T, p; \boldsymbol{N})$. When this system is in chemical equilibrium, along with (9.99), the following condition is also satisfied:

$$\frac{a_{A^-}(T, p; \boldsymbol{N})\, a_{H_3O^+}(T, p; \boldsymbol{N})}{a_{HA}(T, p; \boldsymbol{N})\, a_{H_2O}(T, p; \boldsymbol{N})} = K(T, p) , \tag{9.106}$$

where $K(T, p)$ is the equilibrium constant for the reaction (9.105). Rewriting (9.106) in terms of the activities $a_i^m(T, p; \boldsymbol{N})$ for the solutes, this condition becomes

$$\frac{a_{A^-}^m(T, p; \boldsymbol{N})\, a_{H_3O^+}^m(T, p; \boldsymbol{N})}{a_{HA}^m(T, p; \boldsymbol{N})\, a_{H_2O}(T, p; \boldsymbol{N})} = K(T, p)/(m^{\ominus} M_{H_2O}) = K_a(T, p) . \tag{9.107}$$

The quantity $K_a(T, p)$, the equilibrium constant on the molality basis, is referred to as the *acid dissociation constant* (which depends on the acid considered).

As explained in Section 9.4.8, for a system exhibiting two chemical reactions, the simultaneous equations in (9.97) determine the composition of the system in equilibrium. Below, we demonstrate this concretely, using (9.102) and (9.107).

We represent the extents of reaction corresponding to the reactions (9.98) and (9.105) by ξ and η, respectively, and we choose the state corresponding to $\xi = \eta = 0$ to be characterized by the following (unrealizable) amounts of substance:

$$N_{H_2O} = N - \widehat{N}, \quad N_{HA} = \widehat{N}, \quad N_{H_3O^+} = N_{OH^-} = N_{A^-} = 0 . \tag{9.108}$$

Here, N and $\widehat{N}$ are constants, whose values are fixed when the system is prepared. Note that the relations $N = N_{H_2O} + N_{HA} + N_{H_3O^+} + N_{OH^-} + N_{A^-}$ and $\widehat{N} = N_{HA} + N_{A^-}$ hold for all ξ and η. Equilibrium is defined by the following relations:

$$\widetilde{N}_{H_2O} = N - \widehat{N} - 2\tilde{\xi} - \tilde{\eta} , \quad \widetilde{N}_{HA} = \widehat{N} - \tilde{\eta} , \quad \widetilde{N}_{H_3O^+} = \tilde{\xi} + \tilde{\eta} ,$$

$$\widetilde{N}_{OH^-} = \tilde{\xi} , \quad \widetilde{N}_{A^-} = \tilde{\eta} . \tag{9.109}$$

Now, let us assume that the solution is dilute, in which case we have $a_i^m \simeq m_i/m^{\ominus}$ for each solute, and we can approximate the density of the solution as 1 kg/L. Then introducing the quantities $\tilde{y} = \tilde{\xi}/(VM^{\ominus})$, $\tilde{z} = \tilde{\eta}/(VM^{\ominus})$ and $\hat{n} = \widehat{N}/(VM^{\ominus})$, where $M^{\ominus} = 1\,\mathrm{mol/L}$ is the standard-state molarity, we obtain

$$\tilde{a}^{\mathrm{m}}_{\mathrm{HA}} \simeq \hat{n} - \tilde{z}\,, \quad \tilde{a}^{\mathrm{m}}_{\mathrm{H_3O^+}} \simeq \tilde{y} + \tilde{z}\,, \quad \tilde{a}^{\mathrm{m}}_{\mathrm{OH^-}} \simeq \tilde{y}\,, \quad \tilde{a}^{\mathrm{m}}_{\mathrm{A^-}} \simeq \tilde{z}\,. \tag{9.110}$$

Using these expressions and the additional dilute-solution approximation $\tilde{a}_{\mathrm{H_2O}} \simeq 1$, we obtain

$$(\tilde{y} + \tilde{z})\tilde{y} \simeq K_{\mathrm{W}}(T, p) \tag{9.111}$$

and

$$\frac{\tilde{z}(\tilde{y} + \tilde{z})}{\hat{n} - \tilde{z}} \simeq K_{\mathrm{a}}(T, p)\,. \tag{9.112}$$

With numerical values for $\hat{n}$, $K_{\mathrm{W}}(T, p)$ and $K_{\mathrm{a}}(T, p)$, these conditions can be easily solved as simultaneous equations.

For weak acids, like acetic acid ($\mathrm{CH_3COOH}$), $K_{\mathrm{a}}(T, p)$ is extremely small. For this reason, acids of this type exhibit very little ionization in water, and are characterized by the relation $\hat{n} \gg \tilde{z}$.[50] Considering this case and replacing $\hat{n} - \tilde{z}$ with $\hat{n}$ in (9.112), we obtain the simultaneous equations

$$(\tilde{y} + \tilde{z})\tilde{y} \simeq K_{\mathrm{W}}(T, p)\,, \quad \frac{\tilde{z}(\tilde{y} + \tilde{z})}{\hat{n}} \simeq K_{\mathrm{a}}(T, p)\,. \tag{9.113}$$

These equations have the following solution:[51]

$$\tilde{y} \simeq \frac{K_{\mathrm{W}}(T, p)}{\sqrt{K_{\mathrm{W}}(T, p) + \hat{n}K_{\mathrm{a}}(T, p)}}\,, \quad \tilde{z} \simeq \frac{\hat{n}K_{\mathrm{a}}(T, p)}{\sqrt{K_{\mathrm{W}}(T, p) + \hat{n}K_{\mathrm{a}}(T, p)}}\,. \tag{9.114}$$

With this solution, we can compute the pH for a dilute aqueous solution of a weak acid in equilibrium as follows:

$$\begin{aligned}
\mathrm{pH}(T, p; \boldsymbol{N}) &= -\log_{10} \tilde{a}^{\mathrm{m}}_{\mathrm{H_3O^+}} \\
&\simeq -\log_{10}(\tilde{y} + \tilde{z}) \\
&\simeq -\log_{10} \sqrt{K_{\mathrm{W}}(T, p) + \hat{n}K_{\mathrm{a}}(T, p)}\,.
\end{aligned} \tag{9.115}$$

For example, for a solution of HCN at 1 atm and $25^{\circ}\mathrm{C}$, we have $K_{\mathrm{a}}(T, p) \approx 6.2 \times 10^{-10}$. Using this value, for a solution with $\hat{n} = 0.1$, we obtain[52]

$$\mathrm{pH}(T, p; \boldsymbol{N}) \simeq -\frac{1}{2}\log_{10}(6.2 \times 10^{-11}) \simeq 5.1\,. \tag{9.116}$$

---

[50] Contrastingly, in the case of strong acids, like hydrochloric acid (HCl), $K_{\mathrm{a}}(T, p)$ is very large, implying that $\hat{n} - \tilde{z}$ is small.

[51] There is a second solution, but it is unphysical.

[52] In actual systems, there are many situations in which the relation $K_{\mathrm{W}} \ll \hat{n}K_{\mathrm{a}}$ holds, and we can therefore use the approximation $\mathrm{pH}(T, p; \boldsymbol{N}) \simeq -\log_{10} \sqrt{\hat{n}K_{\mathrm{a}}(T, p)}$, as here. In such a situation, of course, we can ignore the self-ionization of water altogether, setting $\tilde{y} = 0$ at the outset. Doing so, the result can be derived very simply. However, the purpose of the present treatment is not merely to derive the result (9.116), but also to demonstrate the general method that can be applied to more complicated systems exhibiting multiple interconnected reactions.

9.6 The thermodynamics of concentration cells

In this section, we present an elementary introduction to the thermodynamic treatment of chemical batteries, which employ oxidation-reduction ("redox") reactions to generate electricity. The ordinary batteries that we use in our everyday lives are chemical batteries, but the theory needed to carry out a thorough analysis of that particular type of battery is beyond the scope of this book. Here, we instead consider the type of battery for which the theoretical principle of batteries is most clearly manifested, concentration cells.

9.6.1 The functioning of a concentration cell

A concentration cell is a type of galvanic electrochemical cell consisting of two aqueous solutions contained in separate compartments (termed "half-cells") with different concentrations of the same metallic salt, MA. (Here, as above, we use MA to represent an arbitrary metallic salt, with M^+ and A^- the products of its dissociation in aqueous solution.) The two half-cells are separated by a semipermeable wall that allows only A^- to pass through, and an electrode composed of M is immersed in each. To simplify the treatment, we assume that we can ignore the self-ionization of H_2O and that in aqueous solution, MA dissociates completely into M^+ and A^-. Figure 9.6 presents a schematic depiction of such a concentration cell. There, Ag plays the role of M, and NO_3 plays the role of A. We assume that the temperature and pressure of the system are fixed at some values T and p.

Let us first investigate the equilibrium state of this system without the electrodes. To begin, we consider the situation in which the two compartments are separated by an ordinary wall. Each compartment contains an aqueous solution of MA, with the concentration of that in the left-hand compartment higher. We write the chemical potentials of A^- in the left-hand and right-hand compartments as μ_{A-}^L and μ_{A-}^R, respectively. It is reasonable that the chemical potential of the solution

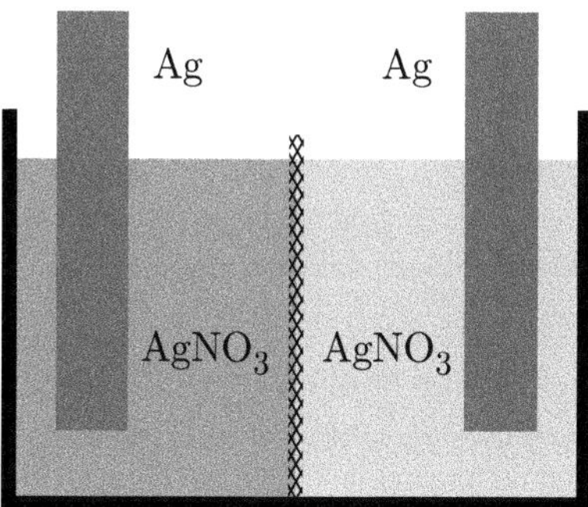

Figure 9.6 A concentration cell consisting of two half-cells containing aqueous solutions with electrodes. Here, as one particular example, we consider the case in which the generic metallic salt MA is $AgNO_3$, which dissociates into Ag^+ and NO_3^- in water. The electrodes are composed of Ag. The half-cells are separated by a semipermeable membrane that is permeable only to NO_3^-. The concentration of the solution in the left half-cell is higher, and as a result, the positive electrode is that on the left.

with higher concentration would be larger, and therefore we assume the relation $\mu_{A^-}^L > \mu_{A^-}^R$. Next, we replace the ordinary wall with a semipermeable wall that allows only A^- to pass through. Due to the difference in chemical potentials, substance will begin to move from the high-concentration side to the low-concentration side. If A^- were an electrically neutral substance, then the equilibrium state to which the system would subsequently evolve would be that for which the balance condition (9.25) holds for A^-, with each compartment containing a spatially uniform solution. In that case, as discussed in Section 9.3.1, osmotic pressure would arise between the compartments. However, for the presently considered system, because the migrating substance possesses electric charge, the situation is quite different. When a small amount of A^- passes through the semipermeable wall, the solution to the right of the wall will become slightly negatively charged, and the solution to the left of the wall will become slightly positively charged. But due to the strong attraction between these oppositely charged portions of the solution, they will be confined to the region very close to the wall, with the rest of the solution remaining electrically neutral. In this way (as shown in Figure 9.7) there will form thin layers of solution on the left-hand and right-hand sides of the wall in which the concentrations of A^- are relatively low and relatively high, respectively. It is in such a configuration that equilibrium is realized. For a system of the type considered here, the migration of A^- will take place only in the neighborhood very close to the wall, and the amount of A^- that moves from the left-hand side to the right-hand side as the electrically charged thin layers form and the system approaches equilibrium will be so small that it could not be detected using the methods of chemical analysis. Therefore, it can be assumed that the composition of each fluid outside of this narrow neighborhood of the wall will be unchanged by the migration and that no measurable osmotic pressure will arise. Also, in the homogeneous region within each fluid, there will be no electric field, and hence the electric potential will be constant.[53]

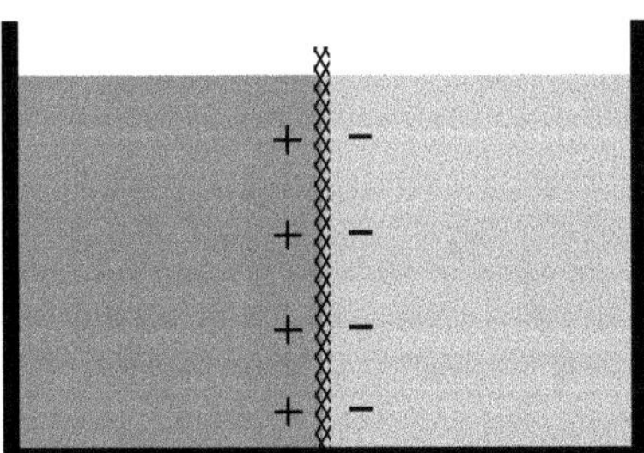

Figure 9.7 As a result of the migration of A^- ions through the semipermeable wall, on the left-hand side, there develops a thin, positively charged layer in which the concentration of A^- is relatively low, and on the right-hand side, there develops a thin, negatively charged layer in which the concentration of A^- is relatively high. However, in the internal region of each solution, the composition is essentially unchanged. Due to the formation of the charged layers on either side of the wall, a potential difference (membrane potential) develops across the wall.

---

[53] This is an example of the general principle in electrostatics that inside a conductor, in equilibrium, the electric field will vanish and thus the electric potential will be constant.

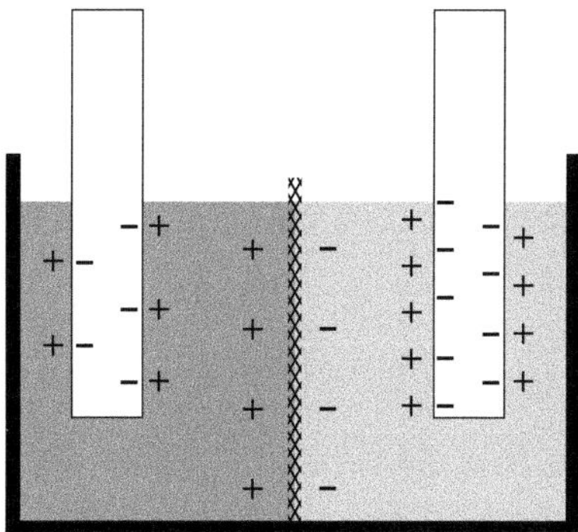

Figure 9.8 The equilibrium state of a battery with unconnected electrodes. The situation in the neighborhood of the semipermeable wall is the same as in the case of Figure 9.7. Upon immersion, the oxidation half-reaction $M \to M^+ + e^-$ will take place at the interface between each electrode and its solution. Eventually, the concentration of M^+ ions in the region surrounding each electrode will become sufficiently large that chemical equilibrium between the oxidation half-reaction $M \to M^+ + e^-$ and the reduction half-reaction $M^+ + e^- \to M$ will be realized. The amounts of M dissolved from the two electrodes are different, and as a result, a potential difference develops between them. Because the ions can move freely in the solution, electrostatic screens form around the electrodes and the wall, and for this reason, the electric field vanishes in each solution sufficiently far from these regions. However, there is a nonzero electric field in the space between the electrodes outside of the solution.

As is clear from Figure 9.7, in equilibrium, there exists a potential difference between the internal regions of the solutions in the two half-cells. In general, a potential that develops in this manner between aqueous solutions on the two sides of a semipermeable wall is called a *membrane potential*.[54] As a matter of practical experimental investigation, we point out that in the case of two aqueous solutions with different compositions, as in the present example, it is not possible using the methods of electrochemistry to directly measure the membrane potential.[55] It may be thought that we could simply insert similar electrodes into each solution and then measure the potential difference between them. However, generally, the value measured in such a situation will not be the membrane potential. The reason for this is that, even if the electrodes are identical, because the two solutions have different concentrations, the conditions at the interfaces between the electrodes and their respective solutions will differ, as shown in Figure 9.8. This difference influences the potential difference measured by the electrodes.[56] Indeed, the calculations presented below show explicitly that the potential difference between

---

[54] The concept of the *membrane potential* plays an important role in biochemistry.

[55] This is closely related to the fact that the chemical potential of a single ionic species cannot be measured directly. Of course, in principle, there is no problem here. The potential difference is a well-defined concept, based on that of the electric field. Also, in principle, it is a measurable quantity. In the present case, it can be determined by measuring the work required to move an inert charge between the internal regions of the two solutions. (A formally computed expression for the membrane potential is given in (9.135).)

[56] To overcome this problem, measurement methods employing such devices as reference electrodes and salt bridges have been devised.

the electrodes, expressed by (9.129) and (9.137), does differ from the membrane potential, given in (9.135).

Next, we consider the situation in which two unconnected electrodes are inserted into the two solutions. In the resulting system, at the interface between each electrode and its solution, the metal substance composing the electrode, M, will undergo the oxidation half-reaction

$$M \rightarrow M^+ + e^- . \tag{9.117}$$

The metal ions, M^+, will then diffuse into the solution, while the electrons, e^-, will remain attached to the electrode. However, because these positive and negative charges will strongly attract each other, the M^+ ions in the solution will be concentrated in the region near the electrode, as shown in Figure 9.8. Eventually, equilibrium will be realized between the electrode and the solution when the concentration of M^+ ions dissolved into the solution becomes sufficiently large. Here, as in the case of the migration of A^- ions through the semipermeable wall, we can assume that the composition of the solution anywhere sufficiently far from the electrode will be unaffected by the reaction taking place at the electrode-solution interface.

Because the concentrations of the metallic salt in the two solutions differ, the quantities of M^+ ions existing in the solutions in the equilibrium state will also differ. It is intuitively clear that more M^+ ions will be shed by the electrode in the more dilute solution, i.e., that on the right-hand side. Thus, more free electrons will become attached to the right electrode, and as a result, a potential difference will develop between the two electrodes. Apparently, the electrode on the left-hand side, with fewer attached electrons, will be the plus side.[57] In equilibrium, similarly to the case without electrodes, in the internal region of each solution, sufficiently far from its electrode and the wall, the electric field will vanish, and the electric potential will be constant.

Below, we carry out two independent calculations to determine the potential difference between the electrodes in an equilibrium state like that depicted in Figure 9.8. In the first, we use an intuitive approach in which we focus on the transport of substance, and in the second, we use a more sophisticated approach employing the electrochemical potential.

9.6.2 Computation of the emf based on the Gibbs free energy

The first method for computing the emf is based on the Gibbs free energy.

First, we consider the case depicted in Figure 9.9(a), in which the two solutions are separated by an impermeable wall and the electrodes are situated outside of the solutions. The Gibbs free energy in this case, $G^{(0)}$, is simply the sum of those of the two fluids and the two electrodes regarded as four independent systems.

---

[57] However, this is a tentative conclusion, because the potential difference between the two sides of the semipermeable wall must also be taken into account. In the following, we elucidate the situation by carrying out calculations of the potential difference between the electrodes without first assuming an identification of positive and negative sides (also, see Problem 9.5).

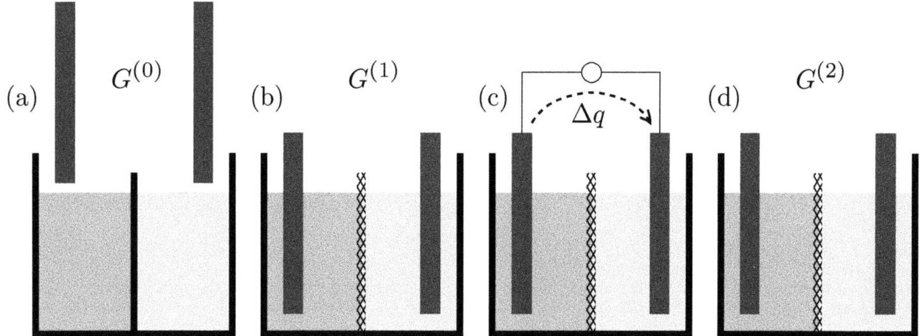

Figure 9.9 (a) The case in which all components of the battery are completely isolated. We write the Gibbs free energy in this case as $G^{(0)}$. (b) The case of a battery with disconnected electrodes. We write the Gibbs free energy in this case as $G^{(1)}$. (c) An infinitesimal amount of charge, Δq, is allowed to move from the left electrode to the right electrode. (d) The same as in (b), but after the migration of the charge Δq. We write the Gibbs free energy in this case as $G^{(2)}$. We can evaluate the emf of the battery by comparing $G^{(1)}$ and $G^{(2)}$.

Let us write the amounts of metallic substance M in the left and right electrodes as $N_{L,1}^{(0)}$ and $N_{R,1}^{(0)}$, respectively. We assume that these electrodes are composed entirely of this single metal, and thus their free energies are given by

$$G_{L,1} = N_{L,1}^{(0)}\,\mu_{\mathrm{M}}, \quad G_{R,1} = N_{R,1}^{(0)}\,\mu_{\mathrm{M}} \ . \tag{9.118}$$

The quantity $\mu_{\mathrm{M}} = \mu_{\mathrm{M}}(T, p)$ here is the chemical potential of a pure sample of M at temperature T and pressure p.

In the aqueous solution on the left-hand side, the amounts of M^+ and A^- are equal. Let us represent this amount by $N_{L,2}^{(0)}$ and the amount of $\mathrm{H_2O}$ by $N_{L,3}^{(0)}$. Then, using the Euler equation for the Gibbs free energy given in (9.21), we obtain the following for the Gibbs free energy of this solution, $G_{L,2}$:

$$G_{L,2} = G[T, p; N_{L,2}^{(0)}, N_{L,2}^{(0)}, N_{L,3}^{(0)}] = N_{L,2}^{(0)} \left(\mu_{\mathrm{M}^+}^{\mathrm{L}} + \mu_{\mathrm{A}^-}^{\mathrm{L}} \right) + N_{L,3}^{(0)}\, \mu_{\mathrm{H_2O}}^{\mathrm{L}} \ . \tag{9.119}$$

Here, $\left(\mu_{\mathrm{M}^+}^{\mathrm{L}} + \mu_{\mathrm{A}^-}^{\mathrm{L}}\right)$ represents the chemical potential corresponding to pairs of A^- and M^+ ions, and $\mu_{\mathrm{H_2O}}^{\mathrm{L}}$ is the chemical potential of $\mathrm{H_2O}$ in this solution. Of course, both chemical potentials are functions of $(T, p; N_{L,2}^{(0)}, N_{L,2}^{(0)}, N_{L,3}^{(0)})$. It is important to note here that we do not consider the chemical potentials of A^- and M^+ independently. This is implicit in our assumption that N_{M^+} and N_{A^-} are equal. The Gibbs free energy of the solution on the right-hand side, which we write $G_{R,2}$, is obtained in the same manner, and hence we have

$$G_{R,2} = N_{R,2}^{(0)} \left(\mu_{\mathrm{M}^+}^{\mathrm{R}} + \mu_{\mathrm{A}^-}^{\mathrm{R}} \right) + N_{R,3}^{(0)}\, \mu_{\mathrm{H_2O}}^{\mathrm{R}} \ . \tag{9.120}$$

Because A^- and M^+ are not treated independently here, $G_{L,2}$ and $G_{R,2}$ each takes the form of the Gibbs free energy of an aqueous solution of a single, electrically neutral substance.

With the above results, we immediately obtain the Gibbs free energy for the system consisting of the individual parts of the battery existing as independent subsystems:

$$
\begin{aligned}
G^{(0)} &= G_{L,1} + G_{R,1} + G_{L,2} + G_{R,2} \\
&= (N_{L,1}^{(0)} + N_{R,1}^{(0)})\mu_{\mathrm{M}} + N_{L,2}^{(0)}(\mu_{\mathrm{M^+}}^{\mathrm{L}} + \mu_{\mathrm{A^-}}^{\mathrm{L}}) + N_{R,2}^{(0)}(\mu_{\mathrm{M^+}}^{\mathrm{R}} + \mu_{\mathrm{A^-}}^{\mathrm{R}}) + G_{\mathrm{H_2O}} .
\end{aligned}
\tag{9.121}
$$

Here, $G_{\mathrm{H_2O}}$ is the total contribution of H_2O, consisting of the sum of the contributions from the two sides.

Next, we consider the case described by Figure 9.9(b), in which the impermeable wall separating the two solutions has been replaced with a semipermeable wall, the two electrodes have been inserted into their respective solutions, and the system has been allowed to reach its new equilibrium state. These changes will be accompanied by changes in the amounts of the various substances composing the system. (For the amount of each substance in the new equilibrium state, depicted in Figure 9.9(b), we use the same notation as above, only replacing the superscript "(0)" with "(1).") However, as discussed previously, the migration of A^- through the wall and the reactions taking place at the electrodes will only affect the compositions of the two solutions in the regions near the wall and the electrodes, respectively. Let us write the resulting Gibbs free energy as

$$
G^{(1)} = G^{(0)} + \delta G .
\tag{9.122}
$$

The quantity δG depends in a complicated manner on the distribution of matter on the surfaces of the wall and the surfaces of the electrodes, and in general, it is extremely difficult to compute. However, because this quantity can be regarded as a mere surface effect, for a sufficiently large system, it will be negligible. We consider this case here, and simply ignore this term in the following. However, we continue to use the notation $G^{(1)}$ to formally represent the Gibbs free energy in the situation described by Figure 9.9(b), while regarding it as being identical to $G^{(0)}$.

Next, we imagine the situation in which a voltmeter is placed in contact with the electrodes in order to measure the potential difference between them. Such a measurement is carried out, for example, by measuring the energy (in the form of work) ΔW transferred to the mechanical world as a small amount of charge $\Delta q > 0$ moves from the left electrode to the right electrode. This process is depicted in Figure 9.9(c). From electrostatic theory, we know that the potential difference, $\mathcal{E}$, can then be obtained as

$$
\mathcal{E} = \frac{\Delta W}{\Delta q} .
\tag{9.123}
$$

Let us next consider in some detail the physical process undergone by the system when charge moves from the left electrode to the right. The movement of a charge $\Delta q > 0$ from left to right is actually realized as the transport of a quantity of electrons with charge $-\Delta q$ from right to left. Due to such a migration, the electron concentrations at the left and right electrodes will change by small amounts. As a result, the state of electrostatic balance in the neighborhood of each electrode too will be upset by a small amount. Because there will now be a surplus of electrons attached to the electrode on the left-hand side, the reduction half-reaction $M^+ + e^- \to M$ will be promoted, and a small amount of M will be deposited on the electrode, while because there will be a shortage of electrons attached to the electrode on the right-hand side, the oxidation half-reaction $M \to M^+ + e^-$ will be promoted, and a small amount of M^+ will diffuse into the solution. These two half-reactions will proceed to such degrees that the original quantities of electrons at the two electrodes are again realized. Thus, when a quantity of charge Δq moves from left to right, the extent of reaction changes by the amount $\Delta \xi = \Delta q / F$, where $F \simeq 9.65 \times 10^4$ C/mol is the Faraday constant. It is thus seen that in a chemical reaction system acting as a battery, we can control the extent of reaction by externally controlling the transfer of charge between electrodes.[58]

Now we determine $G^{(2)}$, the Gibbs free energy for the entire system after the amount of charge Δq has moved from left to right, the electrodes have again been disconnected, and the system is allowed to reach a new equilibrium state. As is clear from the above discussion, under this process, the amounts of M contained in the two electrodes change as follows:

$$N_{L,1}^{(1)} \to N_{L,1}^{(1)} + \Delta \xi , \quad N_{R,1}^{(1)} \to N_{R,1}^{(1)} - \Delta \xi . \tag{9.124}$$

Then, because the metallic ions M^+ cannot pass through the semipermeable wall, as a result of the changes described by (9.124), the amounts of M^+ in the two solutions undergo the following changes:

$$N_{L,2}^{(1)} \to N_{L,2}^{(1)} - \Delta \xi , \quad N_{R,2}^{(1)} \to N_{R,2}^{(1)} + \Delta \xi . \tag{9.125}$$

If only the amounts of M^+ in the solutions changed, the local electric neutrality of each solution would be destroyed. In fact, however, the above changes in the amounts of M^+ are accompanied by nearly identical changes in the amounts of A^-.[59] Thus, ignoring a small error, we regard (9.125) as also describing the changes in the amounts of A^- in the two solutions. With all of the changes in the system thus accounted for, we find that the Gibbs free energy of the system after an amount of charge Δq has moved from the left-hand side to the right-hand side is given by

$$G^{(2)} = G^{(1)} + \Delta \xi \, \mu_M - \Delta \xi \, \mu_M - \Delta \xi (\mu_{M^+}^L + \mu_{A^-}^L) + \Delta \xi (\mu_{M^+}^R + \mu_{A^-}^R) . \tag{9.126}$$

In deriving this expression, we have assumed that the change in G_{H_2O} and the contribution to $G^{(2)}$ resulting from the changes in the quantities $(\mu_{M^+}^L + \mu_{A^-}^L)$ and $(\mu_{M^+}^R + \mu_{A^-}^R)$ can be ignored.[60]

---

[58] Here we have described an operation (discharging) in which energy is extracted from the battery through the migration of charge from the plus side to the minus side. However, if work is performed on the system by the mechanical world, the reverse operation (charging) can also be realized. In this kind of operation, energy is added to the battery through the externally forced migration of charge from the minus side to the plus side.

[59] If we regard the amount of electrons corresponding to Δq to be much larger than the amounts of A^- and M^+ in the narrow regions near the wall and the electrodes where the concentrations differ from the bulk values, then in order to maintain the neutrality of the bulks under the flow of current, it is necessary that the difference between the changes in the amounts of A^- and M^+ be much smaller than these changes themselves.

[60] This assumption is valid only if Δq is sufficiently small. In the case that a large amount of charge is transferred, the concentrations of the two solutions will be altered significantly, and as a result, the

The maximum amount of work that can be performed on the mechanical world when charge moves from the left electrode to the right is, in accordance with (9.76), equal to the decrease in the Gibbs free energy. Therefore, from (9.122) and (9.126), we find

$$\Delta W_{\max} = G^{(1)} - G^{(2)} = \Delta\xi\{(\mu_{\mathrm{M}^+}^{\mathrm{L}} + \mu_{\mathrm{A}^-}^{\mathrm{L}}) - (\mu_{\mathrm{M}^+}^{\mathrm{R}} + \mu_{\mathrm{A}^-}^{\mathrm{R}})\} \ . \tag{9.127}$$

If the operation through which the charge moves between electrodes is carried out quasi-statically,[61] then the above maximum work is identical to the actual work done by the system, ΔW. In this case, from (9.123) and (9.127), we find that the emf can be expressed in terms of the difference between the chemical potentials of the two solutions as

$$\mathcal{E} = \frac{\Delta W_{\max}}{\Delta q} = \frac{\Delta W_{\max}}{F\Delta\xi} = \frac{(\mu_{\mathrm{M}^+}^{\mathrm{L}} + \mu_{\mathrm{A}^-}^{\mathrm{L}}) - (\mu_{\mathrm{M}^+}^{\mathrm{R}} + \mu_{\mathrm{A}^-}^{\mathrm{R}})}{F} \ . \tag{9.128}$$

It is thus seen that while the chemical potential is defined as a quite abstract quantity, this difference of chemical potentials can be determined directly through a simple voltage measurement.

Next, expressing the chemical potential in terms of the activity, in accordance with (9.49), the emf can be written as follows:

$$\mathcal{E} = \frac{RT}{F} \log\left\{ \frac{a_{\mathrm{M}^+}(T,p; N_{L,2}^{(1)}, N_{L,2}^{(1)}, N_{L,3}^{(1)})\, a_{\mathrm{A}^-}(T,p; N_{L,2}^{(1)}, N_{L,2}^{(1)}, N_{L,3}^{(1)})}{a_{\mathrm{M}^+}(T,p; N_{R,2}^{(1)}, N_{R,2}^{(1)}, N_{R,3}^{(1)})\, a_{\mathrm{A}^-}(T,p; N_{R,2}^{(1)}, N_{R,2}^{(1)}, N_{R,3}^{(1)})} \right\} \ . \tag{9.129}$$

In the case that both of the metallic salt solutions can be treated as dilute, with the relations $N_{L,3}^{(1)} \gg N_{L,2}^{(1)}$ and $N_{R,3}^{(1)} \gg N_{R,2}^{(1)}$, we can use the mole fractions of M^+ and A^- on the left-hand and right-hand sides, $x_{\mathrm{L}} = N_{L,2}^{(1)}/(N_{L,3}^{(1)} + 2N_{L,2}^{(1)}) \simeq N_{L,2}^{(1)}/N_{L,3}^{(1)}$ and $x_{\mathrm{R}} = N_{R,2}^{(1)}/(N_{R,3}^{(1)} + 2N_{R,2}^{(1)}) \simeq N_{R,2}^{(1)}/N_{R,3}^{(1)}$, to re-express the above emf as

$$\mathcal{E} \simeq 2\frac{RT}{F} \log\frac{x_{\mathrm{L}}}{x_{\mathrm{R}}} \ . \tag{9.130}$$

For example, if the concentration of the solution on the left is 10 times larger than that of the solution on the right (i.e., $x_{\mathrm{L}}/x_{\mathrm{R}} = 10$), the emf of this dilute concentration cell at 25°C will be[62]

$$\mathcal{E} \simeq 2\frac{RT}{F} \log 10 \simeq 118\,\mathrm{mV} \ . \tag{9.131}$$

changes in $\mu_{\mathrm{M}^+}^{\mathrm{L}}$, $\mu_{\mathrm{A}^-}^{\mathrm{L}}$, $\mu_{\mathrm{M}^+}^{\mathrm{R}}$, $\mu_{\mathrm{A}^-}^{\mathrm{R}}$ and $G_{\mathrm{H_2O}}$ will have non-negligible effects. Specifically, in the case that energy is being extracted from the battery, the difference between the concentrations of the two solutions will decrease as charge flows from left to right, and the battery's emf will resultingly decrease.

   [61] In order to realize this condition, it is necessary to measure the voltage while allowing only a very small amount of current to flow. In the past, various techniques were applied to accomplish this, but today, with digital voltmeters, voltage measurements can be made almost effortlessly with essentially no current passing between electrodes. Batteries for which quasi-static operations of charging and discharging can be realized are sometimes called *reversible batteries*.
   [62] At 25°C, we have $RT/F \simeq 25.7\,\mathrm{mV}$. In electrochemistry, this is an important value.

9.6.3 Computation of the emf based on the electrochemical potential

Electrochemical potential

Here, as preparation for a second derivation of the emf of a concentration cell, we introduce a useful quantity termed the *electrochemical potential*.

Let us consider an object of charge q that is subject to an electric field, whose potential we write $\varphi(\boldsymbol{r})$, where $\boldsymbol{r}$ is the position vector in three-dimensional space. We assume that $\varphi(\boldsymbol{r})$ varies little over the spatial extent of this object. From the theory of electrostatics, we know that when the object is moved from $\boldsymbol{r}_1$ to $\boldsymbol{r}_2$, the work that it performs on the mechanical world (i.e., the agent that controls the motion of the object) is[63]

$$W_{\mathrm{el}} = q\{\varphi(\boldsymbol{r}_1) - \varphi(\boldsymbol{r}_2)\} \,. \tag{9.132}$$

Now, let us suppose that the object considered here is a thermodynamic system. Then, under conditions of fixed temperature T and pressure p, comparing the work given in (9.132) with the relation between the Gibbs free energy and the work appearing in (9.76), we conclude that the Gibbs free energy of this system should include the term $q\varphi(\boldsymbol{r})$ to account for the effect of the electric potential. It is now clear that the Gibbs free energy that we have been using should be regarded as only that for a system that experiences no electric interaction. For a system of the type considered here, the proper free energy is that obtained by adding the term $q\,\varphi(\boldsymbol{r})$. In the case that the system consists of a single (ionic) species, the addition of $q\,\varphi(\boldsymbol{r})$ to the Gibbs free energy corresponds to the addition of $zF\varphi(\boldsymbol{r})$ to the chemical potential (recalling the relation $\mu(T,p) = G[T,p;N]/N$). Here, z is the valence of the ion, and hence $zF = q/N$ is the charge per unit amount of substance. The chemical potential including this electric term is called the *electrochemical potential* and written $\overline{\mu}$. For a system containing electrically charged substances, the role played by $\overline{\mu}$ is analogous to that played by μ for a system containing only electrically neutral substances.

We define the electrochemical potential of the ith substance in a multi-component system as follows:[64]

$$\overline{\mu}_i := \mu_i + z_i F \varphi \,, \tag{9.133}$$

with μ_i and z_i representing the chemical potential and the valence of the ith substance. Because of the spatial variation exhibited by ionic systems, some care is needed to identify the arguments of the quantities appearing here. First, although μ_i is still considered a state function, we now define it locally, with its value at a given position $\boldsymbol{r}$ being determined by $(T, p, \mathbf{n}(\boldsymbol{r}))$, which are regarded as the state variables in the present case. (Here, $\mathbf{n} = (n_1, \ldots, n_m)$ represents the local concentrations of the m components.) Then, although above we have written φ as a function of only $\boldsymbol{r}$, its value at $\boldsymbol{r}$ also depends on the form of $\mathbf{n}(\boldsymbol{r}')$ over the entire system.[65] Therefore, the value of $\overline{\mu}$ at $\boldsymbol{r}$ is specified by the arguments $(T, p, \mathbf{n}(\boldsymbol{r}))$, $\boldsymbol{r}$ and $\mathbf{n}(\boldsymbol{r}')$.

---

[63] The (mechanical) potential energy yielding the force that acts on the object is given by $V(\boldsymbol{r}) = q\varphi(\boldsymbol{r})$, and thus (9.132) is identical to (3.6).

[64] Note that the Gibbs free energy obtained from the Euler equation (9.21) using this definition of the electrochemical potential properly accounts for the work given in (9.132).

[65] Thus, it is a *functional* of $\mathbf{n}(\boldsymbol{r})$.

Investigating the state of balance in a system as we have done previously, we readily derive the result that, under conditions of fixed T and p, balance is realized in a system containing electrically charged substances when each electrochemical potential defined by (9.133) is constant throughout the system. These extended balance conditions form the foundation of equilibrium electrochemistry.

Computation

Let us again consider the situation depicted in Figure 9.8—two aqueous solutions with different concentrations separated by a semipermeable wall and each containing an electrode. Below, we re-derive the potential difference between the electrodes, this time using the electrochemical potential instead of the Gibbs free energy. We find that this derivation is significantly simpler than that given above.

Let us represent the electric potentials at the left and right electrodes by φ_1^{L} and φ_1^{R} and those in the internal regions (i.e., sufficiently far from the wall and electrodes) of the left and right solutions by φ_2^{L} and φ_2^{R}. Because the solutions on either side are separated by a semipermeable wall that allows A^- to pass freely between them, in equilibrium, the solutions must be balanced with respect to A^-, and hence the electrochemical potential of A^- must be constant throughout both solutions. From (9.133), we know that in the internal regions of the solutions, this electrochemical potential is given by $\overline{\mu}_{\mathrm{A}^-} = \mu_{\mathrm{A}^-}^{\mathrm{L}} - F\varphi_2^{\mathrm{L}}$ on the left-hand side and $\overline{\mu}_{\mathrm{A}^-} = \mu_{\mathrm{A}^-}^{\mathrm{R}} - F\varphi_2^{\mathrm{R}}$ on the right-hand side. Thus, the balance condition for A^- is the following:

$$\mu_{\mathrm{A}^-}^{\mathrm{L}} - F\varphi_2^{\mathrm{L}} = \mu_{\mathrm{A}^-}^{\mathrm{R}} - F\varphi_2^{\mathrm{R}} . \tag{9.134}$$

From this relation, we immediately obtain the following expression for the potential difference between the internal regions of the solutions on the left-hand and right-hand sides (i.e., the membrane potential):[66]

$$\varphi_2^{\mathrm{L}} - \varphi_2^{\mathrm{R}} = \frac{\mu_{\mathrm{A}^-}^{\mathrm{L}} - \mu_{\mathrm{A}^-}^{\mathrm{R}}}{F} . \tag{9.135}$$

Similarly, because M^+ is freely exchanged between the electrodes and the solutions, we have the following balance conditions for it:

$$\mu_{\mathrm{M}^+}^{\mathrm{metal}} + F\varphi_1^{\mathrm{L}} = \mu_{\mathrm{M}^+}^{\mathrm{L}} + F\varphi_2^{\mathrm{L}} , \quad \mu_{\mathrm{M}^+}^{\mathrm{metal}} + F\varphi_1^{\mathrm{R}} = \mu_{\mathrm{M}^+}^{\mathrm{R}} + F\varphi_2^{\mathrm{R}} . \tag{9.136}$$

Here, $\mu_{\mathrm{M}^+}^{\mathrm{metal}}$ is the chemical potential of M^+ in a pure sample of the metal composing the electrodes (subject to no electric field). Combining (9.135) and (9.136), we obtain

$$\varphi_1^{\mathrm{L}} - \varphi_1^{\mathrm{R}} = \frac{\mu_{\mathrm{M}^+}^{\mathrm{L}} - \mu_{\mathrm{M}^+}^{\mathrm{R}}}{F} + (\varphi_2^{\mathrm{L}} - \varphi_2^{\mathrm{R}}) = \frac{(\mu_{\mathrm{M}^+}^{\mathrm{L}} + \mu_{\mathrm{A}^-}^{\mathrm{L}}) - (\mu_{\mathrm{M}^+}^{\mathrm{R}} + \mu_{\mathrm{A}^-}^{\mathrm{R}})}{F} , \tag{9.137}$$

confirming the result derived previously, (9.128).

---

[66] With regard to this expression, see Footnote 55 on p. 231 and the discussion in the main text corresponding to that footnote. Referring to (9.129), we see that in terms of the activities, the membrane potential is $\frac{RT}{F} \log \left\{ \dfrac{a_{\mathrm{A}^-}(T,p;N_{L,2}^{(1)},N_{L,2}^{(1)},N_{L,3}^{(1)})}{a_{\mathrm{A}^-}(T,p;N_{R,2}^{(1)},N_{R,2}^{(1)},N_{R,3}^{(1)})} \right\}$.

Problems 9

9.1 (Section 9.1.2) Derive the balance condition appearing in (9.9). The following is perhaps the simplest approach. First, consider a system consisting of m components that is partitioned into two parts by an ordinary wall. Suppose that two arbitrary states are prepared on either side of the wall. Next, suppose that the ordinary wall is replaced with a semipermeable wall that is permeable only to Species 1. Apply analysis similar to that used in Section 7.3 to elucidate the spontaneous change undergone by the system after this operation and derive the fundamental inequality.

9.2 (Sections 9.1 and 9.2) Following Problem 7.2, extend the Gibbs-Duhem equation for a single-component system, (7.62), to a multi-component system. Carry out the derivation twice, once using the Helmholtz free energy and once using the Gibbs free energy. Discuss the aspects of this equation that are unique to multi-component systems.

9.3 (Section 9.3.1) Regarding T, V and $\boldsymbol{N}$ as the controllable quantities, derive the expression for the osmotic pressure in a dilute solution appearing in (9.57) using the approximate form of the Helmholtz free energy given in (9.15). This is a somewhat involved computation, but it provides insight into the physical situation in which osmotic pressure arises. A brief outline is presented below.

First, using (9.15), derive approximate expressions for the pressure and the chemical potential of Species 1 in the two compartments, with T, V and $\boldsymbol{N}$ acting as the independent variables. Write the amounts of Species 1 in the two compartments in equilibrium as follows:

$$\widetilde{N}_1 = \frac{V}{V + V'} N_1^{\text{tot}} - M \ , \quad \widetilde{N}_1' = \frac{V'}{V + V'} N_1^{\text{tot}} + M \ . \tag{9.138}$$

As in the situation considered in Section 9.3.1, we regard Species 2 to be confined to the left compartment, in which its amount is N_2. Obviously, if $N_2 = 0$, equilibrium is realized with $M = 0$, and therefore M represents the amount of Species 1 that moves from the left compartment to the right due to the interaction between the two substances. Because we are employing the dilute solution approximation, we regard N_2/N^{tot} to be a small quantity. The derivations of the chemical potential and pressure should be carried out as expansions to first order in this quantity. The quantity M/N^{tot} should be treated as first order in N_2/N^{tot}.

Next, using the fact that in equilibrium, the values of μ_1 in the two compartments must be equal, derive an expression for M in terms of the chemical potential for a pure sample of Species 1 and the quantity $\partial w_2(T; v)/\partial v$, where $w_2(T; v)$ is the unknown function in (9.15). For a system of this kind, M is not characterized by any particular universality. (This is clear from the fact that M vanishes in the case of an ideal gas).

Finally, from the difference between the values of the pressure in the compartments, obtain an expression for the osmotic pressure. In its initial form, this will not be a simple expression, but it can be simplified by replacing M with the form described above and then using the second relation in (7.61).

9.4 (Section 9.4) Derive the equilibrium condition for a chemical reaction taking place in a system of fixed volume at constant temperature. Give an explicit form of this condition for the case of an ideal gas.

9.5 (Section 9.6) Consider a concentration cell like that studied in the main text, but suppose that the semipermeable wall is permeable only to M^+, instead of A^-. Show that in this case, the emf of the battery is 0. Derive this result using both methods presented in the main text.

10

The Thermodynamics
of Ferromagnetic Materials

In this chapter, we leave the realm of fluids and consider the thermodynamic behavior of ferromagnetic materials. In particular, we investigate phase transitions and critical phenomena in ferromagnets, which have been important topics in condensed matter physics since the nineteenth century. Of course, the structure of thermodynamics itself is independent of the application, and for this reason, after we establish several points of correspondence between ferromagnetic systems and fluid systems, we are able to directly apply the theoretical framework constructed in previous chapters to ferromagnets.[1] After introducing the topics of phase transitions and critical phenomena in ferromagnets, we investigate the method employing the Landau pseudo free energy. Because there is some degree of confusion in the literature regarding this method and its application, we present a detailed explanation of both its fundamental idea and its theoretical structure. Finally, we discuss the scaling hypothesis for critical phenomena.

Before beginning, we note that the content of this chapter is somewhat advanced in comparison with the rest of the book.

10.1 The treatment of ferromagnetic materials

In this section, we formally introduce the thermodynamic theory of ferromagnetic materials while elucidating the theoretical correspondence between ferromagnets and fluids.

10.1.1 Magnetization and the parameterization of equilibrium

Let us consider a ferromagnetic system containing an amount of substance N under isothermal conditions at temperature T. We assume that this ferromagnet is a solid. Intuitively, we can imagine it to be a piece of iron, for example. For such a system, there is a new extensive quantity, unique to the thermodynamic treatment of magnetic systems, the *magnetization*, which we represent by M. The magnetization is the magnetic moment of the ferromagnet as a whole. In the treatment presented here, it is regarded as a scalar.[2] We assume that for the phenomena studied in

---

[1] The treatment given in this chapter assumes an understanding of the material presented up through Chapter 8.

[2] Of course, the magnetization is in general a vector, but for simplicity, here we consider ferromagnets possessing strong uniaxial anisotropy (i.e., magnetic systems that are easily magnetized along one particular axis). In this case, the magnetization can be regarded as a scalar. But it should be noted that the treatment of the general case is essentially the same as that presented here.

Thermodynamics: A Modern Approach. Hal Tasaki and Glenn Paquette, Oxford University Press.
© Hal Tasaki and Glenn Paquette (2026). DOI: 10.1093/9780191878091.003.0010

this chapter, the quantities T, M and N provide a complete parameterization of equilibrium.[3] For this reason, V does not appear as a state variable.

10.1.2 Helmholtz free energy

From the point of view of thermodynamics, the magnetization in ferromagnetic systems is a fundamental extensive quantity that plays a role analogous to that played by the volume in fluid systems. Following this analogy, our analysis is based on the assumption that the Helmholtz free energy $F[T; M, N]$ has been determined through evaluation of the maximum work under operations through which M is varied.[4] Also as in the case of fluid systems, $F[T; M, N]$ is extensive and additive, and it satisfies a variational inequality of the same form as (7.30).

In analogy to the relation (3.31), which expresses the pressure of a fluid in terms of the derivative of $F[T; V, N]$ with respect to V, we have the following relation involving the derivative of $F[T; M, N]$ with respect to M:

$$\frac{\partial F[T; M, N]}{\partial M} = H(T; M, N) \, . \tag{10.1}$$

Just as the pressure can be understood as representing the response felt by an external agent attempting to change the volume of a fluid, the quantity $H(T; M, N)$ can be understood as representing the response felt by an agent attempting to change the magnetization of a ferromagnet. From consideration of the electromagnetic energy, it is seen that $H(T; M, N)$ is the magnitude of the (homogeneous)

---

[3] As explained in Sections 2.4.5 and A.1.3, this means that within any particular context in which we are studying the magnetic behavior of ferromagnetic systems through operations under which we vary M, a parameterization in terms of the state variables T, M and N is sufficient to account for the work performed in all quasi-static operations within that context. For many practical applications, this requires that in such operations, the variation of the volume of the system be negligibly small.

[4] There is a foundational point that slightly complicates the thermodynamic treatment of magnetic systems. In this book, we have constructed the theory of thermodynamics through an operational treatment, in which thermodynamic systems and thermodynamic behavior are understood through consideration of mechanical operations and the work performed in them. In particular, we have relied on the assumption that the operations in question are carried out through variation under which direct control is exerted on the collective extensive variable. (A typical example is an operation under which the volume of a fluid in a container is varied by moving one of the walls in a predetermined manner.) However, in the case of a magnetic system, the magnetization is not generally a directly controllable quantity. This would appear to imply that the construction of thermodynamics carried out in this book does not apply to magnetic systems. However, as we explain below, this is not the case.

In ordinary experiments involving magnetic materials, M is determined in response to the variation of H, the external magnetic field. We thus expect that for a given system and a given time-dependent magnetization $M(t)$, there exists some time-dependent magnetic field $H(t)$ that results in this $M(t)$. Thus, having determined the proper $H(t)$ in advance, we can carry out an operation with any desired $M(t)$. In particular, for a quasi-static operation, in which $H(t)$ and $M(t)$ vary infinitesimally slowly, we need only know the equilibrium dependence of M on H in order to realize the desired $M(t)$, provided that the equilibrium values of M and H are in one-to-one correspondence. (In practical terms, dM/dH must be sufficiently small, because if this quantity is large, experimental errors involved in the control of H will be amplified.) This is always the case unless M is in the phase coexistence region, which corresponds to $H = 0$. (Note, however, that the exclusion of the phase coexistence region is not a problem for the determination of the Helmholtz free energy, because it is constant throughout this region, as seen in Figure 10.2(b).) Thus, we see that although our control of M is indirect, we can carry out operations under which any predetermined time dependence is realized. Considering that even in the case of a fluid system, the assumption that we have direct control over the volume is an idealization, we understand that the situations for magnetic systems and fluid systems are in fact quite similar.

magnetic field in the spatial region containing the ferromagnetic object.[5] Comparing the definition of the pressure given in (3.31) with (10.1), we find that H plays the same role in the case of ferromagnetic materials as $-p$ plays in the case of fluids. With this observation, using the convexity of the Helmholtz free energy, we deduce that $H(T; M, N)$ is a non-decreasing function of M, in analogy to Result 7.1 (p. 146), regarding the volume dependence of the pressure.

Finally, we can express the derivatives of $F[T; M, N]$ in a differential form similar to that for the free energy $F[T; V, N]$, appearing in (7.12):[6]

$$dF = -S\,dT + H\,dM \ . \tag{10.2}$$

10.1.3 Gibbs free energy

The expression for the external magnetic field $H(T; M, N)$, with the independent variable M, connotes the situation in which we stipulate the value of the magnetization, and from this infer the strength of the magnetic field. However, from an experimental point of view, the natural situation is that in which H is controlled and M is determined in response to this. This corresponds to the situation in which the pressure, rather than the volume, is a controllable parameter in the case of a fluid system. With this in mind, following the definition of the Gibbs free energy for a fluid system given in (8.3), we define the Gibbs free energy for a magnetic material using the Legendre transformation as follows:[7]

$$G[T, H; N] := \min_{M} \left\{ F[T; M, N] - HM \right\} \ . \tag{10.3}$$

Because $F[T; M, N]$ is always differentiable with respect to M, choosing some fixed value H_0 of H, the minimization condition here is $\partial F[T; M, N]/\partial M = H_0$. Using (10.1), this can be written as $H(T; M, N) = H_0$. This is a condition on M, and its

---

[5] The derivation of this result is left to Problem 10.1. The relation (10.1) is expressed in CGS units. In MKS units, the proper relation is obtained by replacing M in (10.1) with $J = \mu_0 M$ (the magnetic polarization), where μ_0 is the vacuum permeability. In the treatment of electromagnetic phenomena, it would be best to consistently use MKS units, but because this results in the appearance of μ_0 along with M in the various thermodynamic relations appearing in our discussion, for the sake of simplicity, we use CGS units here. With regard to this point, the reader need only remember that the conversion from CGS units to MKS units is realized by simply replacing M with J. Also, note here that we have ignored the effect of the demagnetizing field created by the magnetization of the magnetic material. In the case of ferromagnetic materials that can act as practically useful magnets, this effect is generally non-negligible.

[6] Considering the formal correspondence between fluid systems and magnetic systems, we should add a term like $\mu\,dN$, representing the variation of F with respect to N. However, in the case of magnetic materials, we do not consider anything that corresponds to a multi-component fluid system, and therefore the chemical potential is essentially meaningless in this context. For this reason, in the thermodynamic treatment of magnetic materials, variation of N is generally not considered.

[7] Explicitly including the volume as an independent variable, the Gibbs free energy here would be a function of T, H, N and V. Interpreting this as a free energy with primary dependence on the fundamental thermodynamic variables T, V and N, to which dependence on H is appended, it would in fact be more natural to regard this as the Helmholtz free energy. This ambiguity is a reflection of the fact that the framework of thermodynamics was constructed primarily through investigation of fluid systems, and when this framework is extended to other types of systems, there is not necessarily a unique correspondence of concepts. In particular, often there exists no definitive identification of types of free energies, and there generally exists no criterion to determine which identification is "best." In statistical mechanics, the dominant convention is to express the quantity $G[T, H; N]$ as $F[T, H; N]$, with the understanding that this represents the function $F[T, H; V, N]$ with the argument V suppressed.

meaning is that the value of M picked out by the Legendre transformation is that realized by the system when H is fixed at the value H_0. Graphically, the situation can be understood by turning the graphs in Figure 10.1 on their sides. (In the case $T < T_c$, $H = 0$, as discussed below, there is a continuous range of values of M that satisfy this condition, and all of these values are realizable.)

Similarly to the case of the Gibbs free energy for a single-component fluid, $G[T, p; N]$, because $G[T, H; N]$ is an extensive quantity whose independent variables include just one extensive quantity, we can write it as $G[T, H; N] = Ng(T, H)$, where $g(T, H)$ is the molar Gibbs free energy, which depends on T and H alone.

Next, carrying out a calculation similar to that presented in (8.14), we obtain the following relation, corresponding to (8.16):

$$M(T, H; N) = -\frac{\partial G[T, H; N]}{\partial H} . \tag{10.4}$$

Note, however, that this equation does not apply to the case $T < T_c$, $H = 0$, because in this case, $G[T; H, N]$ is not differentiable with respect to H. This reflects the fact that below T_c, $M(T; 0, N)$ is not well defined, as discussed in the following section. (This is analogous to the fact that for a fluid system, $V(T, p; N)$ is not well defined in the two-phase coexistence region.)

Finally, including the dependence of G on T, we have the following differential form:[8]

$$dG = -S\,dT - M\,dH . \tag{10.5}$$

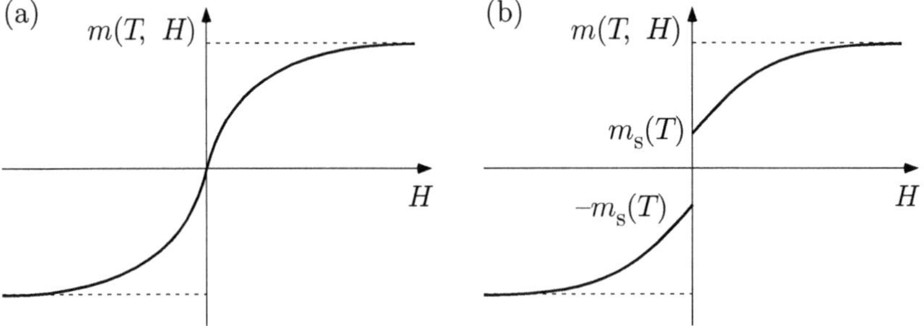

Figure 10.1 Typical behavior exhibited by the magnetization of a ferromagnet. These graphs display the magnetization $m(T, H)$ as a function of the external magnetic field with the temperature fixed. (a) For $T > T_c$, $m(T, H)$ is a smooth, increasing function of H. (b) For $T < T_c$, $m(T, H)$ is a smooth, increasing function of H, except at $H = 0$, where it is not defined. In either case, in the limits $H \to \pm\infty$, the magnetization approaches a fixed value known as the saturation magnetization, which varies among ferromagnetic materials.

---

[8] For a concrete example of the Gibbs free energy in a simple model of magnetic materials, see Problems 10.2 and 10.3.

10.2 Phase transitions and critical phenomena

Like $G[T, H; N]$, the magnetization $M(T, H; N)$ is an extensive quantity whose independent variables include only one extensive quantity. Therefore, we can write it as $M(T, H; N) = Nm(T, H)$, where the magnetization per unit amount of substance, $m(T, H)$, depends on only T and H. In the following, we investigate the behavior of $m(T, H)$. For conciseness, we also refer to this quantity as the *magnetization*.[9]

10.2.1 Ferromagnetic phase transition

Figure 10.1 displays the dependence of the magnetization $m(T, H)$ on H for a typical ferromagnet at two temperatures. As seen in Figure 10.1(a), for sufficiently high temperatures, $m(T, H)$ is a smooth, increasing function of H. In particular, note that $m(T, H)$ is well defined at $H = 0$, with $m(T, 0) = 0$. In other words, at such temperatures, in the absence of an external magnetic field, the material does not act as a magnet. In this situation, the system is said to exhibit *paramagnetism*. Contrastingly, for sufficiently low T, the magnetization is not uniquely defined at $H = 0$, as shown in Figure 10.1(b). For temperatures in this regime, at $H = 0$ the value of the magnetization behaves in accordance with the following:[10]

$$\lim_{H \searrow 0} m(T, H) = m_{\mathrm{s}}(T) > 0 \,, \quad \lim_{H \nearrow 0} m(T, H) = -m_{\mathrm{s}}(T) < 0 \,. \tag{10.6}$$

Roughly stated, the value of m realized at $H = 0$ depends on whether H reached this value from the positive side or the negative side.[11] Here, $m_{\mathrm{s}}(T)$ is an important quantity termed the *spontaneous magnetization* (per unit amount of substance). A system with nonzero spontaneous magnetization can possess finite magnetization even in the absence of an external magnetic field, thus acting as a so-called permanent magnet. In this situation, the system is said to exhibit *ferromagnetism*. A ferromagnetic material undergoes a phase transition at a particular temperature between the high-temperature paramagnetic phase and the low-temperature ferromagnetic phase.

Turning the graphs in Figure 10.1 on their sides, we are able to visualize the behavior of the magnetic field $H(T; M, N)$ as a function of M/N in the two cases

---

[9] To be more precise, we should call m the *molar magnetization* and use *magnetization* only in reference to M, but we trust that no confusion will result from this slightly mistaken terminology.

[10] Experimentally, the behavior described by (10.6) corresponds to isothermal operations in which H is quasi-statically decreased to 0 from the positive side or quasi-statically increased to 0 from the negative side. In general, however, it is not only the values $-m_{\mathrm{s}}(T)$ and $m_{\mathrm{s}}(T)$ that are realizable at $H = 0$. In fact, for $T < T_{\mathrm{c}}$, $H = 0$, in equilibrium the magnetization can take any value between $-m_{\mathrm{s}}(T)$ and $m_{\mathrm{s}}(T)$. This behavior corresponds to that of a fluid system in the liquid-gas coexistence region, where, even with T and p fixed, the volume of the system can take any value between $v_{\mathrm{L}}(T)N$ and $v_{\mathrm{G}}(T)N$.

[11] In actual systems, in the case that the magnetic field is being decreased non-quasi-statically from a finite positive value, the magnetization will remain positive for some time even after the field has become negative. This is an example of hysteresis phenomena. A state of positive magnetization in a negative external field is a type of metastable state. As discussed near the end of Section 7.4 (in particular, in Footnote 27 on p. 149), metastable states cannot be described within the (conventional) thermodynamic framework presented in this book. For further discussion of this point, see Footnote 23 on p. 257.

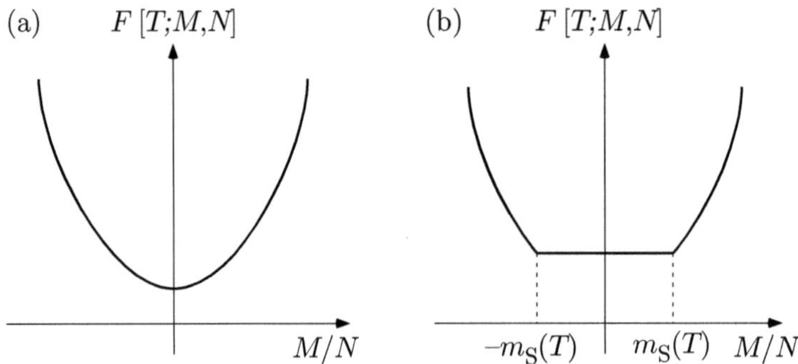

Figure 10.2 Dependence of the Helmholtz free energy $F[T; M, N]$ on M/N in (a) the paramagnetic (high temperature) case and (b) the ferromagnetic (low temperature) case. The appearance of the flat region at the bottom of the graph of $F[T; M, N]$ for low temperatures is a distinctive characteristic of ferromagnetism.

considered. Then, using (10.1), we can obtain a rough idea of the behavior of $F[T; M, N]$ above and below T_{c}. For temperatures at which the system exhibits paramagnetism, corresponding to Figure 10.1(a), $H(T; M, N)$ is a smooth, increasing function of M/N. Thus, in this case, $F[T; M, N]$ is a convex function of M/N possessing a single minimum, located at $M/N = 0$. This behavior is depicted in Figure 10.2(a). Contrastingly, for temperatures at which the system exhibits ferromagnetism, corresponding to Figure 10.1(b), $H(T; M, N)$ is 0 for all values of M/N satisfying $-m_{\mathrm{s}}(T) \leq M/N \leq m_{\mathrm{s}}(T)$. Thus, in this case, $F[T; M, N]$ is constant over this entire range, as seen in Figure 10.2(b). Note that despite this behavior, in this case too, $F[T; M, N]$ is a convex function of M/N. The existence of such a flat region in the graph of $F[T; M, N]$ as a function of M/N is a definitive indication of spontaneous magnetization.

10.2.2 The critical point and critical phenomena

The temperature at which the transition between paramagnetic and ferromagnetic behavior takes place is called the *critical temperature*, written T_{c}. Above, we saw that in the limit of vanishing magnetic field strength, the magnetization behaves as follows:

$$\lim_{H \searrow 0} m(T, H) = \begin{cases} 0 & \text{for } T \geq T_{\mathrm{c}} , \\ m_{\mathrm{s}}(T) > 0 & \text{for } T < T_{\mathrm{c}} , \end{cases}$$

$$\lim_{H \nearrow 0} m(T, H) = \begin{cases} 0 & \text{for } T \geq T_{\mathrm{c}} , \\ -m_{\mathrm{s}}(T) > 0 & \text{for } T < T_{\mathrm{c}} . \end{cases} \tag{10.7}$$

The behavior of $m(T, H)$ over a range of values of T with H fixed is depicted in Figure 10.3. At $H = 0$, for $T > T_{\mathrm{c}}$, we have $m(T, 0) = 0$, while for $T < T_{\mathrm{c}}$, $m(T, 0)$ can realize any value in the gray region. For $H \neq 0$, $m(T, H)$ is a smooth function of T. In this case, there is no clear phase transition, as there is no clear distinction between the paramagnetic and ferromagnetic

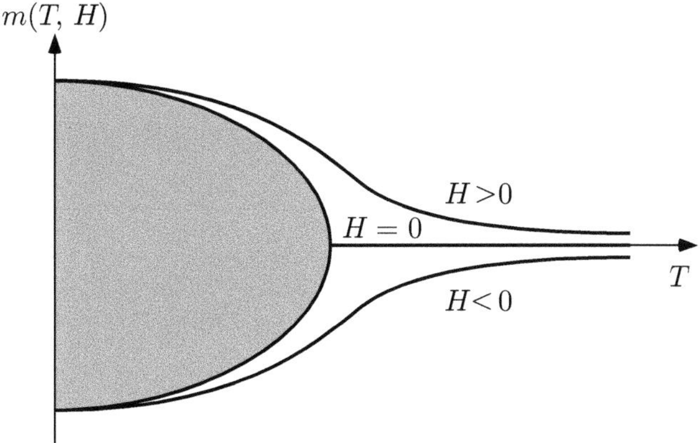

Figure 10.3 Temperature dependence of $m(T, H)$ in the cases that H is fixed with positive, zero and negative values. In the case $H = 0$, for $T < T_{\mathrm{c}}$ the value of m is undetermined, and any value in the gray region can be realized. The boundary of this region coincides with $\pm m_{\mathrm{s}}(T)$.

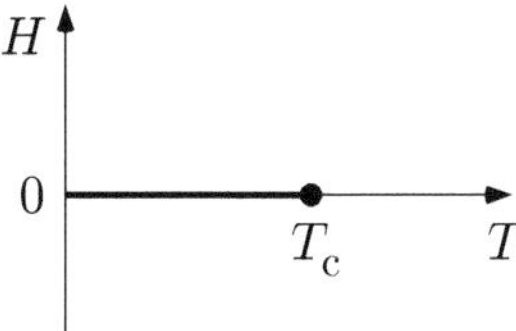

Figure 10.4 Phase diagram for a ferromagnetic material in the T-H plane. The magnetization changes discontinuously only when the line segment connecting the origin, $(0,0)$, and the critical point, $(T_{\mathrm{c}}, 0)$, is crossed. (The critical point $(T_{\mathrm{c}}, 0)$ here is analogous to the critical point $(T_{\mathrm{c}}, p_{\mathrm{c}})$ for a fluid system, discussed in Section 7.5.4.)

phases. Thus, the phase diagram in the T-H plane takes the form displayed in Figure 10.4. The magnetization changes discontinuously when crossing the line segment connecting $(0,0)$ and $(T_{\mathrm{c}}, 0)$. The right endpoint of this line segment, $(T_{\mathrm{c}}, 0)$, is called the *critical point*. This phase diagram is similar to the liquid-gas phase diagram appearing in Figure 7.7. As discussed near the end of Section 10.4, there is in fact a deep connection between these two situations that far exceeds what one might surmise from the mere resemblance of these phase diagrams.

In the neighborhood of the critical point, a ferromagnetic material exhibits a variety of distinctive types of behavior that are collectively known as *critical phenomena*. For example, in the limit $T \nearrow T_{\mathrm{c}}$, the spontaneous magnetization displays the following characteristic type of power-law behavior:[12]

$$m_{\mathrm{s}}(T) \sim (T_{\mathrm{c}} - T)^{\beta} . \tag{10.8}$$

---

[12] The expression $f(x) \sim x^{a}$ for $x \to 0$ means that $f(x)$ possesses the same degree of singularity as x^{a} at 0. More precisely, there exist positive constants C and C' such that the relation $C x^{a} \leq f(x) \leq C' x^{a}$ holds for all sufficiently small x.

The constant β appearing here is one example of a *critical exponent*. Critical exponents play important roles in the quantitative description of critical phenomena. Other examples of power-law behavior characterized by particular critical exponents include the divergence in the limit $T \to T_c$ (with 0 external magnetic field) of the *magnetic susceptibility*,[13,14]

$$\chi(T) := \left.\frac{\partial m(T, H)}{\partial H}\right|_{H=0} \sim |T - T_c|^{-\gamma} , \tag{10.9}$$

and of the specific heat,

$$c(T) := -\frac{T}{N}\frac{\partial^2 G[T, 0; N]}{\partial T^2} = -T\frac{\partial^2 g(T, 0)}{\partial T^2} \sim |T - T_c|^{-\alpha} . \tag{10.10}$$

This specific heat is analogous to the specific heat at constant pressure for a fluid system (see (8.28)).

10.3 A model of ferromagnetic phase transitions based on the Landau pseudo free energy

10.3.1 Introduction

In this section, we construct and analyze a model of ferromagnetic systems based on the Landau pseudo free energy. Through this treatment, we elucidate the characteristics of phase transitions and critical phenomena as described by this model and ascertain the proper place of this model within the theory of thermodynamics. We also attempt to clear up the confusion regarding the interpretation of the Landau pseudo free energy that exists in the literature.

We seek to construct the mathematically simplest thermodynamic model that captures the most essential behavior exhibited by ferromagnetic systems. Although some aspects of this method are rather theoretical, it begins with consideration of empirically observed behavior. The simple mathematical structure obtained from these considerations provides a clear, intuitive understanding of the phase transition phenomena exhibited by ferromagnetic systems. Because this method seeks to model only the essential aspects of phase transitions and does so in a general, mathematically simple manner, its applicability is not limited to magnetic systems. Indeed, this is the standard theoretical approach used to treat phase transitions and critical phenomena quite generally, and essentially the same method has been successfully applied to the study of systems in fields ranging from condensed matter physics to cosmology.[15] In fact, we have already seen an application of this method,

---

[13] The magnetic susceptibility is a measure of the degree to which a material becomes magnetized in response to an external field.

[14] More generally, we could define two critical exponents here, γ and γ', with $\chi(T) \sim (T - T_c)^{-\gamma}$ for $T \searrow T_c$ and $\chi(T) \sim (T_c - T)^{-\gamma'}$ for $T \nearrow T_c$. However, both theoretical and experimental results are consistent with the equality of these exponents, and for this reason, we ignore their distinction. The situation is similar for the exponent α appearing in (10.10).

[15] The fact that the same mathematical framework can be applied to the description of phase transitions over such a wide range of systems is another example of universality in science (see Section 1.2).

namely the treatment of gases based on the van der Waals equation (3.37), studied in Problems 3.3 and 7.8.

10.3.2 Models in thermodynamics

Before proceeding to our treatment of the Landau pseudo free energy, we briefly discuss the concept of a theoretical *model*.

As we have stressed throughout this book, thermodynamics is a universal framework that applies to all macroscopic systems. Therefore, thermodynamics in its general form provides an implicit description of any macroscopic system. However, in order to obtain predictions specific to a particular class of systems or phenomena, we need an explicit description. This kind of description is provided by what we term a "model." In general, we regard a model to be any theoretical system that provides a concrete, explicit description of a particular class of physical systems.[16] Within thermodynamics, the construction of a model is equivalent to the specification of a concrete form of a complete thermodynamic function.

There are various methods for obtaining models. In some cases they are deduced directly from a general theory or theories, while in other cases they are constructed phenomenologically by piecing together physical concepts on the basis of empirical and mathematical considerations. In the latter situation, of course, it is important to consider the consistency of the model with more general theories.

The model of an ideal gas, obtained by specifying the form of the Helmholtz free energy appearing in (7.9), is a typical example of a model in thermodynamics. The simplicity of this specific form has allowed us to obtain a very detailed understanding of the behavior of ideal gases. The primary importance of this model, however, is not that it allows us to understand these gases themselves, but that some of the behavior easily derived from it is in fact observed more universally. The situation is similar for the somewhat more sophisticated phenomenological models provided by the van der Waals theory, defined by the free energy constructed in Problem 7.8, and the Landau theory studied in this chapter. The simplicity of these models allows us to easily derive from them certain essential features of phase transition behavior that are observed over very wide ranges of actual systems.

10.3.3 Construction of the model

For a system with a sufficiently high temperature and a sufficiently small magnetization, it is known empirically that the magnetization and the magnetic field are related as

$$H(T; M, N) \simeq \frac{aTM}{N} \,.$$

(10.11)

The value of the positive constant a here depends on the particular substance in question, but this proportionality relation itself is of universal validity, holding not

---

[16] We should note here that there is no standard definition of the term *model* as used in scientific fields. Although the interpretation of this term presented here is in no way unusual, it should be kept in mind that there are other interpretations that may be more useful in other contexts.

only for ferromagnets but for almost all magnetic materials. This relation is known as *Curie's law.*

Let us first assume that (10.11) holds exactly. Then, using (10.1) and integrating (10.11), we obtain

$$F_0[T; M, N] = N f_0(T) + \frac{aTM^2}{2N} = N \left\{ f_0(T) + \frac{aT}{2} \left(\frac{M}{N} \right)^2 \right\} . \qquad (10.12)$$

Here, the subscript "0" on F_0 indicates that this is the initial form of the free energy that we consider in our construction of the model. The quantity $f_0(T)$ is an unknown function that is not determined by (10.11). We assume that it is a concave, smooth function of T.

From (10.11), we find that the susceptibility (defined in (10.9)) corresponding to the above free energy is

$$\chi_0(T) = \left. \frac{\partial m_0(T, H)}{\partial H} \right|_{H=0} = \left(N \left. \frac{\partial H_0(T; M, N)}{\partial M} \right|_{M=0} \right)^{-1} = \frac{1}{aT} , \qquad (10.13)$$

which is finite for any $T > 0$. This implies that in its present form, this model can only describe paramagnetic behavior, which includes no phase transition. The situation here is analogous to that for an ideal gas, which also exhibits no phase transition. We conclude that to treat ferromagnetic phase transitions, we must extend this approach.

In order to extend the present approach to the description of phase transitions while maintaining the basic manner of thinking behind (10.12), we wish to apply the smallest change to $F_0[T; M, N]$ that makes such a description possible. In this way, we construct the theoretically simplest model that accounts for the most fundamental empirical facts regarding the behavior of ferromagnetic systems, i.e., that there exists a critical temperature, $T_c > 0$, with respect to which the system possesses the following properties: for T sufficiently larger than T_c, the system obeys Curie's law, (10.11); for $T < T_c$, the system exhibits spontaneous magnetization; in the limit $T \searrow T_c$, the susceptibility diverges.[17]

As the first step in our extension of (10.12), in order to account for the divergence of the susceptibility described by (10.9), we replace the term aT in (10.12) with $a(T - T_c)$, where T_c is a positive constant. With this replacement, the relation in (10.11) becomes

$$H(T; M, N) = a(T - T_c)M/N . \qquad (10.14)$$

This is the *Curie-Weiss law.* With this form of $H(T; M, N)$, the susceptibility is given by $\chi_1(T) = \{a(T - T_c)\}^{-1}$, which indeed does diverge as $T \searrow T_c$. However, (10.14) also describes the unphysical behavior that for $T < T_c$, M is a decreasing function of H for all values of H. We next address this point.

As the second step, we add to the free energy a term proportional to $(M/N)^4$. This modification is based on the mathematical reasoning that if we expand an

---

[17] To go beyond a minimal extension, however, it is known that this simplistic approach is insufficient.

even function[18] of (M/N) in a Taylor series, then such a term appears generically.[19] With these considerations, we obtain the following:

$$\widetilde{F}[T; M, N] = N \left\{ f_0(T) + \frac{a(T - T_{\rm c})}{2} \left(\frac{M}{N}\right)^2 + \frac{b}{4} \left(\frac{M}{N}\right)^4 \right\} . \qquad (10.15)$$

This is the Landau pseudo free energy. The quantities a, b and $T_{\rm c}$ are all positive constants, and $f_0(T)$ is a smooth, concave function of T.

For sufficiently high temperatures and sufficiently small magnetizations, the above Landau pseudo free energy is almost identical to the free energy (10.12) obtained from Curie's law. This suggests that at least for systems under such conditions, $\widetilde{F}[T; M, N]$ provides a physically valid description of magnetic phenomena. However, there is a serious problem with $\widetilde{F}[T; M, N]$ as a free energy: For temperatures below $T_{\rm c}$, it is not a convex function of M. This implies that below the critical temperature, there is no variational inequality satisfied by $\widetilde{F}[T; M, N]$. Thus, because the existence of a variational inequality is a necessary property for a proper thermodynamic system, we conclude that $\widetilde{F}[T; M, N]$ cannot be regarded as the Helmholtz free energy of a thermodynamic system. This is the reason that we term this quantity a *pseudo* free energy.

Although the Landau pseudo free energy constructed above is not a convex function, there exists a natural method to derive from it a convex free energy. Briefly stated, this method consists simply of applying the Legendre transformation and then its inverse to $\widetilde{F}[T; M, N]$. Below, we describe this method, and then we present its physical interpretation. This interpretation provides a way to understand why this simple method is able to produce a physically meaningful model of ferromagnetic behavior.

To begin, we introduce the Gibbs free energy corresponding to the pseudo free energy $\widetilde{F}[T; M, N]$, defined through analogy to (10.3) as

$$G[T, H; N] := \min_{M} \left\{ \widetilde{F}[T; M, N] - HM \right\}$$

$$= N \left\{ f_0(T) + \min_{m} \left(\frac{a(T - T_{\rm c})}{2} m^2 + \frac{b}{4} m^4 - Hm \right) \right\} . \qquad (10.16)$$

It is not difficult to show that this $G[T, H; N]$ is a concave function of both H and T (see Problem 8.5). Recalling that for the Gibbs free energy, concavity plays a role equivalent to that played by convexity for the Helmholtz free energy, we see that, at least with regard to its basic form, $G[T, H; N]$ defined above can be regarded as a legitimate Gibbs free energy of a thermodynamic system.

---

[18] For most magnetic systems, microscopic considerations based on quantum mechanics indicate that, in the absence of an external magnetic field, the free energy must be an even function of M. This is a direct result of the fact that for systems of this kind, the system Hamiltonian is invariant under any $180°$ spatial rotation.

[19] However, the fact that the Landau theory does not yield critical exponents that are consistent with their experimentally measured values (see Section 10.4) indicates that this apparently reasonable assumption is of limited validity. Consideration of this interesting situation serves as the starting point for the modern theory of critical phenomena. However, such topics are beyond the scope of this book.

Next, applying the inverse Legendre transformation to $G[T, H; N]$, we make the following definition:

$$F[T; M, N] := \max_{H} \{G[T, H; N] + HM\} . \tag{10.17}$$

Unlike the original $\widetilde{F}[T; M, N]$, the quantity $F[T; M, N]$ is a convex function of M, and hence, with regard to its basic form, this can be regarded as a legitimate Helmholtz free energy of a thermodynamic system.

The free energies $F[T; M, N]$ and $G[T, H; N]$ constitute our model of ferromagnetic systems. Below we present an interpretation of their construction that supports the validity of this model.

10.3.4 Physical interpretation of the construction

Above, we saw how the Legendre transformation removes the pathological behavior of $\widetilde{F}[T; M, N]$ and replaces this function with the theoretically simplest free energy that satisfies the minimum requirements for the description of the ferromagnetic phase transition. However, the physical meaning of this method is still unclear. We now present an interpretation that clarifies this meaning.

The first important observation is that, as discussed in Section 10.2.1, from empirical results, we know that for $T < T_{\mathrm{c}}$, the Helmholtz free energy for a ferromagnetic system must have the general form displayed in Figure 10.2(b), i.e., a convex function whose minimum value is realized along a horizontal line segment. In the following, we demonstrate why the free energy $F[T; M, N]$ obtained from the procedure described above does have this general form, and we also argue that this procedure endows $F[T; M, N]$ with specific properties appropriate for the description of a ferromagnetic system. In particular, from this argument, it is understood how the Legendre transformation is able to extract only the physically meaningful information contained in $\widetilde{F}[T; M, N]$. With this understanding, our identification of $F[T; M, N]$ and $G[T, H; N]$ as providing a valid model of ferromagnetic systems is seen to be quite reasonable. We point out, however, that this interpretation represents merely a plausibility argument. Formulating a more rigorous argument supporting the validity of this model would be an important advance.

We begin by recalling that the construction of the function $\widetilde{F}[T; M, N]$ was carried out in a simple, physically intuitive manner that accounts for the most essential behavior of systems exhibiting ferromagnetic transitions. We can thus assume that, as a free energy, this function does contain physically meaningful information regarding the fundamental nature of such systems. However, as manifested by its non-convex form, it also contains information that is not physically meaningful. We wish to preserve the former and remove the latter. Interestingly, as we now discuss, the Legendre transformation is apparently capable of doing this.

First we study the case $T \geq T_{\mathrm{c}}$. As mentioned above, in this case the Landau pseudo free energy $\widetilde{F}[T; M, N]$ is a convex function of M. For this reason, we consider all information possessed by $\widetilde{F}[T; M, N]$ in this regime to be physically meaningful. The situation regarding the application of the Legendre transformation to a function of this type is depicted in Figure 10.5. As seen there, for each value M_0 of M, there exists a unique value H_0 of H such that the minimum value among

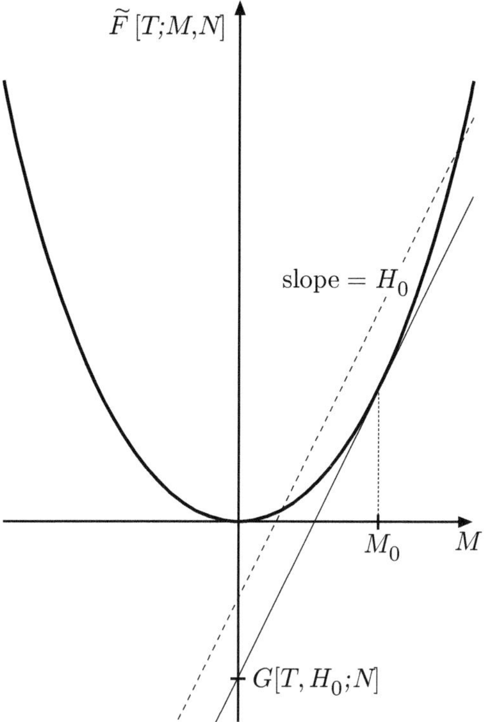

Figure 10.5 Landau pseudo free energy as a function of M (with T and N fixed) at a temperature above T_c. In this case, because $\widetilde{F}[T;M,N]$ is a convex function of M, every value of M participates in the construction of $G[T,H;N]$ under the Legendre transformation: For each value M_0 of M, there exists one value H_0 of H such that the value of $G[T,H_0;N]$ is determined by the value of $\widetilde{F}[T;M_0,N]$ in the manner shown. Explicitly, $G[T,H_0;N]$ is equal to the y-intercept of the line tangent to $\widetilde{F}[T;M,N]$ at $M = M_0$.

the y-intercepts of all lines of slope H_0 that intersect $y = \widetilde{F}[T;M,N]$ is realized by that one of these lines that passes through the point $(M_0, \widetilde{F}[T;M_0,N])$. (The value H_0 is in fact the slope of the curve $\widetilde{F}[T;M,N]$ at M_0, and hence the line in question is that tangent to $\widetilde{F}[T;M,N]$ at M_0.) For this reason, every point in the domain of $\widetilde{F}[T;M,N]$ participates in the construction of $G[T,H;N]$ under the Legendre transformation, and, indeed, all information concerning the form of $\widetilde{F}[T;M,N]$ in the neighborhood of each M_0 is encoded in the form of $G[T,H;N]$ in the neighborhood of the corresponding H_0. Then, because the situation is similar when the inverse Legendre transformation is applied to $G[T,H;N]$, due to its concavity, this transformation returns $\widetilde{F}[T,H;N]$. From these considerations, it is clear that in the $T > T_c$ case considered presently, $\widetilde{F}[T;M,N]$ and $F[T;M,N]$ defined in (10.17) are identical. (See Appendix H for related discussion.)

Next, let us consider a system at a temperature $T < T_c$. In this case, $\widetilde{F}[T;M,N]$ is a non-convex function of M, which qualitatively takes the form depicted in Figure 10.6(a). (The slopes of lines tangent to $\widetilde{F}[T;M,N]$, several of which are plotted in the figure, represent values of H.) First, we consider the regions external to the two minima. In these regions, due to the convexity of $\widetilde{F}[T;M,N]$, the situation is like that considered above in the $T > T_c$ case, and hence we regard all of the

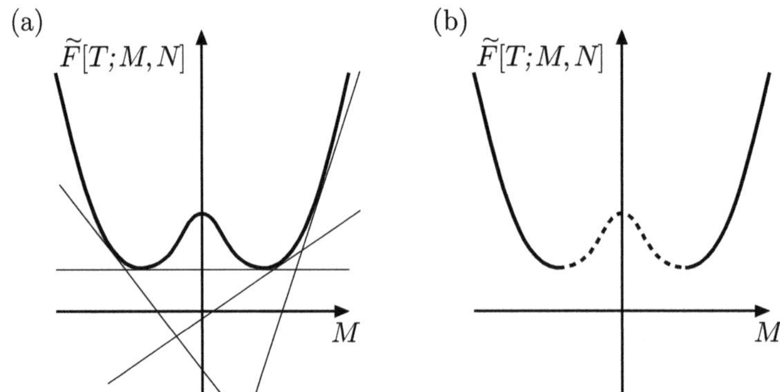

Figure 10.6 (a) Landau pseudo free energy as a function of M (with T and N fixed), along with lines tangent to it, at a temperature below T_c. (b) The dotted part of the graph is not contacted by any tangent lines. This implies that the behavior of $\widetilde{F}[T; M, N]$ in this region, which we postulate to be unphysical, is not reflected in the Legendre transform, $G[T, H; N]$. (Also, see Figure H.2(b) and related discussion in Appendix H.)

information possessed by $\widetilde{F}[T; M, N]$ in these regions to be physically meaningful. Next, we consider the region between the two minima of $\widetilde{F}[T; M, N]$. Because it is the form in this region that is responsible for the unphysical nature of $\widetilde{F}[T; M, N]$ as a free energy, we regard the information possessed by $\widetilde{F}[T; M, N]$ here to contain a physically meaningless component. In other words, we understand the form of $\widetilde{F}[T; M, N]$ here to consist of an underlying physically valid form with some physically meaningless "noise." The simplest assumption is that this physical form is that which can be obtained from the original with the smallest change resulting in a convex function. Apparently, the function obtained in this manner, which we write $\overline{F}[T; M, N]$, is identical to $\widetilde{F}[T; M, N]$ outside of the minima and coincides with the horizontal line connecting these minima in the region between them.[20,21]

Now, let us study how the Legendre transformation treats $\widetilde{F}[T; M, N]$ in the $T < T_c$ case. First, in the regions external to the minima, because the function there is convex, as in the $T > T_c$ case considered above, all information is passed on to the Legendre transform, $G[T, H; N]$. Next, let us consider the region of non-convexity between the minima. Because no point on the graph of $\widetilde{F}[T; M, N]$ in this region is contacted by any tangent line in the construction of $G[T, H; N]$, none of the information concerning the physically meaningless noise contained in $\widetilde{F}[T; M, N]$ is passed on to $G[T, H; N]$. Indeed, we could replace $\widetilde{F}[T; M, N]$ in the region between these minima with any form that maintains its non-convexity (more precisely, any form that remains above $\overline{F}[T; M, N]$ in this region), and the resulting function would be mapped to the same $G[T, H; N]$. Also, there is only one convex function that is mapped to this $G[T, H; N]$, namely, $\overline{F}[T; M, N]$ itself. We thus understand how the Legendre transformation preserves the physically meaningful information possessed by $\widetilde{F}[T; M, N]$ and removes the physically meaningless noise.

---

[20] Mathematically, $\overline{F}[T; M, N]$ is referred to as the convex hull of $\widetilde{F}[T; M, N]$. As described precisely in Theorem H.6 of Appendix H, $\overline{F}[T; M, N]$ is indeed the convex function "closest" to $\widetilde{F}[T; M, N]$.

[21] Recall that in Problem 7.8, we obtained a physical free energy from the van der Waals pseudo free energy in a similar manner.

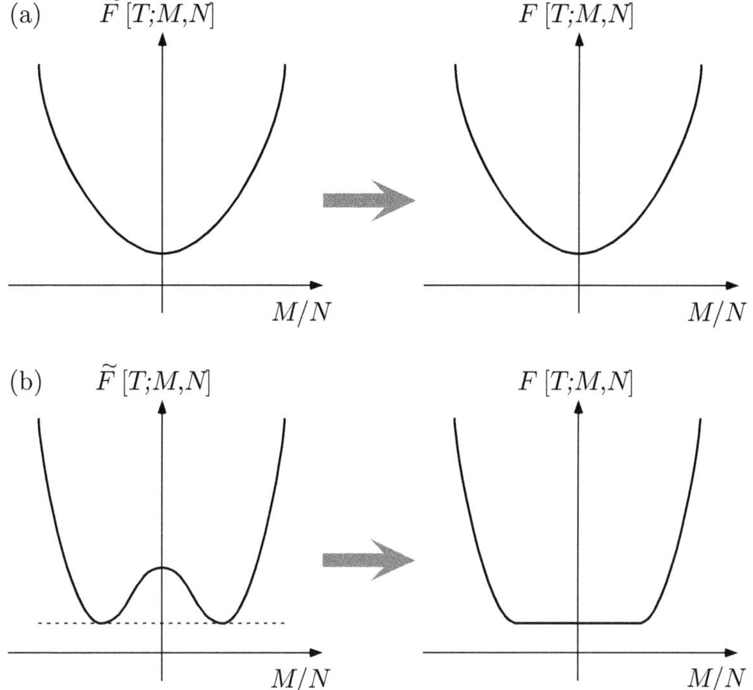

Figure 10.7 Applying the Legendre transformation and then its inverse to the Landau pseudo free energy $\widetilde{F}[T; M, N]$, we obtain a legitimate free energy $F[T; M, N]$, defined by (10.17), which is a convex function of M. (a) For $T > T_{\mathrm{c}}$, $\widetilde{F}[T; M, N]$ itself is convex, and for this reason, here $F[T; M, N]$ is identical to $\widetilde{F}[T; M, N]$. (b) For $T < T_{\mathrm{c}}$, $\widetilde{F}[T; M, N]$ and $F[T; M, N]$ differ. In this case, $F[T; M, N]$ given in (10.17) is obtained from $\widetilde{F}[T; M, N]$ by replacing that portion of $\widetilde{F}[T; M, N]$ between its two minima with the line segment that has these points as its endpoints. The "flat base" of the resulting free energy indicates the existence of spontaneous magnetization.

Furthermore, the facts that $\overline{F}[T; M, N]$ is convex and that its Legendre transform is $G[T, H; N]$ imply that for $T < T_{\mathrm{c}}$, the function $\overline{F}[T; M, N]$ is identical to $F[T; M, N]$ defined in (10.17). (See Appendix H for related discussion.)

From the above considerations, it is natural to interpret $F[T; M, N]$ as the functional form obtained when $\widetilde{F}[T; M, N]$ is stripped of its physically meaningless information. The relationship between $\widetilde{F}[T; M, N]$ and $F[T; M, N]$ both above and below T_{c} is depicted in Figure 10.7.

10.3.5 Analysis of the model

The role of the Landau theory

We now investigate the model of ferromagnetic phase transitions provided by the free energies defined above. Specifically, this is done by analyzing the Gibbs free energy given in (10.16). Before doing so, however, we give some discussion of the role played by the Landau theory in the modern theory of phase transitions and critical phenomena.

As stated above, the model of ferromagnetism based on the Landau pseudo free energy represents an attempt to obtain a basic understanding of ferromagnetic phase transitions and associated critical phenomena through treatment of the simplest possible model. As seen from the investigation presented in this section, this model does indeed provide a faithful qualitative description of the general features exhibited by ferromagnetic systems. However, as a quantitative model, the limitations of the Landau theory have been clearly demonstrated. Below, we derive the predictions provided by the Landau theory for the critical exponents β, γ and α, defined in (10.8), (10.9) and (10.10). It is known that these predictions, given in (10.24), are generally inconsistent with experimental results. (The latter depend on the particular conditions characterizing the system under investigation, but some representative examples are given in (10.25) and (10.26).) Although there are certain situations[22] in which the predictions of the Landau theory are quantitatively consistent with currently accepted experimental values, these situations are limited. Despite this fact, however, quantitative investigation of the Landau theory is meaningful because this theory is the foundation on which the more rigorous, more quantitatively successful theories are built.

General behavior

Above, we established that the pseudo free energy $\widetilde{F}[T; M, N]$ is not a proper free energy and proposed a method for obtaining a proper free energy, $F[T; M, N]$, from it. For this reason, keeping strictly with the rigorous framework of thermodynamic theory, we should amend the definition of $G[T, H; M]$ given in (10.16) by replacing $\widetilde{F}[T; M, N]$ with $F[T; M, N]$. However, for the purpose of investigating the Gibbs free energy of a ferromagnetic system, it is sufficient to continue using $\widetilde{F}[T; M, N]$ in (10.16), because, as explained above, the Legendre transforms of $\widetilde{F}[T; M, N]$ and $F[T; M, N]$ are identical. Furthermore, because $\widetilde{F}[T; M, N]$ has a simple functional form, using it for this purpose is advantageous. For this reason, we use it in the following. It should be kept in mind, however, that within this analysis, the status of $\widetilde{F}[T; M, N]$ is merely that of a convenient "stand in" for the more physically meaningful $F[T; M, N]$.

Before beginning a quantitative analysis, we qualitatively investigate how the minimum in (10.16) is realized. For any fixed temperature $T > T_{\mathrm{c}}$, the pseudo free energy $\widetilde{F}[T; M, N]$ behaves like the function plotted in Figure 10.8(a), possessing a single minimum located at $M = 0$. In this case, the quantity $\widetilde{F}[T; M, N] - HM$ has a unique minimum, which it realizes at a positive value of M for $H > 0$

---

[22] **Advanced note:** The discrepancy between the experimentally observed values of the critical exponents and the values predicted by the Landau theory can be attributed to the influence of fluctuations, which are dominant near the critical point. However, as we move away from T_{c}, the effect of fluctuations weakens, and as a result, the behavior of the system approaches that predicted by the Landau theory. Of course, if we move too far away from T_{c}, the perturbative expansion used to construct $\widetilde{F}[T; M, N]$ will break down, and thus the basic form of the Landau pseudo free energy itself will become invalid. The critical behavior observed in the $T \to T_{\mathrm{c}}$ limit is referred to as "asymptotic critical behavior," and the region in which this behavior appears is referred to as the "critical region." The region in which a system behaves in accordance with the Landau theory is termed the "mean-field region." The region between these two is termed the "crossover region." For many systems, in fact, there is no mean-field region, but systems in which mean-field behavior is observed are certainly not rare. For some types of systems (e.g., many materials that exhibit superconducting transitions), the critical and crossover regions are so narrow that these systems can be regarded as exhibiting mean-field critical behavior, and thus for these systems, the description provided by the Landau theory is sufficient to account for the observed behavior even quantitatively.

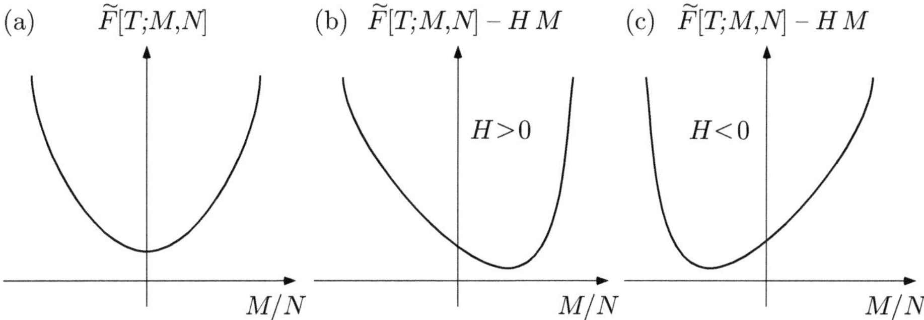

Figure 10.8 (a) The Landau pseudo free energy $\widetilde{F}[T; M, N]$ for $T > T_{\mathrm{c}}$, plotted as a function of M/N. (b) The quantity $\widetilde{F}[T; M, N] - HM$ plotted as a function of M/N in the case $H > 0$. (c) The same as (b), but in the case $H < 0$. In both (b) and (c), the minimum represents $G[T, H; N]$, and the value of M realized there is the magnetization of the system. In the case considered here of temperatures above T_{c}, $\widetilde{F}[T; M, N]$ and $F[T; M, N]$ are identical.

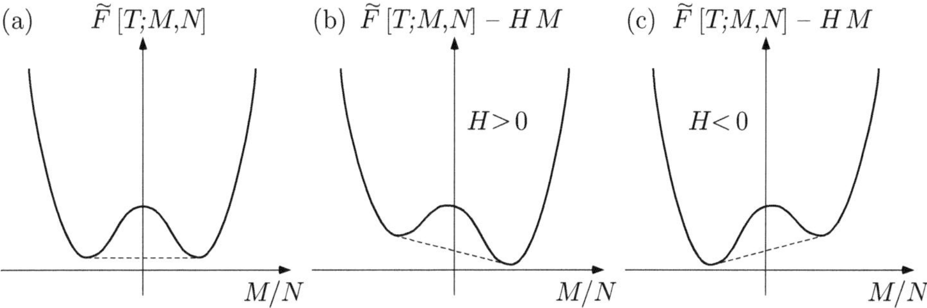

Figure 10.9 The same as Figure 10.8, but in the case $T < T_{\mathrm{c}}$. Here, it is the global minimum of $\widetilde{F}[T; M, N] - HM$ that represents $G[T, H; N]$. In each case, if the non-convex portion of the graph is replaced with the dotted line, the resulting graph is that of $F[T; M, N]$ or $F[T; M, N] - HM$. From these graphs, we see that (10.16) is indeed unchanged if we replace $\widetilde{F}[T; M, N]$ with $F[T; M, N]$.

(Figure 10.8(b)) and at a negative value of M for $H < 0$ (Figure 10.8(c)). Contrastingly, for $T < T_{\mathrm{c}}$, $\widetilde{F}[T; M, N]$ behaves like the function plotted in Figure 10.9(a), possessing two minima, one at a positive value of M and one at a negative value of M. The corresponding minima of $\widetilde{F}[T; M, N] - HM$ in the cases $H > 0$ and $H < 0$ are as depicted in Figures 10.9(b) and (c), respectively.[23]

The quantity $\widetilde{G}(T, H; M, N) = \widetilde{F}[T; M, N] - HM$, plotted in Figures 10.8 and 10.9 above and below the critical temperature, respectively, is also sometimes referred to as the Landau (pseudo) free energy. Note, however, that two of its arguments, H and M, cannot be controlled independently. For this reason, the function $\widetilde{G}(T, H; M, N)$ on its own is not physically meaningful. It has meaning

---

[23] As seen in (b) and (c), one of the minima of $\widetilde{F}[T; M, N] - HM$ is merely a local minimum. In the literature, this local minimum is sometimes interpreted as representing a metastable state. However, as we have argued, within thermodynamics this interpretation is erroneous. We stress that the situation here differs in an essential way from that for the system considered in Problem 7.6. (For related discussion in the context of statistical mechanics, see Footnote 24 on p. 258.)

only when acting as the operand of the min operator in (10.16), and then only to the extent that the Landau theory itself is meaningful.[24]

Phase transitions and critical phenomena

As mentioned in Section 10.1.3, for fixed H, the equilibrium value of M is that at which the minimum in the definition of the Gibbs free energy, (10.3), is realized. Therefore, considering our model based on the Landau pseudo free energy, in order to determine m, it is sufficient to consider the minimization condition $a(T - T_c)m + bm^3 - H = 0$, obtained from (10.16). Although this is a cubic equation and hence can be solved exactly, determining its exact solutions would not help in this analysis, because we seek only to understand the behavior of the magnetization in the neighborhood of $H = 0$.

We begin our analysis by setting $H = 0$ in the above minimization condition and considering the resulting equation, $a(T - T_c)m + bm^3 = 0$. For $T > T_c$, there is just one real root of this equation, $m = 0$, at which the minimum is realized, as seen in Figure 10.8(a). This demonstrates that above the critical temperature, the magnetization necessarily vanishes as $H \to 0$. Next, for $T < T_c$, this equation has the following three roots:

$$m = 0, \pm\sqrt{\frac{a}{b}}\,(T_c - T)^{1/2}\,. \tag{10.18}$$

In this case, as illustrated in Figure 10.9(a), the minimum in (10.16) is realized at the two nonzero roots. As discussed in Section 10.2.1 (and can be understood by considering Figures 10.9(b) and (c)), the positive root is realized in the case that H is slowly decreased to 0 from the positive side, and the negative root is realized in the case that H is slowly increased to 0 from the negative side. Thus, considering (10.7), we see that the spontaneous magnetization is given by

$$m_s(T) = \sqrt{\frac{a}{b}}\,(T_c - T)^{1/2}\,. \tag{10.19}$$

Comparing this relation with the definition of the critical exponent β given in (10.8), we find that for this system, we have $\beta = 1/2$.

Next, let us consider the case of small but nonzero H with $T > T_c$. In this case, m will also be small, and if we choose H sufficiently small, the third-order term in the equation $a(T - T_c)m + bm^3 - H = 0$ can be ignored. Under these conditions, this equation possesses the following approximate solution:

$$m(T, H) = \frac{H}{a(T - T_c)} + O(H^3)\,. \tag{10.20}$$

---

[24] **Advanced note:** In statistical mechanical studies, non-convex pseudo free energies $\widetilde{F}[T; M, N]$ and $\widetilde{G}(T, H; M, N)$ are often derived using a method called the "mean-field approximation." It is important to realize that such derivations always rely on ad hoc calculations employing the special nature of the approximation, and they do not fit within the standard, universal framework of statistical mechanics. As far as we know, there are no proper statistical mechanical definitions of $\widetilde{F}[T; M, N]$ or $\widetilde{G}(T, H; M, N)$ that apply universally to well-defined microscopic models of ferromagnets. Therefore, our assertion that $\widetilde{F}[T; M, N]$ and $\widetilde{G}(T, H; M, N)$ should not be regarded as legitimate free energies also applies within statistical mechanics.

With this, for the magnetic susceptibility we obtain

$$\chi(T) = \left.\frac{\partial m(T, H)}{\partial H}\right|_{H=0} = \frac{1}{a}(T - T_{\mathrm{c}})^{-1} \,, \qquad (10.21)$$

and thus, from (10.9), we find $\gamma = 1$.

Then, again considering the case $H = 0$, if we substitute the values of m at which the minima of $a(T - T_{\mathrm{c}})m + bm^3 = 0$ are realized for $T > T_{\mathrm{c}}$ and $T < T_{\mathrm{c}}$ into (10.16), we derive the following explicit form of the Gibbs free energy:

$$G[T, 0; N] = \begin{cases} N f_0(T) & \text{for } T \geq T_{\mathrm{c}} \,, \\ N\left\{f_0(T) - \frac{a^2}{4b}(T_{\mathrm{c}} - T)^2\right\} & \text{for } T \leq T_{\mathrm{c}} \,. \end{cases} \qquad (10.22)$$

With this expression, for the specific heat we obtain

$$c(T) = -\frac{T}{N}\frac{\partial^2 G[T, 0; N]}{\partial T^2} = \begin{cases} c_0(T) & \text{for } T > T_{\mathrm{c}} \,, \\ c_0(T) + \frac{a^2}{2b}T & \text{for } T < T_{\mathrm{c}} \,. \end{cases} \qquad (10.23)$$

Here, $c_0(T)$ is a smooth function of T given by $c_0(T) = -T f_0''(T)$. We thus see that the specific heat changes discontinuously at the critical temperature. Finally, comparing the above form of $c(T)$ with (10.10), we find $\alpha = 0$.

Summarizing the present analysis, we have obtained the following values of the critical exponents predicted by the Landau theory:

$$\alpha = 0 \,, \quad \beta = \frac{1}{2} \,, \quad \gamma = 1 \,. \qquad (10.24)$$

Interestingly, these values are independent of both the unknown function $f_0(T)$ and the constants a and b. This can be regarded as a strong point of the Landau theory. The values given in (10.24) are referred to as the *classical values* of these critical exponents. They serve as the starting point for the analysis of critical phenomena.

10.4 The scaling hypothesis and a first step beyond the Landau theory

Although the approach outlined above employing the Landau pseudo free energy does provide an intuitive understanding of phase transitions and critical phenomena, and it is able to capture the important qualitative aspects of such phenomena exhibited by actual systems, it is only a first step in their treatment, because, as noted previously, the values of the critical exponents α, β and γ that it predicts— the "classical" values given in (10.24)—are inconsistent with the values measured in experiments on magnetic materials. In this section, we consider experimental results for these scaling exponents, and we introduce a theoretical approach for their prediction that goes beyond the Landau theory.

Before proceeding, we note that because modern experiments used to determine critical exponents are extremely elaborate, we do not attempt to describe the actual methods involved. Instead, we merely report currently accepted results for two types of ferromagnetic systems.

10.4.1 The scaling relation for critical exponents

It has been found through the investigation of many types of ferromagnetic materials that, as one might expect, the value of the critical temperature varies significantly among them. However, despite this variation in T_c, the values of the critical exponents, which describe the singular behavior of systems near their critical points, exhibit a great degree of uniformity. More precisely, it has been found that all of the magnetic materials investigated to this time fall into a small number of categories, each characterized by a particular set of critical exponents. This remarkable property is known as the *quantitative universality* of critical exponents.[25] For an ordinary ferromagnetic material, in the case of uniaxial anisotropy, this combination is

$$\alpha \approx 0.11 \,, \quad \beta \approx 0.325 \,, \quad \gamma \approx 1.24 \,, \tag{10.25}$$

while in the highly isotropic case,[26] it is[27]

$$\alpha \approx -0.14 \,, \quad \beta \approx 0.38 \,, \quad \gamma \approx 1.375 \,. \tag{10.26}$$

Even more remarkably, the experimentally measured values of the critical exponents in both of these cases satisfy the following equation to a very high degree of precision:

$$\alpha + 2\beta + \gamma = 2 \,. \tag{10.27}$$

This kind of equation is called a *scaling relation*.[28] (Note that the classical exponents satisfy this equation exactly.) In addition, it has been found that the values of the critical exponents measured experimentally for other types of magnetic materials and those computed theoretically (both analytically and numerically) on the basis of statistical mechanical models[29] also all satisfy (10.27) to great precision.

The (quantitative) universality exhibited by the critical exponents hints at the existence of some kind of universal structure near critical points. The mathematical elucidation of this structure is an important topic in modern statistical mechanics

---

[25] The term "universality" is being used here with a somewhat narrower, more concrete meaning than in Section 1.2. Also, because it is being used with regard to quantitative behavior, its present meaning is stronger than that in Section 1.2. The underlying reason for this universality is believed to be something of truly profound significance.

[26] As stated in Footnote 2 on p. 241, although magnetization is a vector-valued quantity, we have been studying the case of materials possessing strong uniaxial anisotropy, for which it can be treated as a scalar. A highly isotropic magnetic material is one in which the magnetization can be along any direction, and thus for materials of this kind, it must be treated as a vector. Despite this slight complication, the thermodynamic treatment of highly isotropic systems is essentially the same as that given to this point for strongly anisotropic systems, and for this reason, we do not treat such systems explicitly.

[27] Because α in this case is negative, the specific heat does not diverge, but instead converges in a singular manner to a finite value at T_c.

[28] If we assume the existence of scaling exponents, the Rushbrooke inequality corresponding to (10.27), $\alpha' + 2\beta + \gamma' \geq 2$, can be derived rigorously using only the ordinary theory of thermodynamics (see Problem 10.5). This is one example of how thermodynamic considerations have played an important role in modern condensed matter physics. (With regard to the definitions of the critical exponents α' and γ', see Footnote 14 on p. 248.)

[29] A famous example is the two-dimensional Ising model, an idealized model of a ferromagnet, for which the exact values $\alpha = 0$, $\beta = 1/8$ and $\gamma = 7/4$ have been derived.

(and quantum field theory), and at the present time, it remains a great unsolved problem.[30]

10.4.2 The scaling hypothesis

Within the context of thermodynamics, the scaling hypothesis clearly characterizes the structure existing in systems near their critical points. In the remainder of this section, we discuss this hypothesis and its implications.[31]

For the purpose of the present investigation, we use the Gibbs free energy

$$\tilde{g}(\tau, H) = \frac{G[T_\mathrm{c} + \tau, H; N] - G[T_\mathrm{c}, 0; N]}{N} , \tag{10.28}$$

defined in reference to the critical point. The scaling hypothesis is the following: With sufficiently small τ and H, for any $\lambda \in [0, 1]$, the above Gibbs free energy satisfies the relation

$$\tilde{g}(\lambda^a \tau, \lambda^b H) \simeq \lambda\, \tilde{g}(\tau, H) . \tag{10.29}$$

Here, a and b are constants that are characteristic of the particular material under consideration. (More precisely, they are characteristic of the universal structure that governs the critical phenomena exhibited by that material.) These constants are called *scaling exponents*. Note that (10.29) is quite similar to relations like (3.24) that express the property of extensivity. Mathematically, (10.29) implies that $\tilde{g}(\tau, H)$ is (asymptotically) a generalized homogeneous function, and thus it is indeed similar in meaning to (3.24). However, while within the context of thermodynamics, extensivity is a completely universal property whose existence is nearly self-evident, the scaling relation (10.29) applies only to systems exhibiting critical phenomena in a very narrow neighborhood of the critical point, and the underlying reason for its (apparent) validity is still unclear.

In the remainder of this section, we simply assume the validity of the scaling hypothesis as expressed by (10.29) and investigate its implications for critical phenomena.

First, we replace τ with $-\tau$ in (10.29), take the derivative of each side with respect to H, and then consider the limit $H \searrow 0$. From (10.4), we have $m(T_\mathrm{c} - \tau, H) = -\partial \tilde{g}(-\tau, H)/\partial H$, and thus, using the definition of the spontaneous magnetization given in (10.7), we obtain the relation

$$\lambda^b\, m_\mathrm{s}(T_\mathrm{c} - \lambda^a \tau) \simeq \lambda\, m_\mathrm{s}(T_\mathrm{c} - \tau) . \tag{10.30}$$

---

[30] The classical exponents correspond to the mathematical structure of Gaussian fields. It is known that these values are realized in (theoretical) ferromagnetic systems of four or more spatial dimensions. Also, it has been demonstrated that the universal structure that governs the critical phenomena exhibited by (theoretical) two-dimensional ferromagnets (see Footnote 29 on p. 260) can be described by conformal field theory, a type of quantum field theory. For three-dimensional magnetic systems, the renormalization group approach has been applied with great success since the 1970s, and, more recently, an approach based on three-dimensional conformal field theory has been developed. However, obtaining a complete theoretical understanding of the critical phenomena exhibited by magnetic systems remains an elusive dream.

[31] For a detailed treatment, see Ref. [12].

Defining the quantity $\varepsilon = \lambda^a \tau$, this can be rewritten as

$$m_{\mathrm{s}}(T_{\mathrm{c}} - \varepsilon) \simeq \tau^{(b-1)/a} \, \varepsilon^{(1-b)/a} \, m_{\mathrm{s}}(T_{\mathrm{c}} - \tau) \, . \tag{10.31}$$

Considering the case of fixed, positive τ and writing $T_{\mathrm{c}} - \varepsilon$ as T, from the above we derive the following temperature dependence:

$$m_{\mathrm{s}}(T) \simeq (\text{constant}) \times (T_{\mathrm{c}} - T)^{(1-b)/a} \, . \tag{10.32}$$

We have thus reproduced the relation (10.8), describing the behavior of the spontaneous magnetization near the critical point, and we find that the critical exponent there is given by $\beta = (1 - b)/a$.

Next, taking the derivative of (10.29) twice with respect to H and then setting $H = 0$, from (10.9) we obtain

$$\lambda^{2b} \chi(T_{\mathrm{c}} + \lambda^a \tau) \simeq \lambda \chi(T_{\mathrm{c}} + \tau) \, . \tag{10.33}$$

This can be rewritten as follows:

$$\chi(T_{\mathrm{c}} + \varepsilon) \simeq \tau^{(2b-1)/a} \, \varepsilon^{(1-2b)/a} \, \chi(T_{\mathrm{c}} + \tau) \, . \tag{10.34}$$

Then, applying the same reasoning as above, from this expression we derive the value $\gamma = (2b - 1)/a$ for the critical exponent characterizing the magnetic susceptibility.

Finally, taking the derivative of (10.29) twice with respect to τ, using (10.10), and then repeating the manipulation applied above in the case of the magnetic susceptibility, we obtain the result $\alpha = (2a - 1)/a$ for the critical exponent characterizing the specific heat.

Collecting all of the above results, we have the following:

$$\alpha = \frac{2a - 1}{a} \, , \quad \beta = \frac{1 - b}{a} \, , \quad \gamma = \frac{2b - 1}{a} \, . \tag{10.35}$$

We are thus able to express all three of the critical exponents in terms of the two scaling exponents, a and b.[32] If indeed these expressions are valid, then the scaling relation (10.27) obviously holds.

The classical values of the critical exponents, appearing in (10.24), are realized with the scaling exponents

$$a = \frac{1}{2} \, , \quad b = \frac{3}{4} \, , \tag{10.36}$$

the experimental values for systems with uniaxial anisotropy, appearing in (10.25), are realized with the scaling exponents

$$a \approx 0.529 \, , \quad b \approx 0.828 \, , \tag{10.37}$$

---

[32] Let us point out here that there are several other critical exponents for thermodynamic quantities, in addition to α, β and γ, and these too can be expressed in terms of a and b alone (see Problem 10.6).

and the experimental values for highly isotropic systems, appearing in (10.26), are realized with the scaling exponents

$$a \approx 0.467 , \quad b \approx 0.822 . \tag{10.38}$$

Elucidating the physical mechanisms underlying the apparent validity of the scaling hypothesis and constructing a theory from which the experimental values of the scaling exponents given in (10.37) and (10.38) can be derived are two important problems in statistical mechanics.

The following is an even more mysterious result. From experimental data regarding the liquid-gas phase transition in fluid systems, it is known that in the neighborhood of the critical point (T_c, p_c), choosing the constants A, B, C and D appropriately, the Gibbs free energy

$$\tilde{g}(\tau, H) = \frac{G[T_c + A\tau + BH, p_c + C\tau + DH; N] - G[T_c, p_c; N]}{N} , \tag{10.39}$$

defined in reference to the critical point, also satisfies the scaling hypothesis expressed by (10.29). In addition, the values of the scaling exponents a and b for fluid systems are (with present-day experiments) indistinguishable from their values for ferromagnetic systems with uniaxial anisotropy, given in (10.37). These experimental results suggest that it may be possible to describe the critical phenomena exhibited by both ferromagnets and fluids in terms of the same universal structure. This is a truly remarkable possibility. Obtaining a theoretical understanding of this deep correspondence, which seemingly touches on the essence of scaling phenomena, is another important problem in statistical mechanics.

Problems 10

10.1 (Section 10.1) In this problem, we demonstrate (10.1), which determines the state function $H(T; M, N)$ in terms of the free energy $F[T; M, N]$, and acts as the starting point for the thermodynamic treatment of magnetic materials. However, rather than treating the general case, here we limit our consideration to a particular simple case.

Because we cannot directly control the magnetization, M, we consider the situation in which we use the magnetic field $\boldsymbol{H}$ created by an electromagnet to control M indirectly. Then, we determine the amount of work needed to carry out an operation in which we vary $\boldsymbol{H}$ in such a manner that M changes by a small amount.

Consider a sample of magnetic material in the form of a solid cylinder of radius r and length L around which is wrapped a solenoid of the same length and radius. Suppose that the solenoid has n loops per unit length. We have the following equations relating the magnetic field, $\boldsymbol{H}(\boldsymbol{r}, t)$, the magnetic flux density, $\boldsymbol{B}(\boldsymbol{r}, t)$, the magnetization per unit volume of magnetic material, $\boldsymbol{m}(\boldsymbol{r}, t)$, the electric field, $\boldsymbol{E}(\boldsymbol{r}, t)$, and the electric current density, $\boldsymbol{J}(\boldsymbol{r}, t)$:[33]

---

[33] Of course, more generally, the first of these relations is $\mathrm{rot}\,\boldsymbol{H} = \boldsymbol{J} + \partial \boldsymbol{D}/\partial t$, but here we consider the case in which the operation is carried out sufficiently slowly that the effect of the displacement current is negligible, and hence we can ignore the second term on the right-hand side. Also note that we are using MKS units.

$$\text{rot } \boldsymbol{H} = \boldsymbol{J} \,, \quad \text{rot } \boldsymbol{E} = -\frac{\partial}{\partial t}\boldsymbol{B} \,, \quad \text{div } \boldsymbol{B} = 0 \,, \quad \boldsymbol{B} = \mu_0(\boldsymbol{H} + \boldsymbol{m}) \,.$$

$$(10.40)$$

We consider the case in which L is much greater than r, and therefore we can ignore end effects. From symmetry considerations, we then conclude that $\boldsymbol{H}(\boldsymbol{r},t)$, $\boldsymbol{B}(\boldsymbol{r},t)$ and $\boldsymbol{m}(\boldsymbol{r},t)$ are all independent of position inside the solenoid and are directed along its axis. Let us write the magnitudes of these quantities as $H(t)$, $B(t)$ and $m(t)$. The magnetization $M(t)$ is given by $M(t) = Vm(t)$, where $V = \pi r^2 L$ is the volume of the sample.

We assume that the system is in an environment with fixed temperature. Then, with a constant current I passing through the solenoid, we write the values of $H(t)$, $B(t)$ and $M(t)$ realized in the equilibrium state as H, B and M.

Derive H for this equilibrium state. Then suppose that the current is subsequently changed slowly from I to $I + \Delta I$ during some time interval $[t_1, t_2]$, where ΔI is regarded as an infinitesimal quantity. Let us write the changes undergone by the other quantities due to this change in the current as ΔH, ΔB and ΔM. Determine the electromotive force (emf) $\mathcal{E}(t)$ induced in the solenoid between the times t_1 and t_2 in terms of $dB(t)/dt$. Then, with $\Delta W_{\text{tot}} = \int \mathcal{E}(t)\, I(t)\, dt$ denoting the work performed by the power source during this time interval to overcome the emf and maintain the current, derive the relation $\Delta W_{\text{tot}} = VH\Delta B + O\{(\Delta H)^2\}$. Next, confirm the relation $\Delta W_{\text{tot}} = \Delta(V\mu_0 H^2/2) + \mu_0 H\Delta M + O\{(\Delta H)^2\}$. The quantity $V\mu_0 H^2/2$ represents the energy of the magnetic field, and thus we find that the energy necessary to change the magnetic field is $\mu_0 H\Delta M$. Hence, noting that this change is induced by a quasi-static isothermal operation, we arrive at (10.1). (For discussion of the coefficient μ_0, see Footnote 5 on p. 243.)

10.2 (Section 10.1) From a statistical mechanical model of a system consisting of a particular type of idealized paramagnetic material,[34] the following Gibbs free energy is derived:

$$G[T, H; N] = -k_{\mathrm{B}} T\, N\, N_{\mathrm{A}} \log\left(2\cosh\frac{\mu H}{k_{\mathrm{B}}T}\right) . \qquad (10.41)$$

Here, μ is the magnetic moment of a spin (*not* the chemical potential), N_{A} is the Avogadro constant, and $k_{\mathrm{B}} = R/N_{\mathrm{A}}$ is the Boltzmann constant.

Derive the entropy $S(T, H; N)$, magnetization $m(T, H)$, and specific heat (at constant magnetic field) $c(T, H)$ for this system. Graph $m(T, H)$ as a function of H for fixed T, and $c(T, H)$ as a function of T for fixed $H > 0$.

10.3 (Section 10.1) Reconsider the situation described in the previous problem, and suppose that the relation $\frac{\mu H}{k_{\mathrm{B}}T} \ll 1$ holds. Show that to second order in $\frac{\mu H}{k_{\mathrm{B}}T}$, the Gibbs free energy is given by

---

[34] This model system consists of N moles of microscopic magnets referred to as "spins." Each of these magnets possesses a magnetic dipole moment that can take two values, $+\mu$ and $-\mu$. The states in which these values are realized are conventionally referred to as the "up" and "down" states of the spin.

$$
G[T, H; N] = -k_{\mathrm{B}} T N N_{\mathrm{A}} \left[\log 2 + \frac{1}{2} \left(\frac{\mu H}{k_{\mathrm{B}} T} \right)^2 \right] . \tag{10.42}
$$

Determine the entropy $S(T, H; N)$ and magnetization $m(T, H)$ for a system described by this Gibbs free energy. Then, derive the Helmholtz free energy $F[T; M, N]$ corresponding to this $G[T, H, N]$, and show that it possesses the form of the free energy given in (10.12).

10.4 (Section 10.1) The kind of quasi-static adiabatic operation considered in Problem 8.8 can be adapted to magnetic systems, in which case there exists a phenomenon known as *adiabatic demagnetization*. (To be precise, this should be referred to as *quasi-static adiabatic demagnetization*.) Here we investigate this phenomenon.

Let us consider a magnetic system in an environment with temperature T and magnetic field H. Suppose that after the system reaches the equilibrium state, it is enclosed within adiabatic walls, and then the magnetic field is slowly changed to some new value, H'. Explain how the final temperature of the magnetic material, T', is determined in the general case. Derive the value of T' for the system considered in Problem 10.3.

10.5 (Section 10.4) Show that by reinterpreting the relation between the heat capacity at constant volume and the heat capacity at constant pressure given in (8.43) as a relation characterizing a magnetic system, we obtain

$$
c(T) - c_{\mathrm{M}}(T; N m_{\mathrm{s}}(T), N) = \frac{T}{\chi(T)} \left\{ \frac{\partial m_{\mathrm{s}}(T)}{\partial T} \right\}^2 \tag{10.43}
$$

in the $H \searrow 0$ limit. Here, $c(T)$ is the specific heat (in the case $H = 0$) defined in (10.10), and $c_{\mathrm{M}}(T; M, N)$, given by

$$
c_{\mathrm{M}}(T; M, N) = \frac{1}{N} \frac{\partial U(T; M, N)}{\partial T} = -\frac{T}{N} \frac{\partial^2 F[T; M, N]}{\partial T^2} , \tag{10.44}
$$

is the specific heat under the condition of fixed magnetization (which cannot be measured experimentally). Because the energy, $U(T; M, N)$, is an increasing function of T, we know that $c_{\mathrm{M}}(T; M, N)$ must be positive. This implies the inequality

$$
c(T) \geq \frac{T}{\chi(T)} \left\{ \frac{\partial m_{\mathrm{s}}(T)}{\partial T} \right\}^2 . \tag{10.45}
$$

Considering the limit $T \nearrow T_{\mathrm{c}}$ and using the definitions of the critical exponents α', β and γ' (see Footnote 14 on p. 248) to rewrite (10.45), show that we obtain the Rushbrooke inequality,

$$
\alpha' + 2\beta + \gamma' \geq 2 . \tag{10.46}
$$

It is noteworthy that this inequality follows from only two assumptions: the existence of the critical exponents and the validity of thermodynamics in the present application.

10.6 (Section 10.4) The critical exponent δ is defined through the relation

$$m(T_{\rm c}, H) \sim H^{1/\delta} \tag{10.47}$$

describing critical phenomena in the $H \to 0$ limit. Show that the scaling hypothesis implies

$$\delta = \frac{b}{1 - b} . \tag{10.48}$$

Then, on the basis of this equality, derive scaling relations that include this critical exponent. (For clarity, let us point out that a "scaling relation" is an equality involving only critical exponents. In particular, it does not contain a or b.)

APPENDIX A

Supplement to Chapter 2

In Chapter 2, we established the foundation of thermodynamics as formulated in this book. In this appendix, we present some supplemental discussion and treat several technical details in order to complete the treatment given in Chapter 2.

A.1 The thermodynamic description

In this section, we present a brief discussion of the conceptual point of view from which we construct the theory of thermodynamics. This somewhat abstract discussion provides background for the concrete steps taken in the construction presented in Chapter 2.

In the broadest terms, the goal of thermodynamics is to construct a theory describing the universal structure observed in macroscopic systems. More specifically, the goal is to obtain a universal description of equilibrium states and transitions among them. Below, we briefly discuss how this goal is realized.

A.1.1 Intensive and extensive quantities

As discussed in Section 1.2, seeking theories that describe the universal structure inherent in physical systems is a fundamental approach to doing science. An important step in constructing such theories is deciding what aspects of the systems of interest to consider and what aspects to ignore. In the study of thermodynamic systems, the former are those that are described by intensive and extensive quantities. As discussed in Section 2.2.2, these are quantities that are characterized by the special ways in which they behave when the system size is varied: Intensive quantities and extensive quantities are those that remain constant and scale in proportion to the size of the system, respectively, when this is done.[1] We seek to construct a theory in which systems are described in terms of and investigated through manipulation of macroscopic intensive and extensive quantities. Quantities of these kinds reflect the nature of a system as a whole and are insensitive to system-specific details. By limiting the scope of our description to phenomena that can be understood through consideration of intensive and extensive quantities, we are able to uncover through the purely macroscopic approach described in this book a very general universality inherent in the behavior exhibited by macroscopic systems. This universality is that of equilibrium states. Employing a theory constructed in terms of intensive and extensive quantities, we develop an understanding of equilibrium states and transitions among them that applies to essentially all macroscopic systems.

A.1.2 The role of work

With our focus on intensive and extensive quantities, we take the stance that the universal description we seek can be constructed solely through consideration of mechanical operations and quantities that can be measured in them—most importantly, work. Thus, what we seek is a theoretical framework written in terms of intensive and extensive quantities that

---

[1] It is important to understand the meaning of varying the system size as imagined here. For example, consider the case in which the system size is increased by a factor of 10. This means that we combine the original system with nine identical systems to form a single system (without the performance of work). Extensive quantities are those that increase by a factor of 10 no matter how the original systems are combined, i.e., for any shape of the resulting system.

can account for the work performed in mechanical operations. In rough terms, this is our strategy. However, as discussed in Section A.2.4, we do not require our theory to be capable of accounting for the work performed in all types of operations, which in general could not be done in terms of only intensive and extensive quantities. Instead, we focus on the particular class of *quasi-static* operations, defined in Section A.2. Thus, in precise terms, we wish to construct a theory written in terms of intensive and extensive quantities characterizing equilibrium states that uniquely determine the work performed in quasi-static operations.[2] That there exists such a set of quantities sufficient for this purpose is a fundamental assumption.

A.1.3 Parameterization of equilibrium

Here we consider how the ideas outlined above are implemented in a concrete manner and how this implementation reflects the universality exhibited by thermodynamic systems.

When considering thermodynamic systems, there is a natural question regarding how the general condition of equilibrium should be partitioned into individual states. For any given equilibrium state, there are numerous measurable properties that could be regarded as relevant. How do we select a "proper" defining set from among these? This question is answered by the above considerations: We parameterize equilibrium states with an independent set of intensive and extensive quantities that uniquely determine the work performed in relevant quasi-static operations. Empirically, it is known that the collective extensive variable, X (introduced in Section 2.2.2), and the temperature, T (defined in Section 2.4.4), form one such set of quantities.[3] This is the parameterization used in this book.

While the above considerations are sufficient to identify the parameterization of equilibrium in a general sense, we have not yet identified the explicit makeup of X. To do this in any given case requires consideration of the system in question and the continuous extensive operations through which it is investigated, as we now explain.

As discussed in Section 2.3.2, continuous extensive operations are of essential importance within our construction of thermodynamics. The work measured in such operations is the fundamental quantity in terms of which we understand thermodynamic systems, as, indeed, this work provides us with complete thermodynamic functions, which incorporate complete thermodynamic information (see Appendix F). As defined in Section 2.3.2, each continuous extensive operation is carried out through the controlled variation of mechanically controllable extensive quantities. For a given system under investigation in a given situation, there will be some class of continuous extensive operations of interest. Then, we can choose a complete set of mechanically controllable extensive quantities that are independently variable under this class of operations.[4] The collective extensive variable, X, for this system in the present situation consists of such a set, along with the corresponding amounts of substance. In Chapters 1–8, we investigate single-component systems through application of continuous extensive operations in which, for an individual system, the total volume, V, is the controlled variable. In this case, the collective extensive variable is $X = (V, N)$, where N is the total amount of substance. The variables composing X in the cases of composite

---

[2] While we explicitly set out to construct a theory that is capable of accounting for the work performed in quasi-static operations (as determined by the choice of the quantities used in the parameterization of equilibrium), the theory that we obtain is in fact capable of much more, as is made clear throughout this book. This fact itself points to the fundamental nature of quasi-static work in the description of thermodynamic phenomena.

[3] Explicitly, this means that for any given system under given external conditions (isothermal or adiabatic), the amount of work performed by the system in a quasi-static operation is uniquely determined by the initial values of T and X and the final value of X. All properties of the system and operation not specified by these values are irrelevant with regard to the amount of work performed by the system.

[4] While, strictly speaking, there will always be some freedom in this choice, because there is redundancy among the extensive quantities characterizing a system, in practice there will generally be an obvious natural choice.

systems, multi-component systems (studied in Chapter 9), and magnetic systems (studied in Chapter 10) are easily determined through straightforward generalizations.

The fact that a parameterization of equilibrium states in terms of just a few intensive and extensive quantities is sufficient to account for the work performed in all quasi-static operations is the essence of the universality exhibited by thermodynamic systems. While operations can be distinguished according to their precise mechanical descriptions, it is an empirical fact uncovered through careful and exhaustive experimentation that in the quasi-static case, when we consider only work, many mechanically distinguishable operations are equivalent, and the quantities determining this equivalence are universal, few in number, and easily measured.

A.2 Quasi-static operations

A.2.1 Introduction

In this section, we introduce the class of operations that play the most fundamental role in thermodynamics, quasi-static operations. Intuitively, we can think of these simply as operations that are carried out very slowly. However, because of their importance, it is worthwhile defining these operations precisely.

The assumption underlying the concept of quasi-static operations is that if an operation is carried out very slowly, the system will effectively be in equilibrium throughout its performance, and for this reason, all behavior can be understood by considering only the equilibrium properties of the system.

While the above description provides a useful intuitive picture, it is too simple, as there exist both slow operations that induce non-slow processes (as discussed below) and slow operations that induce slow processes whose effects cannot be understood through consideration of equilibrium properties alone.[5] Thus, *quasi-static* should not be regarded as merely a synonym of *slow*.

A.2.2 Types of operations for which *slow* implies *quasi-static*

Some operations become quasi-static in the infinitesimally slow limit and some do not. In this section, we carry out a formal treatment that distinguishes these types. Before presenting that treatment, however, let us simply state the conclusion: Of the operations considered in this book (see Section 2.3), continuous extensive operations and operations that constitute the introduction of a constraint (i.e., operations in which a partitioning wall is inserted into a system and operations in which a system is surrounded with adiabatic walls) always become quasi-static in the infinitesimally slow limit, while operations that constitute the release of a constraint (i.e., operations in which a partitioning wall is extracted from a system, operations in which surrounding adiabatic walls are removed from a system, and operations in which a fixed internal wall is released and allowed to move freely) become quasi-static in the infinitesimally slow limit only under certain specific conditions.[6] These conditions are elucidated near the end of this section.

---

[5] One such example is considered in Problem 7.10, which regards the Joule-Thomson experiment. Although the processes induced in that experiment are slow, they yield effects (most notably, irreversible change—see Section 6.2.1) that cannot be described in terms of the equilibrium properties of the system.

[6] **Advanced note:** In this discussion, we have ignored operations under which work is performed without the variation of the extensive variables. Although (at least) some such operations could be regarded as becoming quasi-static in the infinitesimally slow limit, this is only because in this limit they become trivial. For example, consider an operation through which a fluid is stirred by an impeller. Under such an operation, even if the number of rotations of the impeller is fixed, the amount of work performed by the system vanishes in the infinitesimally slow limit, and hence this operation imparts no meaningful change to the system. For this reason, we never consider the infinitesimally slow limit of this class of operations.

We now proceed with the formal treatment that yields the above conclusion.

Whether or not an operation is quasi-static can be established by considering the influence that it has on the results of subsequent experiments (i.e., the work performed in subsequent continuous extensive operations). In fact, we use this as the defining characteristic of a quasi-static operation.

A quasi-static operation can be understood as an operation whose effect on a system is in some sense entirely controlled. There are no "residual," uncontrolled effects, because the behavior of the system under the operation is determined solely by equilibrium properties. For this reason, the influence of a quasi-static operation on subsequent experimental results can be "undone" by its reverse operation.[7] For a quasi-static operation, the combined operation consisting of this operation followed by its reverse has no effect on the results of subsequent experiments.

In order to formally identify the types of operations that become quasi-static in the infinitesimally slow limit, we first define the meaning of this limit. Consider an operation applied to a thermodynamic system.[8] Let us call this operation $\mathcal{O}$. Then, considering multiple realizations of this system, we can imagine a family of operations that consists of all operations identical to $\mathcal{O}$ except that each is carried out at a uniformly slower speed. We can parameterize this family of operations by, for example, the inverse of the completion time, τ^{-1}. Then, for $\mathcal{O}$, the limit of an infinitesimally slow operation is understood as the $\tau^{-1} \to 0$ limit applied to this family of operations. Throughout this book, when we say that we consider an infinitesimally slow operation, formally, we are imagining this kind of limiting procedure. The essential features of infinitesimally slow operations are that, for continuous extensive operations, the rate of work performed by the system vanishes, while for all other types of operations, the amount of work performed by the system vanishes.

Here, we describe an approach for investigating an arbitrary operation, $\mathcal{O}$, to determine whether it becomes quasi-static in the infinitesimally slow limit. In essence, this approach consists of applying $\mathcal{O}$ and then its reverse, $\mathcal{O}^*$, to a thermodynamic system and comparing the equilibrium states realized before and after.[9] Considering the infinitesimally slow limit of this sequence of operations, $\mathcal{O}\,\mathcal{O}^*$, we say that $\mathcal{O}$ becomes quasi-static in this limit if the equilibrium states before and after the application of $\mathcal{O}\,\mathcal{O}^*$ are identical.

With the basic idea as outlined above, let us first study the case of continuous extensive operations. (From this point, we use the notation $\mathcal{E}$ for operations specified as continuous extensive operations and $\mathcal{O}$ for operations of unspecified type.) Consider an arbitrary thermodynamic system in equilibrium under any fixed external conditions. Now, choose an arbitrary collection of an arbitrary number n of continuous extensive operations that can each be carried out on this system as prepared. Among these operations, any individual operation may appear any number of times. Then, assigning an arbitrary order to these operations, we consider a process in which the following sequence is carried out on the system: $\mathcal{E}_1 \mathcal{E}_1^* \mathcal{E}_2 \mathcal{E}_2^* \ldots \mathcal{E}_n \mathcal{E}_n^*$. (Here, it is understood that the operations are applied in the order from left to right.) From the definitions of a continuous extensive operation

---

[7] This idea is also treated in Sections 3.1.2 and 4.1.1.

[8] It is useful here to recall that for a given thermodynamic system, an operation is defined by the corresponding time variation of the mechanical world: Two mechanical manipulations applied to the same system are regarded as identical operations if and only if the degrees of freedom of the mechanical world evolve identically under them.

[9] For the sake of completeness, here we define the meaning of the *reverse* of an operation $\mathcal{O}$. For a continuous extensive operation and an operation under which an internal partitioning wall is removed from or inserted into a system, $\mathcal{O}^*$ is the operation under which the evolution of the mechanical world (and thus, the mechanically controllable degrees of freedom of the system) is the time-reversed counterpart to that under $\mathcal{O}$. For an operation under which the type of walls bounding the system (diathermal or adiabatic) is changed, $\mathcal{O}^*$ is an operation under which the initial and final conditions regarding these walls are reversed, with the stipulation that between the applications of $\mathcal{O}$ and $\mathcal{O}^*$, the (unique) external system with which the system makes thermal contact interacts with no other system. For an operation under which a constraint fixing the position of an internal partitioning wall is released or imposed, $\mathcal{O}^*$ is an operation under which the initial and final conditions regarding this constraint are reversed.

(see Section 2.3.2) and its reverse, it is clear that because each $\mathcal{E}_i$ can be carried out on the system in its original state, it can also be carried out in this sequence. Then, considering the situation in which a given operation appears multiple times, from empirical results, we know that the amounts of work that it performs each time become equal in the infinitesimally slow limit of the sequence. This implies that, in this limit, any change imparted to the system under a combined operation $\mathcal{E}_i \mathcal{E}_i^*$ is undetectable through investigation with any $\mathcal{E}_j$. We interpret this as meaning that for any infinitesimally slow continuous extensive operation $\mathcal{E}_i$ and any thermodynamic system in a state to which it can be applied, the states of the system before and after the combined operation $\mathcal{E}_i \mathcal{E}_i^*$ are identical. We therefore conclude that continuous extensive operations always become quasi-static in the infinitesimally slow limit.

Next, we use the above result for continuous extensive operations to devise a method for investigating the other types of operations. Let us consider an arbitrary thermodynamic system in an arbitrary equilibrium state, along with an operator $\mathcal{O}$ (of any type) and a continuous extensive operator $\mathcal{E}$ that can both be applied to the initial state of the system. Then, we consider an experiment in which the sequence $\mathcal{E} \mathcal{E}^* \, \mathcal{O} \mathcal{O}^* \, \mathcal{E} \mathcal{E}^*$ is applied to the system in the infinitesimally slow limit. The operation $\mathcal{O}$ is regarded as becoming quasi-static in this limit if this sequence of operations can be carried out (i.e., if the second application of $\mathcal{E}$ is possible), if the amounts of work performed in the two applications of $\mathcal{E}$ converge to the same value, and if this holds for any chosen $\mathcal{E}$ that can be applied to the same initial state.

The above procedure is sufficient as a general test to determine whether or not any operation becomes quasi-static in the infinitesimally slow limit. With this procedure, empirical results provide the conclusion stated above: Any continuous extensive operation and any operation constituting the implementation of some kind of constraint (as well as any combination thereof) always become quasi-static in the infinitesimally slow limit, while any operation constituting the release of a constraint becomes quasi-static in the infinitesimally slow limit only for systems prepared under certain specific conditions.[10]

Now, let us consider an arbitrary constraint-releasing operation. Then, let us reconsider the above procedure using this operation as $\mathcal{O}$, but in this case, let us suppose that $\mathcal{O}^*$ is carried out as the final step in the preparation of the system. More precisely, we consider a system whose initial state is prepared by applying $\mathcal{O}^*$ to an arbitrary equilibrium state. Empirically, it is found that in the case of such an initial state (and only in this case), $\mathcal{O}$ is always quasi-static. It is thus concluded that for a constraint-releasing operation, if and only if the initial state of the system can be realized as the final state under its reverse (i.e., an operation that imposes the constraint in question), then this operation becomes quasi-static in the infinitesimally slow limit.

A.2.3 Definition of a quasi-static operation

Having identified the types of operations that become quasi-static in the infinitesimally slow limit, we now present a practical test to determine whether or not a given operation or set of operations is carried out sufficiently slowly to be in the quasi-static regime. This test constitutes our operative definition of quasi-static operations.

Suppose that we wish to carry out an experiment that involves operations of various types, and at least some of these operations are intended to be (and are of types that can be) carried out quasi-statically (the "slow" operations). We imagine conducting this experiment, carrying out each of the slow operations slowly, and obtaining some results. Then, considering the repetition of this experiment,[11] if the results that we obtain would

---

[10] For example, consider the case in which a partitioning wall is removed. It is easy to imagine how if the conditions on the two sides of the wall are very different, even in the case that the wall is removed very slowly, this operation will induce non-slow processes in the system.

[11] For each repetition, the system is prepared identically (as determined by macroscopic mechanical measurements) and undergoes identical non-slow operations (as viewed from the mechanical world).

not change (to the desired precision) if any of the slow operations were carried out more slowly, then all of these slow operations are regarded as quasi-static.

A.2.4 Extensivity of the work performed in quasi-static operations

The following postulate is of fundamental significance.

Postulate A.1 (Extensivity of quasi-static work) *The work performed in any quasi-static operation is an extensive quantity.*[12]

It is important to note that the property described by the above postulate is limited to quasi-static operations; in general, the work performed in a non-quasi-static operation is neither extensive nor intensive. This is a fundamentally important realization in the construction of the theory of thermodynamics: Because we seek to construct a theory in terms of intensive and extensive quantities alone, this postulate leads us to conclude that in this construction, we should focus our attention on quasi-static operations, as discussed in Section A.1.2.[13]

A.3 Quantification of temperature: the Kelvin scale

A.3.1 Introduction

In Section 2.4.4, we defined the temperature as a physical quantity. Here we establish its quantification by defining the Kelvin scale. In fact, we present two definitions, consistent with the former and current definitions of the International System of Units, respectively. In practical terms, these two definitions of the Kelvin scale are equivalent. In essence, their only difference is that in the old definition, the triple point of water is defined with an exact value, while the Boltzmann constant is regarded as an experimentally determined quantity, while in the new definition, these roles are reversed.[14]

Let us note that there are various possible approaches to defining the Kelvin scale.[15] The approach used here, which is based on the ideal gas law, is suited to the construction of thermodynamics carried out in this book.[16]

---

[12] As a simple example, consider an operation carried out on an ordinary fluid system through which the volume is changed quasi-statically by a small amount dV. The amount of work performed in this operation is pdV, where p is the pressure of the system. Then, because p is intensive and V (and hence dV) is extensive, the quantity pdV is extensive.

[13] This postulate is not part of our formal construction of thermodynamics, as nowhere do we use it in the derivation of any results. Instead, it should be regarded as a conceptual guide, helping us to determine how the theory should be constructed. In fact, noting that this postulate is a combination of (3.15) and (the quasi-static case of) (4.18), we see that within the construction of the theory carried out in this book, it can be derived as a result.

[14] The new definitions of the International System of Units went into effect in May 2019.

[15] In addition to the macroscopic approaches considered here, there are also microscopic, statistical mechanical approaches.

[16] Because a true ideal gas is a theoretical construct that cannot be realized physically, it may seem that the definitions given here are not on firm footing. This is not the case, however. While the ideal gas equation of state is not exactly satisfied for any actual physical system, it is not difficult to realize systems whose deviation from the behavior described by this equation is arbitrarily small. For example, consider a system prepared by allowing the free expansion of a gas. For any gas and any given initial conditions, under such a process we can realize a system whose equation of state is arbitrarily close to that of an ideal gas by simply making its final volume sufficiently large. We regard the systems considered in this section to be prepared in such a manner, and the operations considered to be such that the ideal gas properties are maintained (to the desired precision) throughout their investigation.

Finally, note that these definitions do not rely on the *assumption* of any form for the equation of state. Rather, they simply take as input the universal limiting behavior found empirically when systems are prepared in the manner described above.

A.3.2 Definition in terms of the triple point of water

We wish to define a temperature scale that uniquely determines the temperature of any environment. We do this by first defining the ratio of the temperatures of two arbitrary environments and then fixing the numerical value of one particular temperature.

Let us consider an ideal gas system under isothermal conditions that is subject to an operation under which its volume is varied quasi-statically. We compare the amounts of work done by the system under this operation in the cases that its isothermal conditions are provided by two environments, Environment 1 and Environment 2. Let these amounts of work be denoted by W_1 and W_2, respectively. Empirically, we know that for any given Environments 1 and 2, the ratio W_1/W_2 is the same for any ideal gas system (as long as each environment provides isothermal conditions for it) and any quasi-static operation. With this observation, representing the temperature of Environment 1 by T_1 and the temperature of Environment 2 by T_2, we define the ratio T_1/T_2 by $T_1/T_2 = W_1/W_2$. Then, fixing the temperature of an environment at the triple point of water with the exact numerical value 273.16, we obtain a unique value for the temperature of each environment. The scale so obtained is the Kelvin temperature scale, expressed in units of kelvins (K).

A.3.3 Definition in terms of the Boltzmann constant

Let us consider an ideal gas system with amount of substance N (measured in moles) under isothermal conditions and consider a quasi-static operation carried out on this system with initial and final values of the volume V_i and V_f. We write the work performed by the system in the operation as W. Empirically, it is known that, for a given environment, the quantity $W/[N \log(V_f/V_i)]$ is the same for all quasi-static operations (with any V_i and V_f) and all ideal gas systems. With this observation, we define the temperature of an environment to be proportional to this quantity, and define its numerical value in units of kelvins as $T = W/[N N_A k_B \log(V_f/V_i)]$, where N_A is the Avogadro constant, given exactly by $6.02214076 \times 10^{23}$, and k_B is the Boltzmann constant, given exactly by $1.380649 \times 10^{-23}\,\mathrm{JK}^{-1}$.

A.3.4 "Ideal gas temperature" and "thermodynamic temperature"

In some textbooks, the temperature scale is not fixed until the theoretical structure of thermodynamics becomes clear, with the point of view that doing so allows for the most natural/convenient scale to be selected. In the development of the theory presented in this book, this would correspond to keeping the temperature scale undetermined until Carnot's theorem is introduced in Section 5.2.1, and then defining the temperature scale to be such that the simple form $f(T', T) = T'/T$ is realized in (5.20). However, it is our view that nothing is gained by doing this.[17] In this book, rather than pretending that we do not know that the Kelvin scale is the proper choice, we have made this choice at the outset, through the definitions presented in Sections A.3.2 and A.3.3. With this approach, Carnot's theorem simply confirms that our choice is the correct one.

Finally, let us take this opportunity to clear up a common misconception. In some works, definition of the temperature scale through the ideal gas equation of state and through the efficiency of the Carnot cycle are regarded as yielding distinct temperature scales, which are commonly given the names "ideal gas temperature" and "thermodynamic temperature,"

---

[17] The content of the presentation would be changed little if we were actually to use such an approach. To see this, suppose that instead of using T as quantified with the Kelvin scale we introduce a temperature variable τ quantified in an arbitrary manner, for example, in reference to some arbitrary thermometer(s). Then, if we assume that both of these quantifications allow for complete parameterizations of equilibrium states (see Sections 2.4.5 and A.1.3), they are equivalent in the sense that we can express T as $T(\tau)$ and τ as $\tau(T)$. Thus, any statement or mathematical expression written in terms of one of these variables can obviously also be rewritten in terms of the other in a straightforward manner.

respectively. However, there is no actual distinction here. As seen from Carnot's theorem, these are identical scales quantifying the same physical attribute. Furthermore, their identity is in no way uncertain or merely the result of a convenient choice of convention. Within the theory, that these two quantities are identical (up to the choice of a trivial overall constant factor) is a clear logical necessity. Also, it is not the case (as is often claimed) that the definition through reference to the Carnot cycle is somehow more general. Such an erroneous belief reflects confusion of the universal behavior described in Sections A.3.2 and A.3.3 with the behavior of actual devices that approximate it. In fact, the method of definition described in Sections A.3.2 and A.3.3, like that based on the Carnot cycle, is entirely general, employing exact, universal thermodynamic behavior without reference to any particular system or substance.

APPENDIX B

Uniqueness of the Entropy

In this appendix, we prove the uniqueness of the entropy, asserted in Section 6.2.[1]

The following states an important, general property of the entropy of a thermodynamic system.

Result B.1 (Uniqueness of the entropy) *Consider an arbitrary thermodynamic system, and suppose that some state function $\widetilde{S}(T; X)$ for this system satisfies the entropy principle[2] and the extensivity and additivity conditions*

$$\widetilde{S}(T; \lambda X) = \lambda\, \widetilde{S}(T; X) \tag{B.1}$$

and

$$\widetilde{S}(T; X, X') = \widetilde{S}(T; X) + \widetilde{S}(T; X') . \tag{B.2}$$

Then there exist a positive constant a and a constant b such that the quantity $\widetilde{S}(T; X)$ and the entropy $S(T; X)$ are related as

$$\widetilde{S}(T; X) = a\, S(T; X) + b \tag{B.3}$$

for all T and X. Furthermore, a is unchanged and b scales by a factor of λ when the system size is scaled by the same factor.

The meaning of the above result is that, up to a trivial normalization and shift of the zero point, $\widetilde{S}(T; X)$ and $S(T; X)$ are identical quantities. Hence, the entropy is in essence uniquely determined by the entropy principle and the conditions of extensivity and additivity. It can thus be said that this principle and these two conditions together capture the essential nature of the entropy.

In the following, $S(T; X)$ represents the entropy defined in (6.5), and $\widetilde{S}(T; X)$ represents an arbitrary state function that satisfies the conditions of Result B.1.

Suppose that the relation $S(T; X_1) < S(T; X_2)$ holds for some temperature T and two values X_1 and X_2 of the collective extensive variable that can be reached from one another operationally. Then, let us choose some numbers a and b satisfying the relations

$$\widetilde{S}(T; X_1) = aS(T; X_1) + b, \quad \widetilde{S}(T; X_2) = aS(T; X_2) + b . \tag{B.4}$$

By the entropy principle, an adiabatic operation inducing the transition $(T; X_1) \xrightarrow{\text{a}} (T; X_2)$ is irreversible. Thus, because $\widetilde{S}$ also satisfies the entropy principle, we have $\widetilde{S}(T; X_1) < \widetilde{S}(T; X_2)$, which implies $a > 0$.

Next, choose a state $(T; X)$ where X can be reached from X_1 under some operation. Let us first assume that the relation $S(T; X) \leq S(T; X_1)$ holds. We then determine a constant $\lambda \geq 0$ through the equality

$$(\lambda + 1)S(T; X_1) = \lambda S(T; X_2) + S(T; X) . \tag{B.5}$$

---

[1] The derivation given in this appendix consists essentially of the reasoning used by Lieb and Yngvason [9], adapted to the theoretical construction of thermodynamics used in this book. Our derivation is also partly based on an unpublished work by Shin-ichi Sasa and Katsuhiko Sato, in which they carry out an elementary rewriting of the definition of the entropy given in Lieb and Yngvason's work.

[2] In other words, Result 6.5 (p. 117) holds in the case that $S(T; X)$ is replaced with $\widetilde{S}(T; X)$.

Regarding the left-hand side and right-hand side of this equation as the entropies of a composite system in the two states $(T; \lambda X_1, X_1)$ and $(T; \lambda X_2, X)$, and using Result 6.2 (p. 111), we conclude that there exist quasi-static adiabatic operations under which the transitions

$$(T; \lambda X_1, X_1) \xleftrightarrow{\text{qa}} (T; \lambda X_2, X) \tag{B.6}$$

are realizable.

The realizability of the transitions in (B.6) is expressed in terms of the entropy principle in application to the state function $\widetilde{S}$ as

$$(\lambda + 1)\widetilde{S}(T; X_1) = \lambda \widetilde{S}(T; X_2) + \widetilde{S}(T; X) . \tag{B.7}$$

Solving (B.5) and (B.7) for λ and equating the results yields

$$\frac{\widetilde{S}(T; X_1) - \widetilde{S}(T; X)}{\widetilde{S}(T; X_2) - \widetilde{S}(T; X_1)} = \frac{S(T; X_1) - S(T; X)}{S(T; X_2) - S(T; X_1)} . \tag{B.8}$$

Substituting (B.4) into this equation and simplifying, we obtain the relation that we seek, (B.3).

In the case of X for which the relation $S(T; X) > S(T; X_1)$ holds, instead of operations inducing the transitions in (B.6), we consider quasi-static adiabatic operations inducing the transitions

$$(T; \lambda X_2, X_1) \xleftrightarrow{\text{qa}} (T; \lambda X_1, X) . \tag{B.9}$$

Then, repeating the arguments given above with respect to these operations, we again obtain (B.3).

The desired result has thus been obtained for all states with a given value of T. Then, using the implication of the entropy principle that $S(T; X)$ and $\widetilde{S}(T; X)$ are unchanged under any quasi-static adiabatic operation, we immediately obtain the desired result for all values of T. Finally, with the assumption of extensivity, Result B.1 is proved.

Upper Limit on the Efficiency
of a Heat Engine

In Section 5.3, we presented Result 5.4 (p. 101), due to Carnot, which establishes the universal upper limit on the efficiency of a heat engine. Here, using the entropy principle for compound states discussed in Section 6.5, we rigorously prove Carnot's upper limit, asserted in (5.51), for situations much more general than those considered in Section 5.3. The main idea used in this proof is due to Lieb and Yngvason [9].

We study a system with collective extensive variable X acting as a heat engine. We treat its heat baths not as fixed-temperature environments, but as ordinary thermodynamic systems, with collective extensive variables Y and Z, respectively. However, because we can take the Y and Z systems to be as large as we wish, the treatment here obviously also applies to the case in which these systems act as effectively infinitely large, fixed-temperature environments.

Now, let us consider an arbitrary thermodynamic cycle of the following kind. With all three systems under adiabatic conditions, the engine and the heat baths begin in equilibrium states $(T; X)$, $(T_\mathrm{L}; Y)$ and $(T_\mathrm{H}; Z)$. Next, the engine undergoes an operation during which it is in thermal contact with either or both heat baths at arbitrary times and its collective extensive variable is varied in an arbitrary manner. At the end of the cycle, the engine is returned to its original equilibrium state, and again all systems are placed under adiabatic conditions. This cycle, no matter how complicated it may be, can be expressed as a single adiabatic transition:

$$\{(T_\mathrm{L}; Y)|(T; X)|(T_\mathrm{H}; Z)\} \xrightarrow{\text{a}} \{(T_\mathrm{L}'; Y)|(T; X)|(T_\mathrm{H}'; Z)\} \,. \tag{C.1}$$

The vertical bars separating the states in this expression indicate that these are equilibrium states of separate systems, separated by adiabatic walls (see Section 6.5). Note that although the engine returns to its initial state under this operation, the two heat baths do not, due to the exchange of heat with the engine. The heat absorbed by the cold heat bath (the Y system) is $Q_\mathrm{L} = U(T_\mathrm{L}'; Y) - U(T_\mathrm{L}; Y)$, and the heat expelled by the hot heat bath (the Z system) is $Q_\mathrm{H} = U(T_\mathrm{H}; Z) - U(T_\mathrm{H}'; Z)$. Thus, from the law of energy conservation, we conclude that the amount of work done by the system on the mechanical world during this one-cycle adiabatic operation is $W = Q_\mathrm{H} - Q_\mathrm{L}$. In the following, in order to ensure that the X system does indeed function as an engine, we assume the relations $Q_\mathrm{L} > 0$ and $Q_\mathrm{H} > 0$. Hence, we have $T_\mathrm{H} > T_\mathrm{H}'$ and $T_\mathrm{L} < T_\mathrm{L}'$.

From the entropy principle for compound states, Result 6.8 (p. 129), we know that the entropy will not decrease in an adiabatic operation. Therefore we have

$$S((T; X)|(T_\mathrm{L}; Y)|(T_\mathrm{H}; Z)) \leq S((T; X)|(T_\mathrm{L}'; Y)|(T_\mathrm{H}'; Z)) \,. \tag{C.2}$$

Combining this with the additivity of the entropy, expressed by (6.58), we obtain

$$S(T_\mathrm{L}; Y) + S(T_\mathrm{H}; Z) \leq S(T_\mathrm{L}'; Y) + S(T_\mathrm{H}'; Z) \,, \tag{C.3}$$

which we rewrite as

$$S(T_{\mathrm{H}}; Z) - S(T_{\mathrm{H}}'; Z) \leq S(T_{\mathrm{L}}'; Y) - S(T_{\mathrm{L}}; Y) \,. \tag{C.4}$$

Next, assuming that $S(T; Z)$ is differentiable with respect to T for temperatures satisfying $T_{\mathrm{H}}' \leq T \leq T_{\mathrm{H}}$ and using (6.15), the left-hand side of (C.4) can be evaluated as follows:

$$S(T_{\mathrm{H}}; Z) - S(T_{\mathrm{H}}'; Z) = \int_{T_{\mathrm{H}}'}^{T_{\mathrm{H}}} dT\, \frac{\partial S(T; Z)}{\partial T} = \int_{T_{\mathrm{H}}'}^{T_{\mathrm{H}}} dT\, \frac{1}{T}\, \frac{\partial U(T; Z)}{\partial T}$$
$$\geq \frac{1}{T_{\mathrm{H}}} \int_{T_{\mathrm{H}}'}^{T_{\mathrm{H}}} dT\, \frac{\partial U(T; Z)}{\partial T} = \frac{1}{T_{\mathrm{H}}} \{U(T_{\mathrm{H}}; Z) - U(T_{\mathrm{H}}'; Z)\} = \frac{Q_{\mathrm{H}}}{T_{\mathrm{H}}} \,. \tag{C.5}$$

In this derivation, we have used the fact that the relation $T \leq T_{\mathrm{H}}$ holds for the integrand. Similarly, for the right-hand side of (C.4), we obtain

$$S(T_{\mathrm{L}}'; Y) - S(T_{\mathrm{L}}; Y) \leq \frac{Q_{\mathrm{L}}}{T_{\mathrm{L}}} \,. \tag{C.6}$$

Together, (C.4), (C.5) and (C.6) yield the relation

$$\frac{Q_{\mathrm{L}}}{Q_{\mathrm{H}}} \geq \frac{T_{\mathrm{L}}}{T_{\mathrm{H}}} \,. \tag{C.7}$$

Combining this with the definition of the efficiency given in (5.43), we obtain the universal upper limit on the efficiency of a heat engine expressed by (5.51).

We emphasize that in the above derivation, at no point is it assumed that the heat baths are larger than the engine or that the temperatures of the heat baths are fixed. Also, there is no assumption that any of the systems are in equilibrium at any point between the initial and final states. For these reasons, the result derived here is much more powerful than that derived in Section 5.3.

Next, we argue that the result (C.7) also applies to a heat engine undergoing many identical cycles without reaching equilibrium between. Such a continuously running engine is much more realistic than the engines treated in the main text, which all begin in an equilibrium state and reach an equilibrium state after each cycle. To treat a continuously running engine within the present formulation, we imagine that initially the entire system consisting of the engine and the heat baths is in an equilibrium state, then it undergoes many cyclic operations (without reaching equilibrium), and finally it settles into a new equilibrium state. We regard this entire process as constituting a single transition represented by (C.1). Let us assume that the heat baths are sufficiently large that their temperatures remain effectively unchanged during the entire process. Then, within this process, there will generally be initial and final transient periods and some period of steady running between. If the number of cycles is sufficiently large that the effect of the transient periods is negligible, the efficiency for the whole process will be essentially the same as that for each individual cycle of the steady running period. We thus conclude that the upper bound (C.7) also applies to the case of a continuously running engine.

The broad applicability of the rigorous result (C.7) demonstrated here represents one example of the great usefulness of the entropy principle for compound states.

Thermodynamics
of the Triple Point

In Section 7.5.4, we briefly discussed the special point in the phase diagram of a pure substance at which the solid, liquid and gas phases all coexist. This is called the *triple point* (see Figure 7.7). As pointed out several times in the main text, our description of the equilibrium states of a single-component system in terms of the $(T; V, N)$ parameterization is insufficient for specifying states at the triple point.[1]

In this appendix, we investigate the nature of the triple point in some detail, and we show how, with a slight modification, our theory is capable of describing the full set of equilibrium states corresponding to this point. Throughout this appendix, we treat the case of an individual system consisting of a single substance.

D.1 Equilibrium states within the triple-point region

We study a system in which the solid, liquid and gas phases can coexist in equilibrium at some temperature T_3 and pressure p_3, which together specify the triple point. For these three phases at this temperature and pressure, we represent the volume per unit amount of substance by v_S, v_L and v_G and the energy per unit amount of substance by u_S, u_L and u_G.

Let us consider an equilibrium state at the triple point for which the amounts of substance in the solid, liquid and gas phases are given by N_S, N_L and N_G, respectively. We regard the total amount of substance,

$$N = N_S + N_L + N_G \, , \tag{D.1}$$

to be fixed when the system is prepared. Then, from the additivity of the volume and energy, the total volume and energy in this state are given by[2]

$$V = v_S N_S + v_L N_L + v_G N_G \, , \quad U = u_S N_S + u_L N_L + u_G N_G \, . \tag{D.2}$$

Now, in the V-U plane, the set of all states characterized by the pressure p_3, temperature T_3, and amount of substance N form the triangular region displayed in Figure D.1. We refer to this region in the V-U plane as the "triple-point region." In this region, the values of V and U are determined as in (D.2) by the values of N_S, N_L and N_G, with which they vary continuously. As seen from the figure, choosing some volume V_0 between the extreme points on the left-hand and right-hand sides (corresponding to $N_S = N$, $N_L = N_G = 0$ and $N_G = N$, $N_S = N_L = 0$), there exists a range of values of U falling within the triple-point region. Clearly, with the parameterization in terms of $(T; V, N)$, all states within this range correspond to $(T_3; V_0, N)$. It is thus seen that in the triple-point region, the parameterization in terms of $(T; V, N)$ is indeed insufficient.

---

[1] To understand the precise meaning of this statement, see Sections 2.4.5 and A.1.3.
[2] These relations can be determined empirically from the work performed in adiabatic operations.

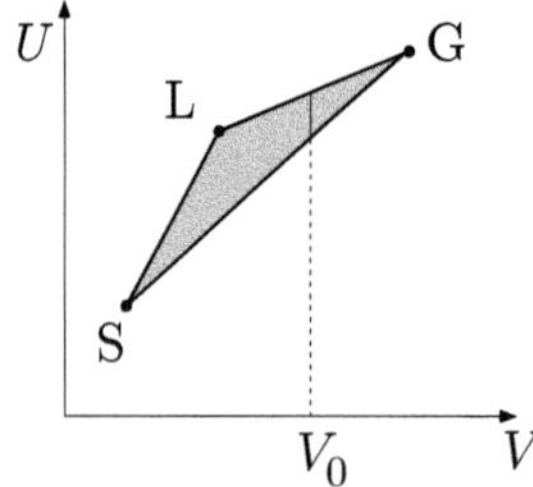

Figure D.1 In the V-U plane, the triple point corresponds to a triangular region, as shown here. The points labeled "S," "L" and "G" indicate the states on the boundary of this region in which the system consists entirely of a solid, liquid and gas, respectively. The coordinates of these three extremal points are $(Nv_{\mathrm{S}}, Nu_{\mathrm{S}})$, $(Nv_{\mathrm{L}}, Nu_{\mathrm{L}})$ and $(Nv_{\mathrm{G}}, Nu_{\mathrm{G}})$. All states lying on the vertical line drawn in the graph have volume V_0, and thus they all correspond to $(T_3; V_0, N)$. This demonstrates that the $(T; V, N)$ parameterization does not uniquely specify equilibrium states at the triple point.

D.2 Phase diagrams

Figure D.2 describes the neighborhood of the triple point through three schematic phase diagrams corresponding to different parameterizations of equilibrium states.

The phase diagram in Figure D.2(a) employs the $(T, p; N)$ parameterization, with the horizontal and vertical axes representing $1/p$ and T.[3] (This figure is essentially the same as Figure 7.7.) With this parameterization, the two-phase coexistence regions are reduced to curves, which in the figure are represented by line segments (labeled "S+L," "S+G" and "L+G") for simplicity. The three-phase coexistence region is reduced to the single point constituting the common endpoint of these three curves. The solid-phase, liquid-phase and gas-phase regions (indicated by "S," "L" and "G") are separated by the phase-coexistence curves. This is the most conventional type of phase diagram, but in it, only states characterized by a pure solid, pure liquid or pure gas phase are represented unambiguously. Every point on the phase-coexistence curves represents multiple equilibrium states.

The phase diagram in Figure D.2(b) employs the $(T; V, N)$ parameterization, with the horizontal and vertical axes representing V and T. Here, the positions of the solid-phase, liquid-phase and gas-phase regions are similar to those in (a), but in this case, the two-phase coexistence regions separating these single-phase regions are no longer represented by curves, but instead by two-dimensional regions. With this parameterization, all equilibrium states other than those in the three-phase coexistence region are unambiguously represented by distinct points in the phase diagram. The three-phase coexistence region is represented by a line segment, on which each point corresponds to multiple equilibrium states.[4] Hence, as pointed out above, the specification of states provided by this parameterization is also incomplete.

The phase diagram in Figure D.2(c) employs the (U, V, N) parameterization, with the horizontal and vertical axes representing V and U. In this case, there is a one-to-one correspondence between equilibrium states and points in the phase diagram. The important difference between this phase diagram and that in (b) is that in the present case, the three-phase coexistence region is faithfully represented by a two-dimensional region, the triangular region indicated by "S+L+G."[5]

---

[3] We have used $1/p$ instead of p here so that the positions of the phase regions within the phase diagram are similar to those of the corresponding regions in the other phase diagrams, in which the horizontal axis corresponds to V.

[4] In the V-T plane, this necessarily is a horizontal line segment.

[5] That this is actually a triangular region in the V-U plane follows from (D.2).

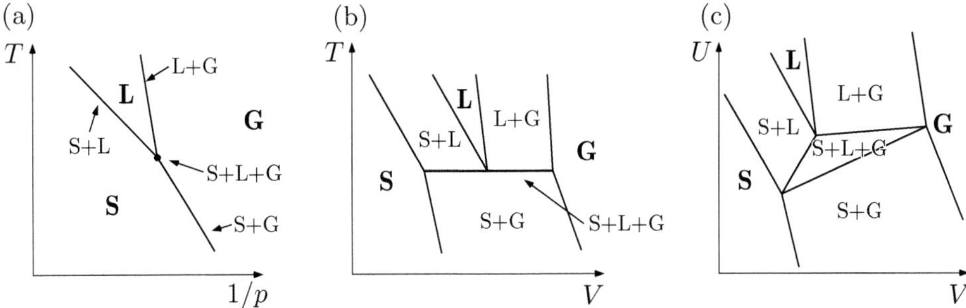

Figure D.2 Schematic phase diagrams in the neighborhood of the triple point. (a) In the $1/p$-T plane, the phase diagram is essentially the same as the more familiar diagram appearing in Figure 7.7. With this parameterization, all two-phase coexistence regions are represented by curves (here, line segments for simplicity), and the three-phase coexistence region is represented by a single point. (b) In the V-T plane, the two-phase coexistence regions are faithfully depicted as two-dimensional regions in the phase diagram, but the three-phase coexistence region is represented by a line segment. (c) In the V-U plane, there exists a one-to-one correspondence between equilibrium states and points in the phase diagram. Thus, in this case, we have a completely unambiguous representation of all equilibrium states.

D.3 Operational thermodynamics in the triple-point region

As discussed above, the standard parameterization in terms of T, V and N does not distinguish individual states within the triple-point region. In order to address this point, we introduce the following parameterization, which accounts for all states of a system consisting of a single substance:

$$
(T; X) := \begin{cases} (T; V, N, N_S) & \text{for } T = T_3 \text{ and } \dfrac{V}{N} \in (v_S, v_G), \\ (T; V, N) & \text{otherwise}. \end{cases} \tag{D.3}
$$

The quantity N_S here is the amount of substance in the solid phase. Thus, in order to specify all equilibrium states under all conditions, N_S is added as a parameter when the system is inside the triple-point region. Although this parameterization is in some sense ad hoc, and mathematically somewhat singular, it is the simplest way within our operational approach to realize a complete thermodynamic description of a state space that includes the triple point.[6] Of course, as discussed above, we know that the parameterization in terms of just the three quantities U, V and N is also generally sufficient to specify all states, even within the triple-point region. It is thus apparent that the treatment of the triple point is theoretically simpler when using the (U, V, N) parameterization.

Now, let us reconsider the construction of thermodynamics presented in this book, this time using the complete parameterization (D.3). Obviously, outside of the triple-point region, there is no change, and we therefore only need to investigate the construction within the triple-point region.

Consider a system in an arbitrary state inside the triple-point region. Then, suppose that with the system surrounded by adiabatic walls, it is subject to an operation through which the volume is varied quasi-statically by an amount ΔV. Recalling that the pressure takes the constant value p_3 throughout the triple-point region, we see that the work performed by the

---

[6] In general, a treatment of this kind is necessary for any system whose state space contains a region in which three or more phases coexist.

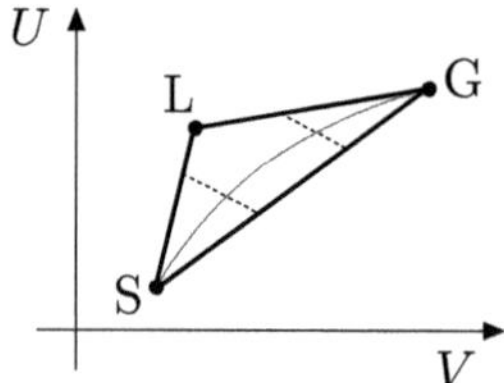

Figure D.3 Schematic depiction of paths in the V-U plane followed by the state of the system under quasi-static processes within the triple-point region. Under a quasi-static adiabatic operation, the state follows a path of constant slope $-p_3$. Under a quasi-static isothermal operation, the state follows a curve connecting the corner points S and G that border the gas-phase and solid-phase regions, respectively.

system in this process is simply $W = p_3 \Delta V$ for any ΔV, as long as the system remains within the triple-point region throughout the operation. Therefore, because this process is adiabatic, the change in the energy of the system is simply $\Delta U = -W = -p_3 \Delta V$. Viewed in the V-U plane, during this operation the state of the system traces out a trajectory along a line of constant slope $-p_3$. We thus see how through the variation of V, the state can be made to traverse the entire triple-point region along this line in either direction. The dotted line segments in Figure D.3 represent two such trajectories.[7]

Let us next suppose that the system is in contact with an environment at temperature T_3.[8] We again start from an arbitrary state within the triple-point region and vary the volume quasi-statically. Considering the phase diagram in Figure D.2(b), we see that the corresponding trajectory of the state in the V-T plane lies along the horizontal segment labeled "S + L + G." Then, considering Figure D.2(c), it is understood that in the V-U plane, this trajectory follows a curve that connects the left and right corners of the triple-point region bordering the solid-phase and gas-phase regions, respectively. A curve of this type is drawn in Figure D.3.[9] Clearly, the work performed by the system in this operation is again $W = p_3 \Delta V$.

From the above discussion, it is clear that by appropriately combining quasi-static adiabatic and isothermal operations, we can cause the system to make a transition between any two states in the triple-point region while following a trajectory that remains entirely inside this region. Because the temperature of the system is always T_3 during such operations, we can use them to determine the Helmholtz free energy within the triple-point region. From the fact that the work performed by the system in an operation of the type under consideration is always $p_3 \Delta V$, it follows that the change in the Helmholtz free energy is $\Delta F = -p_3 \Delta V$ in all cases.[10] This implies that within the triple-point region, the Helmholtz

---

[7] Let us note that although for visualization purposes, we are considering these processes in the V-U plane, the (U, V, N) parameterization of equilibrium states is not necessary to carry out the treatment discussed here. Obviously, this entire treatment could also be carried out in terms of the parameterization (D.3).

[8] Because the temperature of the system is constant for both types of operations considered here, it may seem that the characterizations "adiabatic" and "isothermal" do not apply in the triple-point region as elsewhere. However, we emphasize that the definitions of adiabatic and isothermal conditions presented in Chapter 2 are entirely general, unambiguously distinguishing these two types of operations even within the triple-point region.

[9] The precise forms of these isothermal curves cannot be determined from the principles of thermodynamics alone. In fact, as far as the authors are aware, these forms are not known.

[10] **Advanced note**: To be completely rigorous, this construction requires some extra care. The definition of the Helmholtz free energy (see Section 3.6.1) is based on the principle of maximum work, Result 3.3 (p. 52), whose proof relies on the fact that a quasi-static isothermal transition can be exactly reversed. Within the triple-point region, however, this is generally not possible, because in this case, we cannot precisely control $X = (V, N, N_S)$. (Specifically, we do not have control over N_S during such an operation.) Nevertheless, it can be proved that the principle of maximum work holds even within

free energy is determined uniquely by V and N, and is independent of N_S.[11] For this reason, we write it simply as $F[T_3; V, N]$, rather than $F[T_3; V, N, N_S]$.

Using the extensivity of the Helmholtz free energy (see (3.24)) and the fact that the relation $\partial F[T_3; V, N]/\partial V = -p_3$ holds for any V and N satisfying $v_S \leq V/N \leq v_G$, we find that within the triple-point region, $F[T_3; V, N]$ is given by

$$F[T_3; V, N] = -p_3 V + \mu_3 N \quad \text{for } \frac{V}{N} \in [v_S, v_G] \,, \tag{D.4}$$

where the (undetermined) quantity μ_3 is interpreted as the chemical potential, which (like the pressure) is constant within the triple-point region.

With the expression for the energy within the triple-point region, $U(T_3; V, N, N_S)$, given by (D.2),[12] from (D.1), (D.2) and (D.4), along with the definition (6.5), we find that the entropy within the triple-point region is given by

$$S(T_3; V, N, N_S) = \frac{U(T_3; V, N, N_S) - F[T_3; V, N]}{T_3}$$

$$= s_S N_S + s_L N_L + s_G N_G \,, \tag{D.5}$$

where the quantities s_S, s_L and s_G are defined as[13]

$$s_S = \frac{u_S + p_3 v_S - \mu_3}{T_3} \,, \quad s_L = \frac{u_L + p_3 v_L - \mu_3}{T_3} \,, \quad s_G = \frac{u_G + p_3 v_G - \mu_3}{T_3} \,. \tag{D.6}$$

From the forms of the energy and entropy given by (D.2) and (D.5), it is seen that, when regarded as functions of $(T; V, N)$, these quantities generally change discontinuously as the triple point is crossed.

D.4 Continuity of state functions

Finally, in the table below, we consider the continuity of several state functions as the system passes through the various two-phase and three-phase coexistence regions in the cases of the three parameterizations of equilibrium states studied in this appendix. Some of these results were derived above, while the rest can be derived in similar manners. In each case described as "continuous," the function is continuous along all paths as the system passes through the region in question, while in each case described as "discontinuous," the function is generically discontinuous as the system crosses the region in question, although there may be particular paths along which it is continuous.[14]

It is particularly noteworthy here that even though the parameterizations in terms of the variables T, V and N and the variables T, p and N result in incomplete specifications of

the triple-point region by using the fact that any two states within this region can be connected by some combination of quasi-static adiabatic and isothermal operations and that the corresponding work is always equal to $W = p_3 \Delta V$, where ΔV is the total change in the volume.

[11] It can therefore be concluded that the variational principle (7.30) is valid in the entire state space, including the triple-point region.

[12] Note that there is a one-to-one correspondence between (N_S, N_L, N_G) and (V, N, N_S).

[13] The expressions in (D.6) follow directly from the Euler equation for the entropy, $S[U, V, N] = \{U + p(U, V, N)V - \mu(U, V, N)N\}/T(U, V, N)$, which can be derived from (F.1) analogously to the manner in which $F[T; V, N]$ is derived in (7.8).

[14] In the V-U plane, let us consider a quasi-static process through which a system approaches, enters, traverses, and eventually exits the triple point-region (see Figure D.2(c)). If this is done under adiabatic conditions, then, because the entropy is constant during such a process, when considering it in the V-T and p-T planes, both $S(T; V, N)$ and $S(T, p; N)$ will be seen as varying continuously as the system crosses the triple point. The situation is similar for the energy, although somewhat more complicated.

the equilibrium states of a system, the complete thermodynamic functions $F[T; V, N]$ and $G[T, p; N]$ are always continuous. This important fact can be regarded as a manifestation of the properties of complete thermodynamic functions and the Legendre transformation (see Appendices F and H).

Parameterization	State function	Two-phase coexistence region (phase boundary)	Three-phase coexistence region (triple point)
	$S[U, V, N]$	continuous	continuous
U, V, N	$T(U, V, N)$	continuous	continuous
	$p(U, V, N)$	continuous	continuous
	$F[T; V, N]$	continuous	continuous
T, V, N	$p(T; V, N)$	continuous	continuous
	$U(T; V, N)$	continuous	discontinuous
	$S(T; V, N)$	continuous	discontinuous
	$G[T, p; N]$	continuous	continuous
T, p, N	$U(T, p; N)$	discontinuous	discontinuous
	$S(T, p; N)$	discontinuous	discontinuous

On the Treatment of Mechanical Potential Energy

Consider a thermodynamic system subject to an external field. Obviously, in general, the anisotropic influence imparted by the field will have some effect on the behavior of the system. In our treatment of fluid systems in the main text, we have ignored this effect, because for most systems and behavior of practical interest, it is negligible. In this appendix, considering the case of gravity, we present a more generally valid treatment. First we construct a generalized Helmholtz free energy that fully accounts for the effect of gravity. Next, we obtain the balance condition implied by this free energy and derive from this condition a relation determining the equilibrium height dependence of the pressure in a fluid.

E.1 Generalized Helmholtz free energy

Let us consider a fluid system in a gravitational field. The effect of this field will influence the work performed in mechanical operations, and thus, in principle, it must be accounted for in the Helmholtz free energy. Noting the similarity between the Helmholtz free energy and the gravitational potential energy with regard to the performance of work, it is not unreasonable to postulate that the effect of the gravitational field can be accounted for by simply appending the "ordinary" Helmholtz free energy with a potential energy term, writing the generalized Helmholtz free energy for an individual system in the form $F(T; V, N) + \mathcal{U}(z)$, as a function of the temperature T, total volume V, total amount of substance N, and a spatial coordinate z parameterizing the gravitational potential energy. However, this naive construction yields a valid free energy only if the system is sufficiently short along the direction of gravity that the inhomogeneity caused by gravity is negligible.[1] Here we construct a generalized Helmholtz free energy that is valid in the case that this inhomogeneity cannot be ignored.

For simplicity, we consider a single-component fluid with amount of substance N, confined to a connected finite region $\mathcal{V} \subset \mathbb{R}^3$ with no internal walls and subject to a constant gravitational force pointing in the negative z direction. We assume that the fluid is under isothermal conditions and is in an equilibrium state characterized by a single temperature T.[2] The equilibrium state is uniquely determined by specifying T, $\mathcal{V}$ and N (as long as the system is not at the triple point), and therefore we express it as $(T; \mathcal{V}, N)$. But it is important to keep in mind here that $\mathcal{V}$ represents a specific region of space, characterized by a particular shape and position, not merely a volume.

We can perform isothermal operations on this system by varying the volume, shape and position of the region $\mathcal{V}$ in an arbitrary manner. With the assumption that a perpetual motion machine of the second kind cannot be realized even under the influence of gravity

---

[1] In the main text, we have implicitly assumed that all of the systems considered are sufficiently small in this sense (except in the situation described by Figure 3.4). In addition, we have assumed that changes in the potential energy of the system are negligible (except in the situations described by Figures 3.4 and 8.1).

[2] This assumption is valid in ordinary situations of practical interest, but let us point out that this is not an entirely trivial point. It is known that general relativistic effects introduce a spatial dependence of the temperature, and therefore in situations that such effects cannot be ignored, the system will not be characterized by a single temperature in equilibrium. However, even systems of this type can be properly described by thermodynamics, as long as the gravitational field is time independent; it is only necessary to use energy instead of temperature in the parameterization of equilibrium states (see Section 27 of L.D. Landau and E.M. Lifshitz, *Statistical Physics*, Vol. 5 (3rd Ed.), Butterworth-Heinemann (1980), ISBN 978-0-7506-3372-7).

(see Problem 3.1), we postulate that Kelvin's principle (Postulate 3.1 (p. 47)) is also valid for this larger class of isothermal operations. Let $W_{\max}(T; (\mathcal{V}, N) \to (\mathcal{V}', N))$ denote the maximum work performed by the system on the mechanical world under isothermal operations through which the region occupied by the system is varied from $\mathcal{V}$ to some arbitrary region $\mathcal{V}'$. Then, repeating the logic employed in Section 3.5, we conclude that there exists a function $F[T; \mathcal{V}, N]$ with the property

$$W_{\max}(T; (\mathcal{V}, N) \to (\mathcal{V}', N)) = F[T; \mathcal{V}, N] - F[T; \mathcal{V}', N] \,, \tag{E.1}$$

for any T, N, $\mathcal{V}$ and $\mathcal{V}'$. We can regard the function $F[T; \mathcal{V}, N]$ as a generalization of the ordinary Helmholtz free energy. However, because the concentration of the fluid composing such a system will not necessarily be spatially uniform even in equilibrium, $F[T; \mathcal{V}, N]$ does not generally possess extensivity, which is one of the essential properties of the Helmholtz free energy in the case that the effect of external fields is negligible.

We now investigate how the generalized free energy $F[T; \mathcal{V}, N]$ is related to the ordinary free energy for a fluid system studied to this point. For this purpose, we first divide the region $\mathcal{V}$ into n small subregions with volumes $V_1, V_2, \ldots, V_n$. We assume that the subregions are sufficiently short along the direction of gravity that the inhomogeneity of the fluid due to gravity is negligible within each. (The precise meaning of this assumption is explained in Footnote 3 on p. 286.)

With the above assumptions, the system can be treated using the formalism developed in the main text. Writing the amount of substance in the ith subregion as $\widetilde{N}_i$, we can show that the relation (E.1) for the maximum work is satisfied if we choose the generalized Helmholtz free energy as

$$F[T; \mathcal{V}, N] = \sum_{i=1}^{n} \left\{ F[T; V_i, \widetilde{N}_i] + gM\widetilde{N}_i z_i \right\} \,. \tag{E.2}$$

Here, z_i is the z-coordinate of (the center of mass of) the ith subregion, M is the molar mass of the substance, and g is the acceleration due to gravity. Note that $(T; V_i, \widetilde{N}_i)$ is the equilibrium state of the system obtained from the ith subregion when it is surrounded with walls (without the performance of work).

To demonstrate (E.1), we observe that any transition $(T; \mathcal{V}, N) \to (T; \mathcal{V}', N)$ can be realized in the following steps. (i) Given the equilibrium state $(T; \mathcal{V}, N)$, the n subregions are converted into n subsystems through the insertion of diathermal walls. (It is assumed that the only change undergone by the system in this process is that each subregion has become surrounded by impermeable barriers.) (ii) The subsystems are spatially separated so that each can be individually manipulated. (iii) The volume and/or shape of each subsystem is quasi-statically varied in a predetermined manner, under isothermal conditions at temperature T. (iv) The positions of the subsystems are arranged in such a manner that together they occupy the desired final region $\mathcal{V}'$, with all neighboring systems balanced. (Given the desired $\mathcal{V}'$, this places a strong constraint on the operations carried out in (iii).) (v) All of the walls are removed, and the subsystems thereby become subregions again. (Because neighboring subsystems are balanced before the removal of the walls, this step can be carried out quasi-statically.)

The total work W performed in the above quasi-static procedure is given by[3]

$$W = \sum_{i=1}^{m} \left\{ F[T; V_i, \widetilde{N}_i] - F[T; V_i', \widetilde{N}_i] + gM\widetilde{N}_i(z_i - z_i') \right\} \,, \tag{E.3}$$

---

[3] Let us clarify the assumption implicit in this expression. In the above procedure, the work performed by the ith subsystem when its shape and/or volume is varied is assumed to be given by $F[T; V_i, \widetilde{N}_i] - F[T; V_i', \widetilde{N}_i]$. Of course, this introduces an error, as the influence of the spatial variation of the concentration of the fluid within this subsystem is ignored. The fundamental assumption here is that the total error consisting of the sum of these errors over all subsystems can be made negligibly small by choosing the subregions to be sufficiently small. While the treatment presented in this appendix rests on this assumption, it is known to be valid quite generally.

where V_i and V_i' are the original and final volumes and z_i and z_i' are the original and final heights of the ith subsystem. Therefore, with the generalized free energy given by (E.2), we have $W = F[T; \mathcal{V}, N] - F[T; \mathcal{V}', N]$. The validity of (E.1) thus follows from the principle of maximum work (Result 3.3 (p. 52)).[4]

Finally, let us point out that there is no unique proper division of the region $\mathcal{V}$ into subregions. In fact, as seen from (E.1), any division for which each subregion is sufficiently small yields the same generalized free energy through the formula (E.2).

E.2 Balance conditions and the spatial variation of pressure

As discussed in Section 7.4, two systems are regarded as *balanced* if no observable change results when they are merged into a single system through the removal of a wall. With a trivial generalization, we also regard any two systems that are both balanced with a third system to be balanced with each other. Obviously, for a fluid system in equilibrium, the individual systems obtained when its subregions are separated by walls are all balanced. Below we derive conditions that follow from this observation.

Using the (extended) principle of maximum work and the relation (E.1) for the maximum work, we can repeat the argument given in Section 7.3 to derive a direct generalization of the variational principle (7.31) regarding the transport of substance between two subsystems. With the form (E.2), this variational principle reads

$$F[T; \mathcal{V}, N] = \min_{\substack{N_1,\dots,N_n \\ (\sum_{i=1}^n N_i = N)}} \sum_{i=1}^n \{F[T; V_i, N_i] + gMN_i z_i\} \; . \tag{E.4}$$

We now proceed as in Section 7.4 to derive the corresponding balance condition. With $\widetilde{N}_i$ representing the amount of substance in the ith subregion in the equilibrium state $(T; \mathcal{V}, N)$, we have

$$\frac{\partial}{\partial \xi} \sum_{i=1}^n \left\{ F[T; V_i, \widetilde{N}_i + \xi \Delta_i] + gM(\widetilde{N}_i + \xi \Delta_i) z_i \right\} \bigg|_{\xi=0} = 0 \tag{E.5}$$

for any $\Delta_1, \dots, \Delta_n$ satisfying $\sum_{i=1}^n \Delta_i = 0$. Recalling (7.6), we thus obtain

$$\sum_{i=1}^n \left\{ \mu[T; V_i, \widetilde{N}_i] + gMz_i \right\} \Delta_i = 0 \; . \tag{E.6}$$

Then, because the quantities $\Delta_1, \dots, \Delta_n$ are arbitrary, this reduces to the condition

$$\mu[T; V_i, \widetilde{N}_i] + gMz_i = \mu[T; V_j, \widetilde{N}_j] + gMz_j \; , \tag{E.7}$$

for all i and j. While, interpreted strictly, this result applies only to the case in which the subregions are separated by walls, extending the meaning of $\mu[T; V_i, \widetilde{N}_i]$ to also represent the chemical potential in the ith subregion of the original system, (E.7) naturally describes this system as well. We thus find that the chemical potential plus the potential energy (rather than the chemical potential alone) is constant throughout a fluid system in equilibrium under the influence of gravity.[5]

As an application of the balance condition (E.7), here we derive a relation determining the equilibrium height dependence of the pressure in a fluid. Consider a fluid system under

---

[4] Recall that the principle of maximum work is derived without assuming extensivity. It follows from only Kelvin's principle and the reversibility of a quasi-static isothermal operation.

[5] This condition is essentially the same as the condition that the electrochemical potential given in (9.133) be constant.

the influence of a constant gravitational field in the equilibrium state $(T; \mathcal{V}, N)$. Because the chemical potential depends on z, so does the pressure. Let us represent the pressure at a position of height z by $p(z)$ (as an abbreviated form of $p(T; \mathcal{V}, N; z)$). Then, expressing the chemical potential as a function of T and p, as in (8.18), we rewrite the condition (E.7) as

$$\mu(T, p(z)) + Mgz = \text{(constant)} . \tag{E.8}$$

Differentiating this expression with respect to z, we obtain

$$p'(z) = -M\, n(T, p(z))\, g , \tag{E.9}$$

where $n(T, p(z))$ is the molarity (moles per unit volume). Here we have used the relations (8.19) and (8.16), which together imply $\partial\mu(T, p)/\partial p = 1/n(T, p)$. The differential equation (E.9) is a standard formula in fluid statics. It can be solved to obtain an expression for $p(z)$ when the functional form of $n(T, p)$ is given. Let us examine two simple cases.

First we consider liquids. Using the very good approximation that liquids are incompressible, we drop the p dependence of $n(T, p)$, writing it as $n(T)$. Then from (E.9), we obtain

$$p(z) = p(0) - Mn(T)gz . \tag{E.10}$$

Of course, this relation can also be obtained directly from the mechanical definition of pressure. In the case of water, with an ambient pressure of 1 atm, the pressure will vary by approximately 10% in a system of 1 meter depth.

Next, we consider an ideal gas. In this case, we have $n(T, p(z)) = p(z)/RT$, and thus (E.9) becomes

$$p'(z) = -\frac{Mg}{RT}\, p(z) . \tag{E.11}$$

Solving, we obtain

$$p(z) = p(0)\, \exp\!\left[-\frac{Mg}{RT}z\right] . \tag{E.12}$$

On Earth and at ordinary temperatures, Mg/RT is so small that the variation of $p(z)$ displayed by any practically controllable system would be entirely negligible. However, (E.12) does provide an approximate description of the altitude dependence of the pressure in Earth's atmosphere that is reasonably accurate over sufficiently narrow ranges. (For this reason, it is sometimes referred to as the *barometric formula.*) It breaks down mainly because, due to non-equilibrium effects, the temperature also varies with altitude.

Summary of Complete Thermodynamic Functions

In this appendix, we carry out a systematic survey of complete thermodynamic functions in the context of multi-component systems, parameterized by $(T; V, \boldsymbol{N})$, elucidating the main properties of these functions and the relations among them.[1] We also present the differential forms for these functions, which constitute the *fundamental equations* of thermodynamics (as named by Gibbs).

The background necessary to fully understand the treatment given in this appendix is provided by the discussion given in Chapter 7 regarding differential forms[2] and in Chapter 8 regarding the relationship between the Helmholtz free energy and the Gibbs free energy. With a firm understanding of these topics, the reader should be able to appreciate the elegant mathematical structure of complete thermodynamic functions described in this appendix.

Below, we treat five quantities: the entropy, S, the energy, U, the Helmholtz free energy, F, the enthalpy, H, and the Gibbs free energy, G. All of these are extensive quantities possessing both extensivity and additivity, as expressed by (3.24) and (3.25) for the particular case of F.

Among the quantities considered here, the Helmholtz free energy and the Gibbs free energy were treated in the main text as complete thermodynamic functions. The remaining three—the energy, entropy and enthalpy—were treated merely as state functions. This reflects the fact that whether or not any given one of the quantities considered here acts as a complete thermodynamic function depends on the choice of its independent variables. For example, if the enthalpy is regarded as a function of T, V and $\boldsymbol{N}$ or of T, p and $\boldsymbol{N}$, then it (i.e., the function $H(T; V, \boldsymbol{N})$ or $H(T, p; \boldsymbol{N})$) is a state function, but not a complete thermodynamic function, while if it is regarded as a function of p, S and $\boldsymbol{N}$, then it (i.e., the function $H[p; S, \boldsymbol{N}]$) is a complete thermodynamic function, because in this case, it incorporates all of the information about the system. This illustrates how a single thermodynamic quantity can play various physical roles. The fact that quantities of this kind exist in thermodynamics is one of the most interesting features of its theoretical structure. The independent variables with which a quantity acts as a complete thermodynamic function are referred to as the *natural independent variables* for that quantity. The above five quantities as functions of their natural independent variables are as follows: $S[U, V, \boldsymbol{N}]$, $U[S, V, \boldsymbol{N}]$, $F[T; V, \boldsymbol{N}]$, $H[p; S, \boldsymbol{N}]$, $G[T, p; \boldsymbol{N}]$.

As mentioned in Appendix D, the parameterization of equilibrium states in terms of the three extensive variables U, V and $\boldsymbol{N}$ allows for the unique specification of each equilibrium state in the most general situation, even with the triple point included. In this case, the entropy, $S[U, V, \boldsymbol{N}]$, is a complete thermodynamic function. The function $S[U, V, \boldsymbol{N}]$ is once differentiable and concave with respect to $(U, V, \boldsymbol{N})$. With V and $\boldsymbol{N}$ fixed, it is an increasing function of U. Its derivatives with respect to these variables are conveniently expressed in the differential form[3]

$$dS = \frac{dU}{T} + \frac{p}{T}\,dV - \sum_{i=1}^{m} \frac{\mu_i}{T}\,dN_i \; . \tag{F.1}$$

---

[1] This appendix should be studied in conjunction with Problems 8.1–8.5.

[2] Specifically, it is important to understand how the differential form of a complete thermodynamic function identifies its natural independent variables and expresses its derivatives with respect to these variables.

[3] As explained in Section 7.1.4, differential forms are convenient tools for remembering the independent variables and derivatives of complete thermodynamic functions. However, it is important to keep in mind that these expressions are not mathematically meaningful at points where the function in question is not differentiable.

The variational inequality for the entropy is the following:

$$S[U + U', V + V', \boldsymbol{N} + \boldsymbol{N}'] \geq S[U, V, \boldsymbol{N}] + S[U', V', \boldsymbol{N}'] \ . \tag{F.2}$$

Imagine that we fix V and $\boldsymbol{N}$. Then, because $S[U, V, \boldsymbol{N}]$ is an increasing function of U, there exists a corresponding inverse function $U[S, V, \boldsymbol{N}]$. Noting that the inverse of a function (if it exists) contains all the information of the original function, we thus see that $U[S, V, \boldsymbol{N}]$ is also a complete thermodynamic function.[4] The function $U[S, V, \boldsymbol{N}]$ is once differentiable and convex with respect to $(S, V, \boldsymbol{N})$. Its derivatives with respect to these variables can be read off of the differential form

$$dU = T\, dS - p\, dV + \sum_{i=1}^{m} \mu_i dN_i \ . \tag{F.3}$$

The variational inequality for the energy is the following:

$$U[S + S', V + V', \boldsymbol{N} + \boldsymbol{N}'] \leq U[S, V, \boldsymbol{N}] + U[S', V', \boldsymbol{N}'] \ . \tag{F.4}$$

Taking the Legendre transformation[5] of $U[S, V, \boldsymbol{N}]$ with respect to S, we obtain $F[T; V, \boldsymbol{N}]$, with respect to V, we obtain $H[p; S, \boldsymbol{N}]$, and with respect to both S and V, we obtain $G[T, p; \boldsymbol{N}]$. Because all of the information possessed by a convex function is preserved under the Legendre transformation, it follows that $F[T; V, \boldsymbol{N}]$, $H[p; S, \boldsymbol{N}]$ and $G[T, p; \boldsymbol{N}]$ are also complete thermodynamic functions.

Expressing the Helmholtz free energy $F[T; V, \boldsymbol{N}]$ as the Legendre transform of $U[S, V, \boldsymbol{N}]$, we have

$$F[T; V, \boldsymbol{N}] = \min_{S}\{U[S, V, \boldsymbol{N}] - TS\} \ . \tag{F.5}$$

The inverse transformation is given by

$$U[S, V, \boldsymbol{N}] = \max_{T}\{F[T; V, \boldsymbol{N}] + TS\} \ . \tag{F.6}$$

The function $F[T; V, \boldsymbol{N}]$ is once differentiable and convex with respect to $(V, \boldsymbol{N})$, while it is continuous and concave with respect to T. Its derivatives with respect to these variables can be read off of the differential form[6]

$$dF = -S\, dT - p\, dV + \sum_{i=1}^{m} \mu_i dN_i \ . \tag{F.7}$$

The variational inequality for the Helmholtz free energy is the following:

$$F[T; V + V', \boldsymbol{N} + \boldsymbol{N}'] \leq F[T; V, \boldsymbol{N}] + F[T; V', \boldsymbol{N}'] \ . \tag{F.8}$$

It is also interesting (and sometimes useful) to directly relate $F[T; V, \boldsymbol{N}]$ to the entropy as a complete thermodynamic function, $S[U, V, \boldsymbol{N}]$. Because $U[S, V, \boldsymbol{N}]$ is defined as the inverse function of $S[U, V, \boldsymbol{N}]$ (with fixed V and $\boldsymbol{N}$), from (F.5) we readily obtain

---

[4] Once the concept of adiabatic operations is established, energy is probably the most easily understood state function. However, in general, that which is directly obtained in experiments is not energy as a complete thermodynamic function, $U[S, V, \boldsymbol{N}]$, but instead energy as a mere state function, $U(T; V, \boldsymbol{N})$. For this reason, even if we were to carry out exhaustive and precise measurements of the energy, we would not be able to obtain a complete thermodynamic description of the system. In order to obtain a complete description of a system through the energy as a complete thermodynamic function, it is necessary to first obtain the Helmholtz free energy or the entropy.

[5] This important mathematical tool is used repeatedly in this appendix. The theory of Legendre transformations is treated in Appendices G and H.

[6] The relationship between (F.7) and (F.3) can be found by writing $F = U - TS$ and then using the computational method for differential forms demonstrated in (8.22). The situation is similar for (F.14) and (F.18).

$$F[T; V, \boldsymbol{N}] = \min_U \{U - TS[U, V, \boldsymbol{N}]\} \,, \tag{F.9}$$

which can be rewritten as

$$-\frac{F[T; V, \boldsymbol{N}]}{T} = \max_U \left\{ S[U, V, \boldsymbol{N}] - \frac{U}{T} \right\} . \tag{F.10}$$

This relation implies that $-F[T; V, N]/T$ (which is called the *Massieu function*) is a convex function of $1/T$. The inverse transformation of (F.10) is given by

$$S[U, V, \boldsymbol{N}] = \min_T \left\{ \frac{U - F[T; V, \boldsymbol{N}]}{T} \right\} . \tag{F.11}$$

The enthalpy $H[p; S, \boldsymbol{N}]$ is obtained from $U[S, V, \boldsymbol{N}]$ as follows:

$$H[p; S, \boldsymbol{N}] = \min_V \{U[S, V, \boldsymbol{N}] + pV\} . \tag{F.12}$$

The inverse transformation is given by

$$U[S, V, \boldsymbol{N}] = \max_p \{H[p; S, \boldsymbol{N}] - pV\} . \tag{F.13}$$

The function $H[p; S, \boldsymbol{N}]$ is once differentiable and convex with respect to S and $\boldsymbol{N}$, while it is continuous and concave with respect to p. Its derivatives can be read off of the differential form

$$dH = T\,dS + V\,dp + \sum_{i=1}^{m} \mu_i dN_i . \tag{F.14}$$

The variational inequality for the enthalpy is the following:

$$H[p; S + S', \boldsymbol{N} + \boldsymbol{N'}] \leq H[p; S, \boldsymbol{N}] + H[p; S', \boldsymbol{N'}] . \tag{F.15}$$

The Gibbs free energy $G[T, p; \boldsymbol{N}]$ is obtained from $U[S, V, \boldsymbol{N}]$ as follows:

$$\begin{aligned}
G[T, p; \boldsymbol{N}] &= \min_{S,V} \{U[S, V, \boldsymbol{N}] - TS + pV\} \\
&= \min_V \{F[T; V, \boldsymbol{N}] + pV\} \\
&= \min_S \{H[p; S, \boldsymbol{N}] - TS\} .
\end{aligned} \tag{F.16}$$

The inverse transformation is given by

$$U[S, V, \boldsymbol{N}] = \max_{T,p} \{G[T, p; \boldsymbol{N}] + TS - pV\} . \tag{F.17}$$

The function $G[T, p; \boldsymbol{N}]$ is once differentiable and convex with respect to $\boldsymbol{N}$, while it is continuous and concave with respect to (T, p). Its derivatives can be read off of the differential form

$$dG = -S\,dT + V\,dp + \sum_{i=1}^{m} \mu_i dN_i . \tag{F.18}$$

The variational inequality for the Gibbs free energy is the following:

$$G[T, p; \boldsymbol{N} + \boldsymbol{N'}] \leq G[T, p; \boldsymbol{N}] + G[T, p; \boldsymbol{N'}] . \tag{F.19}$$

Finally, we note that by taking the Legendre transformation of the quantities considered above with respect to $\boldsymbol{N}$, we can obtain complete thermodynamic functions depending on $(\mu_1, \ldots, \mu_m)$ rather than $\boldsymbol{N}$.

APPENDIX G

Convex Functions

The mathematical concepts of convex functions and the Legendre transformation are very natural and useful in application to thermodynamics and statistical mechanics. Despite this fact, there are very few physics textbooks in which these concepts are treated in a systematic manner. The treatments presented in this and the following appendix provide the minimal understanding of convex functions and the Legendre transformation needed in physics. These are self-contained treatments, and for this reason, there is some overlap between them and the more concrete considerations given in the main text.

A command of introductory real analysis is sufficient to understand the content of these appendices. The proofs presented are mathematically rigorous.

G.1 Convex functions of a single variable

G.1.1 Definition

Consider a real-valued function $f(x)$ defined on an interval I.[1] This interval may be open, closed or half-open, as well as bounded, unbounded or half-bounded.[2] The function $f(x)$ is said to be *convex* if the relation

$$f(\lambda x_1 + (1 - \lambda)x_2) \le \lambda f(x_1) + (1 - \lambda)f(x_2) \tag{G.1}$$

holds for all $x_1, x_2 \in I$ and all $\lambda \in [0, 1]$. The meaning of this definition is easily understood from Figure G.1.

A function $f(x)$ is called *concave* if the function $-f(x)$ is convex. In this appendix, we treat only convex functions, but all of the discussion and results given here apply also to the case of concave functions, with only obvious changes of sign and reversals of inequalities.

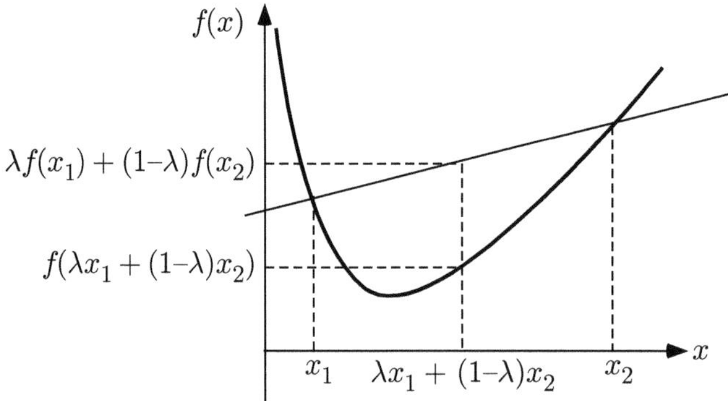

Figure G.1 Graphical depiction of convexity. For any two points $(x_1, f(x_1))$ and $(x_2, f(x_2))$ lying on the graph of $f(x)$, the line connecting these points lies above (or coincides with) this graph for all $x \in (x_1, x_2)$. This is the situation described by the definition, (G.1).

---

[1] An interval is a connected set of real numbers. An interval consisting of a single real number is called a *degenerate interval*, while any interval that is not degenerate is called a *proper interval*.

[2] In applications to thermodynamics, however, the functions we consider are generally defined on open intervals, usually $I = \mathbb{R}$ or $I = \mathbb{R}_+$.

Now, consider arbitrary values $y_1, y_2, y_3 \in I$ with $y_1 < y_2 < y_3$, and set $\lambda = (y_3 - y_2)/(y_3 - y_1)$. Rearranging, this gives $y_2 = \lambda y_1 + (1 - \lambda)y_3$. Then, substituting y_1 for x_1 and y_3 for x_2 in (G.1), we obtain

$$(y_3 - y_2)f(y_1) - (y_3 - y_1)f(y_2) + (y_2 - y_1)f(y_3) \geq 0 \, . \tag{G.2}$$

Thus, that the inequality (G.2) hold for arbitrary y_1, y_2 and y_3 under the conditions specified above is a necessary and sufficient condition for $f(x)$ to be convex. This inequality is useful for proving various properties of convex functions.

G.1.2 Theorems

In this subsection, we present several theorems expressing some of the important properties of convex functions. Proofs are given in Section G.3.

First, let us compare the definition given above with the characterization provided by the following theorem, which is often used in physics textbooks as the definition of a convex function.

Theorem G.1 (Convexity and the second derivative) *For a convex function $f(x)$ defined on an interval I, the relation $f''(x) \geq 0$ holds for any x at which $f''(x)$ is defined. Conversely, if a function $f(x)$ defined on an interval I is twice differentiable and satisfies $f''(x) \geq 0$ for all $x \in I$, then $f(x)$ is convex.*

Considering the simplicity of the above statements, it may seem that it would be better to avoid the somewhat complicated definition given in (G.1) and simply use the condition $f''(x) \geq 0$ to define a convex function. However, as stated several times in the main text, at phase transition points, many thermodynamic functions are not twice differentiable. For this reason, the mathematically more general definition (G.1) must be regarded as the proper definition even from the physics point of view.

The theorem below describes two methods of constructing convex functions from other convex functions. Because this theorem follows directly from the definition, we do not present a proof.

Theorem G.2 (Construction of convex functions from convex functions) *Let $f(x)$ and $g(x)$ be convex functions defined on an interval I. Then, for arbitrary real numbers a and b, $f(ay + b)$ is a convex function of y defined on the interval $\{y \mid ay + b \in I\}$, while for arbitrary positive real numbers a and b, $a f(x) + b g(x)$ is a convex function of x defined on I.*

The following is a powerful theorem.

Theorem G.3 (Continuity of a convex function) *A convex function $f(x)$ defined on an open interval I is continuous at all values of $x \in I$.*

A proof of this theorem is given in Section G.3, but its validity can be understood intuitively by considering the graph of a discontinuous function, as illustrated in Figure G.2.

The above theorem, which applies to convex functions on open intervals, provides information about convex functions on non-open intervals as well. Consider a convex function $f(x)$ on a (proper) non-open interval I. Because the restriction of $f(x)$ to the interior of I is also convex, Theorem G.3 implies that $f(x)$ can be discontinuous only at the endpoint(s) of I.[3,4]

---

[3] The interior of an interval I is the largest open subinterval of I. For example, the interiors of $[a, b]$ and $[a, \infty)$ are (a, b) and (a, ∞).

[4] For functions $g(x)$ with domain A and $\tilde{g}(x)$ with domain B satisfying $B \subset A$ and $\tilde{g}(x) = g(x)$ for all $x \in B$, $\tilde{g}(x)$ is called the *restriction* of $g(x)$ to B.

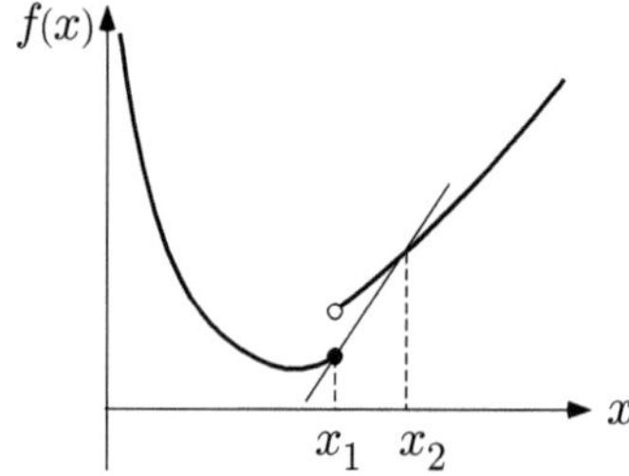

Figure G.2 Demonstration of the non-convexity of a discontinuous function. Choosing x_1 and x_2 as in the figure, it is readily seen that the inequality in (G.1) does not hold.

Next, we introduce the concepts of the right and left derivatives, which are very useful in application to convex functions. Consider a function $f(x)$ defined on an interval I and a point $x_0 \in I$ that is not the right endpoint of I. The limit

$$f'_+(x_0) = \lim_{\varepsilon \searrow 0} \frac{f(x_0 + \varepsilon) - f(x_0)}{\varepsilon} \tag{G.3}$$

is called the *right derivative* of $f(x)$ at x_0. If this limit exists, $f(x)$ is said to be *right differentiable* at x_0. The meaning expressed by "right" here is that in this limit, ε approaches 0 from the positive side. Similarly, for $x_0 \in I$ that is not the left endpoint of I, the limit

$$f'_-(x_0) = \lim_{\varepsilon \searrow 0} \frac{f(x_0) - f(x_0 - \varepsilon)}{\varepsilon} \tag{G.4}$$

is called the *left derivative* of $f(x)$ at x_0. If this limit exists, $f(x)$ is said to be *left differentiable* at x_0. A function $f(x)$ is differentiable at a point x_0 if and only if $f'_+(x_0)$ and $f'_-(x_0)$ exist and are equal.

The left and right derivatives are easily understood intuitively. For example, in the case of the function

$$f(x) = \begin{cases} x & \text{for } x \leq 0 \,, \\ 2x & \text{for } x \geq 0 \,, \end{cases} \tag{G.5}$$

we have $f'_-(0) = 1$ and $f'_+(0) = 2$.

The following theorem expresses an important property of convex functions.

Theorem G.4 (Left and right derivatives and lines of support) *Let $f(x)$ be a convex function defined on an open interval I. Then both the right derivative, defined in (G.3), and the left derivative, defined in (G.4), exist at all values of $x \in I$, and the relations*

$$f'_-(x_1) \leq f'_+(x_1) \leq f'_-(x_2) \leq f'_+(x_2) \tag{G.6}$$

hold for all $x_1, x_2 \in I$ satisfying $x_1 < x_2$. Also, for arbitrary $x_0 \in I$, if we choose α in accordance with

$$f'_-(x_0) \leq \alpha \leq f'_+(x_0) \,, \tag{G.7}$$

then the relation

$$f(x) \geq f(x_0) + \alpha(x - x_0) \tag{G.8}$$

holds for all $x \in I$.

It is thus seen that for a convex function, there exist quantities that are very similar to the derivative even at points of non-differentiability. For example, although the Gibbs free

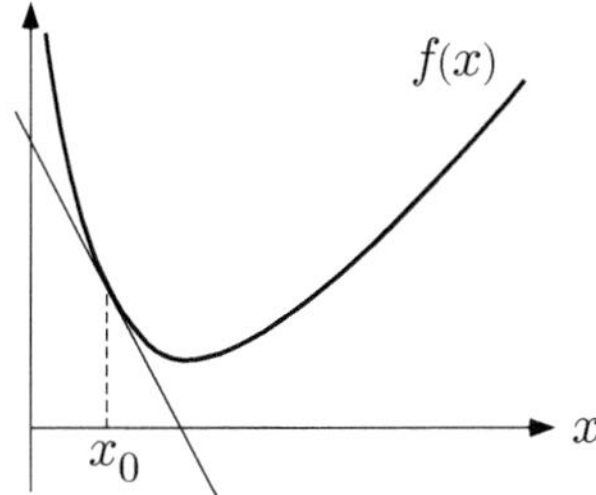

Figure G.3 Demonstration that a convex function lies above any line tangent to it. Here, the curve represents $f(x)$, and the line tangent to it is the line of slope $f'(x_0)$ passing through the point $(x_0, f(x_0))$, whose formula appears on the right-hand side of (G.10).

energy is non-differentiable with respect to T at a phase transition point, we can rigorously define the right and the left derivatives with respect to T at such a point.

Applying the above theorem in the case that $f'(x)$ exists, we obtain the following useful result. (We omit the proof, because it is quite straightforward.)

Theorem G.5 (The derivative and lines of tangency) *Let $f(x)$ be a convex function defined on an interval I. Then, for any $x_1, x_2 \in I$ at which $f(x)$ is once differentiable, $x_1 < x_2$ implies*

$$f'(x_1) \leq f'(x_2) \,. \tag{G.9}$$

In other words, for values of x where it is defined, $f'(x)$ is a non-decreasing function of x. Also, for any $x_0 \in I$ at which $f(x)$ is once differentiable, the relation

$$f(x) \geq f(x_0) + f'(x_0)\,(x - x_0) \tag{G.10}$$

holds for all $x \in I$.

The right-hand side of (G.10) is the formula for the line of slope $f'(x_0)$ that passes through the point $(x_0, f(x_0))$, in other words, the line of tangency to $f(x)$ at x_0.[5] The inequality in (G.10) expresses the fact that $f(x)$ lies entirely above any line tangent to it, except at the point (or points) of tangency. This situation is depicted in Figure G.3.

The theorem below follows directly from (G.10).

Theorem G.6 (Minimum as the unique critical point) *For a convex function $f(x)$ defined on an interval I, if we have $f'(x_0) = 0$ at some $x_0 \in I$, then $f(x_0)$ is the minimum value of $f(x)$.*

This theorem (and its generalization, Theorem G.10) is of fundamental importance in the derivation of balance conditions for thermodynamic systems.

G.2 Convex functions of multiple variables

In this section, we consider convex functions on $\mathbb{R}^n$. First we present their definition, and then we briefly discuss their fundamental properties.

A subset C of the n-dimensional Euclidean space, $\mathbb{R}^n$, is said to be a *convex set* if the relation $\lambda \boldsymbol{x}_1 + (1 - \lambda)\boldsymbol{x}_2 \in C$ holds for all $\boldsymbol{x}_1, \boldsymbol{x}_2 \in C$ and all $\lambda \in [0, 1]$. The meaning of

---

[5] An analogous line with slope α satisfying $f'_-(x_0) \leq \alpha \leq f'_+(x_0)$ in the case that $f(x)$ is not differentiable at x_0 is referred to as a *line of support*.

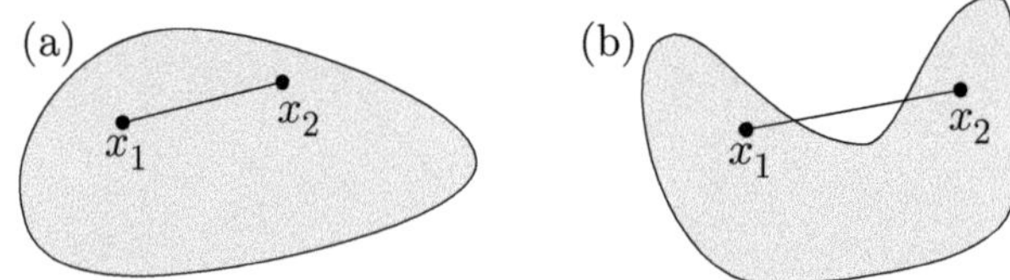

Figure G.4 Two subsets of the two-dimensional Euclidean space, $\mathbb{R}^2$. (a) A convex subset. (b) A non-convex subset.

this definition can be understood from Figure G.4. Note that both $\mathbb{R}^n$ itself and $\mathbb{R}^n_+$ are convex sets.[6]

A real-valued function $f(\boldsymbol{x})$ defined on a convex set $C \subseteq \mathbb{R}^n$ is said to be a *convex function* if the relation

$$f(\lambda \boldsymbol{x}_1 + (1 - \lambda)\boldsymbol{x}_2) \leq \lambda f(\boldsymbol{x}_1) + (1 - \lambda)f(\boldsymbol{x}_2) \tag{G.11}$$

holds for all $\boldsymbol{x}_1, \boldsymbol{x}_2 \in C$ and all $\lambda \in [0, 1]$.

The following theorem, which corresponds to Theorem G.2 in the single-variable case, also follows directly from the definition.

Theorem G.7 (Construction of convex functions from convex functions) *Let $f(\boldsymbol{x})$ and $g(\boldsymbol{x})$ be convex functions each with (convex) domain $C \subseteq \mathbb{R}^n$. Then, for any matrix $\mathsf{M} \in \mathbb{R}^{n \times n}$ and any vector $\boldsymbol{a} \in \mathbb{R}^n$, $f(\mathsf{M}\boldsymbol{y} + \boldsymbol{a})$ is a convex function of $\boldsymbol{y}$ with domain $\{\boldsymbol{y} \mid \mathsf{M}\boldsymbol{y} + \boldsymbol{a} \in C\}$. Also, for any positive real numbers a and b, $a f(\boldsymbol{x}) + b g(\boldsymbol{x})$ is a convex function of $\boldsymbol{x}$ with domain C.*

The following theorem elucidates one of the relationships between multiple-variable convex functions and single-variable convex functions.[7]

Theorem G.8 (Construction of single-variable convex functions) *Let $f(\boldsymbol{x})$ be a convex function with domain $C \subseteq \mathbb{R}^n$. Then, for any $\boldsymbol{x}_1, \boldsymbol{x}_2 \in C$, the function $\tilde{f}(y)$ of $y \in \mathbb{R}$ defined as*

$$\tilde{f}(y) = f(y\,\boldsymbol{x}_1 + (1 - y)\boldsymbol{x}_2)\,, \tag{G.12}$$

whose domain is the interval $\{y \mid y\,\boldsymbol{x}_1 + (1 - y)\boldsymbol{x}_2 \in C\}$, is also convex.

This theorem follows directly from the definitions, and we therefore omit its proof. As a special case, it implies that for a convex function $f(\boldsymbol{x})$, if we fix all components of $\boldsymbol{x} = (x_1, \ldots, x_n)$ except one arbitrary x_i and consider $f(\boldsymbol{x})$ to be a function of x_i alone, then this function too is convex.

The following generalizes Theorem G.3.[8]

Theorem G.9 (Continuity of a convex function of multiple variables) *Any convex function $f(\boldsymbol{x})$ with an open domain $C \subseteq \mathbb{R}^n$ is continuous at all $\boldsymbol{x} \in C$.*

---

[6] It is natural to regard the parameter space for a collection of n extensive thermodynamic variables as $\mathbb{R}^n_+$.

[7] From Theorem G.3, we know that the function $\tilde{f}(y)$ appearing in this theorem is necessarily continuous if its domain is an open interval.

[8] Because, as far as the authors are aware, there is no concise proof of this theorem, and because we do not use this theorem in this book, we simply present it without proof.

Finally, we present the following generalization of Theorem G.6, which establishes the fact that a critical point of a convex function of multiple variables must be its global minimum point.

Theorem G.10 (Minimum as the unique critical point) *Let $f(\boldsymbol{x})$ be a convex function with domain $C \subseteq \mathbb{R}^n$, and suppose that for some $\boldsymbol{x}_0 \in C$ we have*

$$\mathrm{grad} f(\boldsymbol{x}_0) = (0, \ldots, 0) \ . \tag{G.13}$$

Then $f(\boldsymbol{x})$ realizes its minimum value at $\boldsymbol{x}_0$.

The proof of this theorem is quite simple. First, assume that there exists $\boldsymbol{x}_1 \in C$ for which $f(\boldsymbol{x}_1) < f(\boldsymbol{x}_0)$ holds. Then note that the function

$$\tilde{f}(y) = f(y\,\boldsymbol{x}_1 + (1 - y)\boldsymbol{x}_0) \tag{G.14}$$

is convex and satisfies $\tilde{f}'(0) = (\boldsymbol{x}_1 - \boldsymbol{x}_0) \cdot \mathrm{grad} f(\boldsymbol{x}_0) = 0$. Hence, Theorem G.6 implies the relation $\tilde{f}(y) \geq \tilde{f}(0)$. However, this contradicts the relation $\tilde{f}(1) < \tilde{f}(0)$, which follows directly from the assumption.

G.3 Proofs

In this section, we present proofs of Theorems G.1, G.3 and G.4. (The remaining theorems— with the exception of Theorem G.9—were proved above or can be proved quite easily.) As preparation, we first prove the following lemma, which elucidates a useful property of convex functions.

Lemma G.11 *Let $f(x)$ be a convex function defined on an interval I. Choose arbitrary $x_1, x_2 \in I$ satisfying $x_1 < x_2$, and let $g(x)$ be the linear function that coincides with $f(x)$ at the points x_1 and x_2:*

$$g(x) = f(x_1) + \frac{f(x_2) - f(x_1)}{x_2 - x_1}(x - x_1) = f(x_2) + \frac{f(x_2) - f(x_1)}{x_2 - x_1}(x - x_2) \ . \tag{G.15}$$

We then have

$$f(x) \begin{cases} \geq g(x) & \text{for } x \leq x_1 \ , \\ \leq g(x) & \text{for } x_1 \leq x \leq x_2 \ , \\ \geq g(x) & \text{for } x_2 \leq x \ . \end{cases} \tag{G.16}$$

The meaning of these inequalities can be understood from Figure G.1, where the diagonal line represents $g(x)$.

Proof *Considering the case $x_1 \leq x \leq x_2$, and setting $y_1 = x_1$, $y_2 = x$ and $y_3 = x_2$ in (G.2), we obtain*

$$f(x) - g(x) = f(x) - f(x_1) - \frac{f(x_2) - f(x_1)}{x_2 - x_1}(x - x_1)$$

$$= \frac{1}{x_2 - x_1}\{(x_2 - x_1)f(x) - (x_2 - x)f(x_1) - (x - x_1)f(x_2)\}$$

$$= -\frac{1}{x_2 - x_1}\{(y_3 - y_2)f(y_1) - (y_3 - y_1)f(y_2) + (y_2 - y_1)f(y_3)\} \leq 0 \ . \tag{G.17}$$

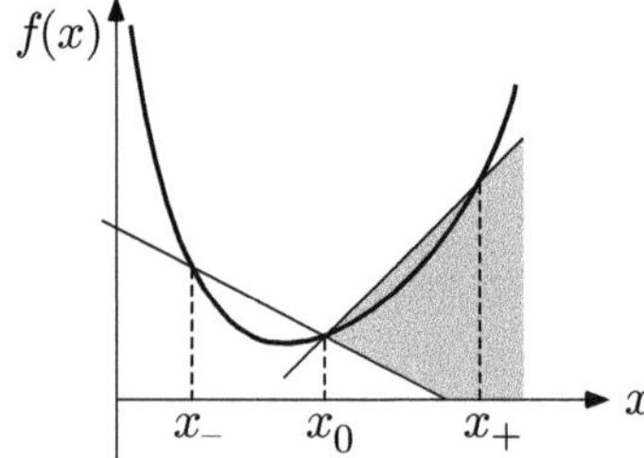

Figure G.5 Approach used for proving the continuity of a convex function. For x satisfying $x_0 < x < x_+$, the function $f(x)$ must be contained in the shaded region bounded by the two lines. Therefore, in the limit $x \searrow x_0$, $f(x)$ is constrained in such a manner that $f(x)$ converges to $f(x_0)$.

The remaining cases can be demonstrated through similar application of (G.2) by properly accounting for the order relations among x, x_1 and x_2. ∎

Proof of Theorem G.3 *Below, we directly prove continuity, without using the intuitive understanding provided by Figure G.2 (which indeed is not easy to formalize). The method of proof used here can be understood from Figure G.5.*

Choose arbitrary $x_-, x_0, x_+ \in I$ satisfying $x_- < x_0 < x_+$. We will prove that $f(x)$ is continuous at $x = x_0$. First, note that for any x satisfying $x_0 < x < x_+$, with the identifications $x_1 = x_0$ and $x_2 = x_+$, the second inequality in (G.16) becomes

$$f(x) \le f(x_0) + \frac{f(x_+) - f(x_0)}{x_+ - x_0}(x - x_0) . \tag{G.18}$$

Also, with the identifications $x_1 = x_-$ and $x_2 = x_0$, the third inequality in (G.16) becomes

$$f(x) \ge f(x_0) + \frac{f(x_0) - f(x_-)}{x_0 - x_-}(x - x_0) . \tag{G.19}$$

Then, defining the constants C and C' as $C = \{f(x_+) - f(x_0)\}/(x_+ - x_0)$ and $C' = \{f(x_0) - f(x_-)\}/(x_0 - x_-)$, the above inequalities imply

$$C(x - x_0) \ge f(x) - f(x_0) \ge C'(x - x_0) . \tag{G.20}$$

It is thus shown that in the limit $x \searrow x_0$, we have $f(x) \to f(x_0)$. Hence, $f(x)$ is right-continuous at x_0. Next, exchanging the roles of x_+ and x_- and repeating the above argument, we find that $f(x)$ is also left-continuous at x_0. We therefore conclude that $f(x)$ is continuous at x_0. Because x_0 is arbitrary, this completes the proof. ∎

Proof of Theorem G.4 *Let $f(x)$ be a convex function defined on an open interval I. Define the function $\gamma(x, y)$ with domain $\{(x, y) \mid x \in I, y \in I, x < y\}$ as*

$$\gamma(x, y) = \frac{f(y) - f(x)}{y - x} . \tag{G.21}$$

Obviously, $\gamma(x, y)$ is the slope of the line passing through $(x, f(x))$ and $(y, f(y))$. Also, $\gamma(x, y)$ is non-decreasing in both x and y, as we now show. First, note that for any x, y and y' satisfying $x < y < y'$, (G.2) implies

$$\gamma(x, y') - \gamma(x, y) = \frac{(y' - y)f(x) - (y' - x)f(y) + (y - x)f(y')}{(y' - x)(y - x)} \ge 0 . \tag{G.22}$$

Similarly, it can be shown that for any x, x' and y satisfying $x < x' < y$, we have $\gamma(x', y) - \gamma(x, y) \geq 0$.

Fix arbitrary $x \in I$ and choose $\delta > 0$ such that $x - \delta \in I$. Next, with $x + \varepsilon \in I$, regard $\gamma(x, x + \varepsilon)$ as a function of $\varepsilon > 0$. Clearly, $\gamma(x, x + \varepsilon)$ is non-decreasing in ε and satisfies $\gamma(x, x + \varepsilon) \geq \gamma(x - \delta, x)$ for every ε. Then, because a non-decreasing function bounded from below on an open interval converges (to its infimum) in the limit that the left endpoint of the interval is approached, it follows that $\lim_{\varepsilon \searrow 0} \gamma(x, x + \varepsilon)$ exists. This is the right derivative $f'_+(x)$. The existence of the left derivative can be demonstrated similarly.

Choose arbitrary x_1 and x_2 with $x_1 < x_2$. For any ε satisfying $0 < \varepsilon < x_2 - x_1$, the non-decreasing nature of $\gamma(x, y)$ implies

$$\gamma(x_1 - \varepsilon, x_1) \leq \gamma(x_1, x_1 + \varepsilon) \leq \gamma(x_2 - \varepsilon, x_2) \leq \gamma(x_2, x_2 + \varepsilon) \,. \tag{G.23}$$

Taking $\varepsilon \searrow 0$, we obtain (G.6).

Again, choose arbitrary x_1 and x_2 with $x_1 < x_2$. For any x satisfying $x < x_1$ or $x > x_2$, (G.16) implies

$$f(x) \geq f(x_1) + \gamma(x_1, x_2)(x - x_1) \,. \tag{G.24}$$

Setting $x_1 = x_0$ and taking the limit $x_2 \searrow x_0$ in (G.24), we obtain (G.8) with $\alpha = f'_+(x_0)$, and setting $x_2 = x_0$ and taking the limit $x_1 \nearrow x_0$ in (G.24), we obtain (G.8) with $\alpha = f'_-(x_0)$. Obviously, (G.8) also holds for values of α between $f'_-(x_0)$ and $f'_+(x_0)$. ∎

Proof of Theorem G.1 *Consider a convex function $f(x)$ and suppose that it is twice differentiable at some value x_0. Then, from (G.9), we obtain*

$$f''(x_0) = \lim_{\varepsilon \searrow 0} \frac{f'(x_0 + \varepsilon) - f'(x_0)}{\varepsilon} \geq 0 \,. \tag{G.25}$$

This completes the proof of the first assertion.

Next, suppose that $f(x)$ is twice differentiable and satisfies $f''(x) \geq 0$ at all x. Integrating $f''(x)$, we have

$$f'(x) = f'(x_0) + \int_{x_0}^{x} dy\, f''(y) \,. \tag{G.26}$$

Then, integrating this expression for $f'(x)$, we obtain

$$\begin{aligned}
f(x) &= f(x_0) + \int_{x_0}^{x} dz\, f'(z) \\
&= f(x_0) + (x - x_0)f'(x_0) + \int_{x_0}^{x} dz \int_{x_0}^{z} dy\, f''(y) \\
&\geq f(x_0) + (x - x_0)f'(x_0) \,, \tag{G.27}
\end{aligned}$$

where we have used the inequality $f''(y) \geq 0$. Thus, we find that $f(x)$ lies above the line tangent to $(x_0, f(x_0))$ everywhere except at the point(s) of tangency, as in Figure G.3. The convexity of $f(x)$ follows almost directly from this fact, as we show below.

With arbitrary x_1 and x_2 satisfying $x_1 < x_2$ and arbitrary $\lambda \in [1, 0]$, setting $x_0 = \lambda x_1 + (1 - \lambda)x_2$, we can rewrite (G.27) as

$$f(x) \geq f(\lambda x_1 + (1 - \lambda)x_2) + \{x - (\lambda x_1 + (1 - \lambda)x_2)\}\, f'(x_0) \,. \tag{G.28}$$

Then, setting $x = x_1$ in (G.28) and multiplying both sides by λ, we find

$$\lambda f(x_1) \geq \lambda f(\lambda x_1 + (1 - \lambda)x_2) + \lambda(1 - \lambda)(x_1 - x_2)f'(x_0) , \tag{G.29}$$

while setting $x = x_2$ in (G.28) and multiplying both sides by $(1 - \lambda)$, we find

$$(1 - \lambda)f(x_2) \geq (1 - \lambda)f(\lambda x_1 + (1 - \lambda)x_2) - \lambda(1 - \lambda)(x_1 - x_2)f'(x_0) . \tag{G.30}$$

Adding these two expressions, the two terms containing $f'(x_0)$ cancel, and we are left with the condition defining a convex function, (G.1). ∎

APPENDIX H

Legendre Transformation

In this appendix, we present a general, systematic treatment of the Legendre transformation.[1] In Sections H.1–H.3, we focus mainly on the application of the Legendre transformation to convex functions. In this case, the Legendre transformation possesses simple properties, which allow it to serve as a powerful tool within the theory of thermodynamics. In Section H.4, we briefly consider the application of the Legendre transformation to non-convex functions. Proofs for the theorems presented in this appendix are given in Section H.5.

H.1 Definition of the Legendre transformation

H.1.1 Definition

Let $f(x)$ be a real-valued function defined on an interval $I \subseteq \mathbb{R}$. As in the previous appendix, I may be open, closed or half-open, bounded, unbounded or half-bounded. The Legendre transform, $f^*(\alpha)$, of $f(x)$ is defined as follows:[2]

$$f^*(\alpha) := \sup_{x \in I} \{\alpha x - f(x)\} = - \inf_{x \in I} \{f(x) - \alpha x\} \ . \tag{H.1}$$

The domain of $f^*(\alpha)$, which we write I^*, is the set of all α for which the above supremum exists:

$$I^* := \{\alpha \in \mathbb{R} \,|\, \sup_{x \in I} \{\alpha x - f(x)\} \in \mathbb{R}\} \ . \tag{H.2}$$

Thus, a given α is contained in I^* if and only if the function $\alpha x - f(x)$ is bounded from above (i.e., there exists $M \in \mathbb{R}$ such that $\alpha x - f(x) < M$ holds for all $x \in I$). More intuitively, α is contained in I^* if we can draw a line of slope α that is below the graph of $f(x)$ for all $x \in I$.

We say that the Legendre transform $f^*(\alpha)$ exists when I^* is non-empty. As seen in Theorem H.1, the Legendre transform of a convex function always exists.

---

[1] As pointed out in Chapter 8, the form of the Legendre transformation treated here is slightly different from that used in the main text. Comparing (H.1) with (8.3), it is seen that the two forms employ different sign conventions. That employed in this appendix is regarded as the standard convention within mathematics. There are two important implications of this difference. First, with the standard convention, a Legendre transform is always convex, while with the thermodynamic convention, it is always concave. Second, with the standard convention, the Legendre transformation is its own inverse, while with the thermodynamic convention (as seen in the main text), the Legendre transformation and its inverse are distinct. We also note that in the mathematical literature, the Legendre transformation as defined here is generally referred to as the *Legendre-Fenchel transformation*. This is a generalization of the transformation introduced by Legendre.

[2] In this definition, sup and inf represent the *supremum* and *infimum*. For a set of real numbers, the supremum is the smallest number greater than or equal to all elements of the set, and the infimum is the largest number less than or equal to all elements of the set. (In other words, they are the least upper bound and greatest lower bound of the set, respectively.) The supremum and infimum are generalizations of the maximum and minimum: While the supremum and infimum exist more generally, if the maximum/minimum exists, it is identical to the supremum/infimum. In the physics literature, the Legendre transformation is often defined in terms of max or min instead of sup or inf, because for practical applications, the difference is usually of little consequence.

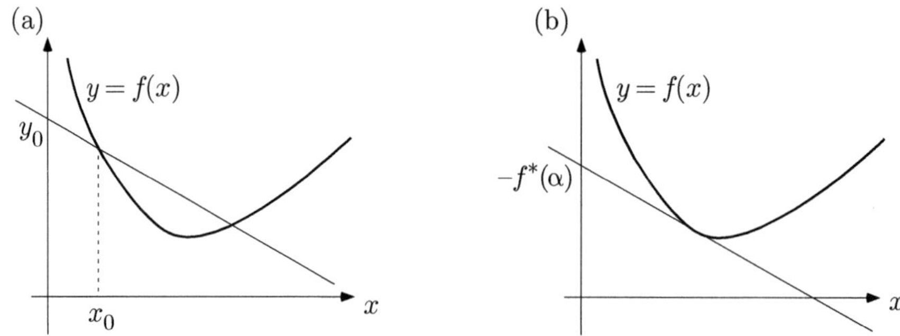

Figure H.1 Graphical representation of (H.1) for a convex function. (a) y_0 is the y-intercept of the line of slope α passing through the point $(x_0, f(x_0))$. (b) $-f^*(\alpha)$ represents the minimum value of y_0 realized when x_0 is varied over all possible values.

H.1.2 Graphical representation of (H.1)

While the above definition may seem somewhat abstract, it can be understood in a simple, intuitive manner, as we now describe. In the following discussion, we first consider convex functions and then non-convex functions, in order to elucidate the role played by convexity in the Legendre transformation.

Consider the convex function $f(x)$ plotted in Figure H.1. Then, for a given value of α, imagine the family of lines with slope α that intersect the graph of $f(x)$. The y-intercept for any such line, y_0, is equal to $f(x_0) - \alpha x_0$, where x_0 is a point of intersection. From among these lines, choose that one for which y_0 is smallest. (Clearly this minimum value exists in the case of the function $f(x)$ and the value of α considered in Figure H.1.) Identifying this minimum value of y_0 as $-f^*(\alpha)$, we have $f^*(\alpha) = -\min_x \{f(x) - \alpha x\}$.

To understand the fundamental role played by convexity, let us next consider the two non-convex functions depicted in Figures H.2(a) and (b). As above, we consider a line of slope α that passes through some point $(x_0, f(x_0))$ on the graph of $f(x)$.

As a typical example of a concave function, in Figure H.2(a) we consider $f(x) = -x^2$, with $I = \mathbb{R}$. As shown in the figure, for a line with any given slope $\alpha \in \mathbb{R}$, moving x_0 far enough to either the left or right, the y-intercept decreases without bound. We thus conclude that for $f(x) = -x^2$ with $I = \mathbb{R}$, the domain I^* is empty.

Next, let us consider the function plotted in Figure H.2(b), where it is understood that I is $\mathbb{R}$ and that $f(x)$ is convex except in the region of the bump. In this case, $f^*(\alpha)$ is defined for some range of values of α, but the region between the two minima of the bump plays no role in the determination of the function $f^*(\alpha)$. To see this, let α_0 denote the slope of the line that is tangent to both "dips" in this function, as shown in the figure. It is then found that for $\alpha < \alpha_0$, $f^*(\alpha)$ is determined by the behavior of $f(x)$ in the region $x < x_1$, while for $\alpha > \alpha_0$, $f^*(\alpha)$ is determined by the behavior of $f(x)$ in the region $x > x_2$. (Obviously, $f^*(\alpha_0)$ is determined by $f(x_1)$ and $f(x_2)$.) Hence, no information concerning the behavior of $f(x)$ for $x \in (x_1, x_2)$ is reflected by $f^*(\alpha)$. In this way, information is lost when the Legendre transformation is applied to a function of this kind.

H.1.3 Simplified definition of the Legendre transform

In the special case that $f(x)$ is once-differentiable and $f'(x)$ is a continuous, increasing function,[3] to determine $f^*(\alpha)$ we need only find the unique solution of $\frac{d}{dx}(\alpha x - f(x)) = 0$.

---

[3] This is a stronger condition than convexity.

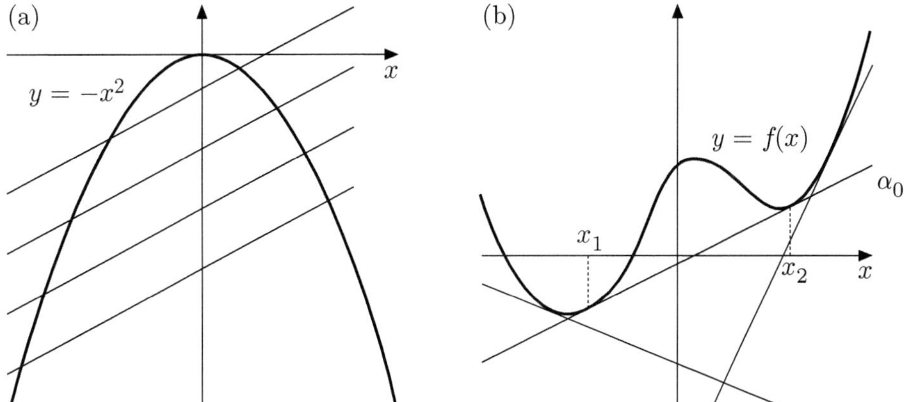

Figure H.2 Graphical representation of (H.1) for non-convex functions. (a) Here we consider $f(x) = -x^2$, with $I = \mathbb{R}$, as a typical example of a concave function. In this case, clearly, it is not possible to draw a line of any slope that remains below $f(x)$ for all $x \in I$. Therefore, for $f(x) = -x^2$ with domain $\mathbb{R}$, the Legendre transform does not exist. (b) For the function $f(x)$ considered here, $f^*(\alpha)$ is defined for some range of values of α. However, for $\alpha < \alpha_0$, the function $f^*(\alpha)$ reflects information regarding the form of $f(x)$ only for values of x satisfying $x < x_1$, while for $\alpha > \alpha_0$, it reflects such information only for values of x satisfying $x > x_2$. Thus, information regarding the form of $f(x)$ on the interval (x_1, x_2) is not reflected by $f^*(\alpha)$. This information has been lost under the Legendre transformation.

Writing this solution as $x^*(\alpha)$ and substituting this into (H.1), we obtain

$$f^*(\alpha) = \alpha\, x^*(\alpha) - f(x^*(\alpha))\,. \tag{H.3}$$

In many physics textbooks, this relation is regarded as the definition of the Legendre transform. However, in cases that $f'(x)$ does not exist or $x^*(\alpha)$ is not unique, this relation is inapplicable.

H.1.4 Fundamental properties of the Legendre transform

Below, we present four theorems stating some fundamental properties of the Legendre transform defined above. Their proofs are given in Section H.5.

The following two theorems reflect the close connection between convexity and the Legendre transformation.

Theorem H.1 (Existence of $f^*(\alpha)$ for a convex function) *For any convex function $f(x)$, the Legendre transform $f^*(\alpha)$ exists (i.e., the domain I^* defined in (H.2) is non-empty).*

Theorem H.2 (Convexity of the Legendre transform) *For any function $f(x)$, if $f^*(\alpha)$ exists, then it is a convex function.*

The simple inequality given in the theorem below expresses an essential property of the Legendre transformation. For this reason, it is quite useful for demonstrating various other properties.

Theorem H.3 (The Legendre transformation and Young's inequality) *For any function $f(x)$ (with domain I) whose Legendre transform $f^*(\alpha)$ (with domain I^*)*

exists, the following Young's inequality necessarily holds for all $x \in I$ and all $\alpha \in I^*$:

$$f(x) + f^*(\alpha) \geq x\,\alpha \,. \tag{H.4}$$

The following theorem asserts a kind of insensitivity of the Legendre transform of a convex function $f(x)$ to the variation of $f(x)$ at finite endpoints of I.

Theorem H.4 (Convex functions with open and non-open domains) *For a convex function $f(x)$ defined on a non-open proper interval I, let $\tilde{I}$ denote the interior of I and $\tilde{f}(x)$ denote the restriction of $f(x)$ to $\tilde{I}$. Then, the Legendre transforms of $f(x)$ and $\tilde{f}(x)$ are identical (including their domains).*

H.2 Reconstructing $f(x)$ from $f^*(\alpha)$

In this section, we present an intuitive description of how a convex function $f(x)$ can be reconstructed from its Legendre transform, $f^*(\alpha)$. For this purpose, we consider only the simplest case of a convex function on an open interval possessing no linear regions and for which the slope tends to $\pm\infty$ as the limits of I are approached. The general case is considered in Theorem H.5.

Let us reconsider the graphical representation of the definition of $f^*(\alpha)$ displayed in Figure H.1. When the slope of the line, α, is varied over I^*, the value(s) of x_0 corresponding to the minimum of y_0 varies over I. In the simple situation considered here, there is a one-to-one correspondence between the values of α and the values of x_0. Due to this property, all information concerning the form of $f(x)$ is incorporated into $f^*(\alpha)$. It is therefore reasonable to conjecture that the former could be reconstructed from the latter. Here, we investigate a method for doing this.

Given a function $f^*(\alpha)$, let us choose a value of α and construct the line with slope α and y-intercept $-f^*(\alpha)$, defined by the equation

$$y = \alpha\,x - f^*(\alpha) \,. \tag{H.5}$$

Recalling the manner in which $f^*(\alpha)$ was obtained, we know that $f(x)$ lies above this line, except at the point where the two are tangent. Next, choose a second value of α, and construct another line in the same manner as the first. We then have two lines lying below $f(x)$ everywhere except at their points of tangency. Repeating this process indefinitely, the function $f(x)$ is reconstructed as the envelope of the set of lines so obtained, as illustrated in Figure H.3.

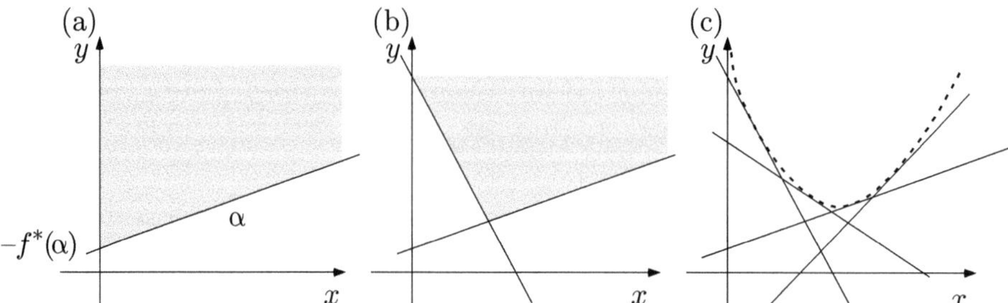

Figure H.3 Method for reconstructing $f(x)$ from known $f^*(\alpha)$. (a) The function $f(x)$ lies entirely in the shaded region above the line of slope α and y-intercept $-f^*(\alpha)$, except at their point of tangency. (b) Adding a second line constructed like the first, the region in which $f(x)$ can exist is narrowed. (c) Repeating this process indefinitely, the graph of $f(x)$ takes form as the envelope of the set of all lines constructed in this manner.

Let us now formulate the above procedure mathematically. Suppose that we wish to determine the value of $f(x)$ for one particular value of $x \in I$. We know that $f(x)$ can nowhere be below the line given in (H.5), and hence for every $\alpha \in I^*$, we have the following relation:

$$f(x) \geq \alpha\, x - f^*(\alpha) \; . \tag{H.6}$$

This is Young's inequality (see (H.4)). This inequality holds for all values of α, and the equality holds for one value. We therefore conclude that $f(x)$ can be expressed as[4]

$$f(x) = \sup_{\alpha \in I^*} \{x\alpha - f^*(\alpha)\} \; . \tag{H.7}$$

Comparing (H.7) with (H.1), we see that these two expressions have identical forms with the roles of $f(x)$ and $f^*(\alpha)$ reversed. We have thus found that if we apply the Legendre transformation twice to a convex function $f(x)$, the original function is returned.

The following theorem formalizes the intuitive procedure considered above, while being entirely general.

Theorem H.5 (Inverting the Legendre transformation) *For any convex function $f(x)$ defined on an open interval I, writing the Legendre transform of its Legendre transform as $f^{**}(x)$, the following relation holds:*

$$f^{**}(x) = f(x) \text{ for all } x \in I \; . \tag{H.8}$$

H.3 Examples

Helmholtz free energy and Gibbs free energy

As discussed in Section 8.1, the Helmholtz free energy $F[T; V, N]$ and the Gibbs free energy $G[T, p; N]$ are connected through the Legendre transformation. Although the Legendre transformation used in that context and the standard Legendre transformation treated here have different sign conventions, there is a simple relation between these transformations, as we now see.

For a given Helmholtz free energy $F[T; V, N]$, fixing T and N and representing the volume by the variable x, let us write

$$f(x) = F[T; x, N] \; , \tag{H.9}$$

with the domain of $f(x)$ understood to be $(0, \infty)$. Because $f(x)$ is a convex function, its Legendre transform, $f^*(\alpha)$, exists. Roughly speaking, the domain of $f^*(\alpha)$ corresponds to the range of $f'(x)$, and hence (because $-f'(x)$ is the pressure), α must be negative. Thus, with T and N possessing the values fixed above, we define the Gibbs free energy as

$$G[T, p; N] = -f^*(-p) \; . \tag{H.10}$$

It is readily seen that defined in this way, $G[T, p; N]$ is a concave function of p with domain $(0, \infty)$. Note that (H.10) is identical to (8.3). Furthermore, the validity of the expression in (8.8), representing the inverse Legendre transformation of $G[T, p; N]$ in that context, is guaranteed by Theorem H.5.

---

[4] In the simple situation considered presently, obviously sup could be replaced with max.

Analytical mechanics

The Legendre transformation also plays an important role in analytical mechanics. There, the Hamiltonian, $H(q, p, t)$, is obtained as the Legendre transform of the Lagrangian, $L(q, u, t)$:[5]

$$H(q, p, t) = \max_{u}\{up - L(q, u, t)\} . \tag{H.11}$$

Function on a degenerate interval

Let us consider a function $f(x)$ defined on a degenerate interval I, i.e., an interval consisting of a single point $a \in \mathbb{R}$. Obviously, any such function is convex, as it trivially satisfies (G.1). Then, because for any $\alpha \in \mathbb{R}$, we also have $a\alpha - f(a) \in \mathbb{R}$, it is seen that the Legendre transform is given by $f^*(\alpha) = a\alpha - f(a)$ with $I^* = \mathbb{R}$. It is easily confirmed that $f^{**}(x)$ coincides with $f(x)$.

Monomial functions

Next, consider the function

$$f(x) = \frac{x^p}{p} , \tag{H.12}$$

with constant $p > 1$. We take the domain of $f(x)$ to be $I = \mathbb{R}_+$. Because $f'(x)$ is a continuous, increasing function for all $x > 0$, the Legendre transform of this function can be obtained from (H.3). With the solution of the equation $\alpha = f'(x) = x^{p-1}$ given by $x^*(\alpha) = \alpha^{1/(p-1)}$, we have

$$f^*(\alpha) = \alpha \cdot \alpha^{1/(p-1)} - \frac{(\alpha^{1/(p-1)})^p}{p} = \frac{\alpha^q}{q} . \tag{H.13}$$

Here, q is a constant determined by the equation

$$\frac{1}{p} + \frac{1}{q} = 1 . \tag{H.14}$$

Clearly, q is necessarily greater than 1. Interestingly, in the case $p = 2$, $f(x)$ and $f^*(\alpha)$ have identical functional forms.

Applying Young's inequality (H.4) to this example, we obtain

$$\frac{x^p}{p} + \frac{\alpha^q}{q} \geq x\alpha \tag{H.15}$$

for arbitrary $x > 0$ and $\alpha > 0$ and arbitrary $p > 1$ and $q > 1$ satisfying (H.14). Defining the variables $X = x^p$ and $Y = \alpha^q$, we can write (H.15) as

$$\frac{X}{p} + \frac{Y}{q} \geq X^{1/p} Y^{1/q} . \tag{H.16}$$

This is a generalization of the well-known inequality relating the arithmetic and geometric means, which corresponds to the special case $p = q = 2$.

---

[5] In this expression, q represents the generalized coordinates, and u is the velocity variable corresponding to q (which is usually expressed as $\dot{q}$). In the Hamiltonian, p is the momentum variable conjugate to q, and together they form a set of canonical coordinates.

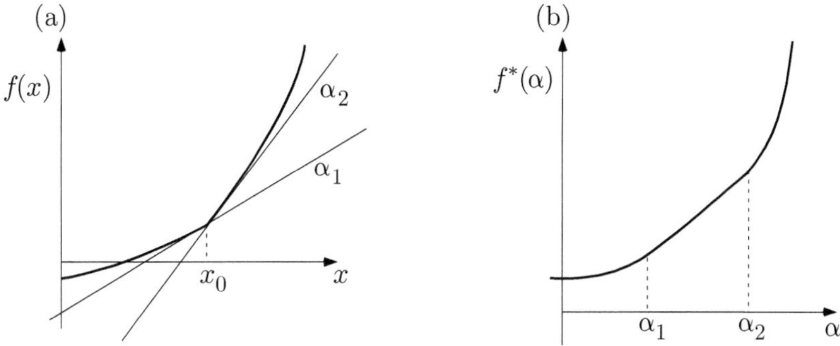

Figure H.4 (a) The function $f(x)$ is non-differentiable at x_0. (b) The Legendre transform of $f(x)$.

Exponential functions

As another simple example, let us consider the exponential form $f(x) = A\,\mathrm{e}^{Bx}$ (with $I = \mathbb{R}$). In the case $A < 0$ (for both $B > 0$ and $B < 0$), with considerations similar to those applied to the function in Figure H.2(a), it is easily seen that the Legendre transform of $f(x)$ does not exist. We therefore consider the case $A > 0$. Then, because $f'(x)$ is a continuous, increasing function, the simplified definition (H.3) is valid in this case. Applying this definition, we readily obtain $f^*(\alpha) = \frac{\alpha}{B}\left(\log\frac{\alpha}{AB} - 1\right)$ (with $I^* = \mathbb{R}_+$ for $B > 0$ and $I^* = \mathbb{R}_-$ for $B < 0$).

Function with a point of non-differentiability

Finally, we consider the Legendre transformation applied to the function $f(x)$ plotted in Figure H.4(a), which is non-differentiable at $x = x_0$. As shown in the figure, the derivative of $f(x)$ changes discontinuously at this point. The left and right derivatives at x_0 are related as $f'_-(x_0) = \alpha_1 < f'_+(x_0) = \alpha_2$. For this reason, as can be understood from the definition (H.1) and from Figure H.3, for all values of the slope α satisfying $\alpha_1 \leq \alpha \leq \alpha_2$, the line that minimizes the y-intercept passes through $(x_0, f(x_0))$. Thus, in this range, the value of this minimum y-intercept is a linear function of α. The resulting Legendre transform, $f^*(\alpha)$, is an everywhere differentiable function that varies linearly for $\alpha \in (\alpha_1, \alpha_2)$, as seen in Figure H.4(b). The reader should confirm the converse—that the Legendre transform of a function possessing a linear region, like that in Figure H.4(b), contains a corresponding point of non-differentiability. (Such a situation was considered in Section 8.4.)

H.4 The Legendre transformation applied to non-convex functions

In this section, we briefly study the Legendre transformation applied to non-convex functions. With Theorem H.2, we established that if it exists, the Legendre transform of even a non-convex function $f(x)$ is convex. Here we consider the interesting properties of the function biconjugate to $f(x)$ under the Legendre transformation (i.e., the Legendre transform of the Legendre transform of $f(x)$). The theorem given below is relevant to the model of ferromagnetic phase transitions based on the Landau pseudo free energy (Section 10.3) and the van der Waals theory of fluids (Problems 3.3 and 7.8).

Theorem H.6 (Non-convex functions and their biconjugate functions) *Let $f(x)$ be a non-convex function with domain I whose Legendre transform $f^*(\alpha)$ has non-empty domain I^*. The Legendre transform of $f^*(\alpha)$, $f^{**}(x)$, satisfies $f^{**}(x) \leq f(x)$ for all $x \in I$. Furthermore, for any convex function $g(x)$ with domain I that satisfies*

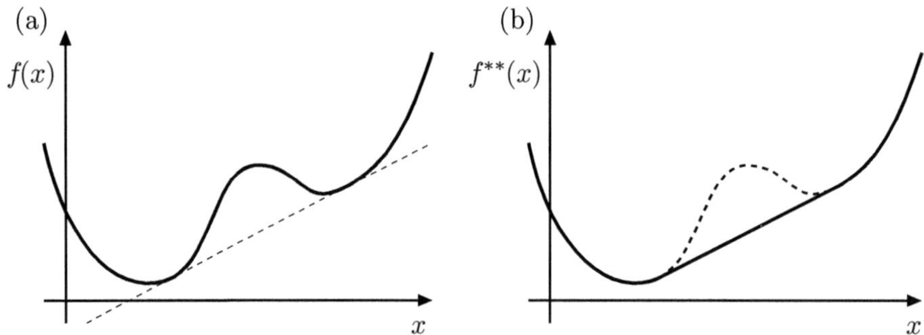

Figure H.5 (a) A non-convex function $f(x)$ with two local minima. (b) Its biconjugate function, $f^{**}(x)$.

$g(x) \leq f(x)$ *for all* $x \in I$, *we have the relation* $g(x) \leq f^{**}(x)$ *for all* $x \in I$. *In this sense,* $f^{**}(x)$ *is the largest convex function that nowhere exceeds* $f(x)$.

With these results, it is natural to interpret $f^{**}(x)$ as the convex function "closest" to $f(x)$. The function $f^{**}(x)$ on the interval I is called the *convex hull* of $f(x)$.

Figure H.5 depicts a typical example. There, $f(x)$ is a non-convex function with two local minima. The function biconjugate to $f(x)$ can be obtained by first drawing the unique line tangent to the graph of $f(x)$ at two points, and then replacing the part of the graph between these two points with the line segment connecting them (also see Figure 10.7). To understand that this simple construction does indeed yield $f^{**}(x)$, it suffices to consider the graphical method described by Figure H.2 and confirm that the two graphs appearing in Figure H.5 have the same Legendre transform.

H.5 Proofs

In this section, we prove the theorems presented in this appendix.

Proof of Theorem H.1 *For a convex function defined on an open interval, the result follows from (G.8), as this relation implies that* I^* *contains each value of* α *satisfying (G.7). With this, Theorem H.4 provides the result for a convex function defined on a non-open proper interval. The case of a degenerate interval was treated explicitly in Section H.3.* ∎

Proof of Theorem H.3 *From (H.1), for arbitrary* $x \in I$ *and* $\alpha \in I^*$, *we immediately obtain*

$$f(x) + f^*(\alpha) = \sup_{y \in I} \{f(x) + \alpha y - f(y)\} \geq \alpha x \ . \quad \blacksquare \tag{H.17}$$

Proof of Theorem H.2 *For any* $\alpha_1, \alpha_2 \in I^*$ *and any* $\lambda \in [0,1]$, *Young's inequality implies*

$$\lambda f^*(\alpha_1) \geq \lambda x \alpha_1 - \lambda f(x) \ , \tag{H.18}$$

$$(1-\lambda)f^*(\alpha_2) \geq (1-\lambda)x\alpha_2 - (1-\lambda)f(x) \tag{H.19}$$

for all $x \in I$. *Adding these inequalities, we obtain*

$$\lambda f^*(\alpha_1) + (1-\lambda)f^*(\alpha_2) \geq x\{\lambda\alpha_1 + (1-\lambda)\alpha_2\} - f(x) \ . \tag{H.20}$$

Therefore, $\lambda f^*(\alpha_1) + (1-\lambda)f^*(\alpha_2)$ is an upper bound of $x\{\lambda\alpha_1 + (1-\lambda)\alpha_2\} - f(x)$ on I. Because the supremum is the minimum upper bound, this implies the following:

$$\lambda f^*(\alpha_1) + (1-\lambda)f^*(\alpha_2) \geq \sup_{x\in I}\{x\{\lambda\alpha_1 + (1-\lambda)\alpha_2\} - f(x)\} \ . \tag{H.21}$$

From (H.1), we thus obtain

$$\lambda f^*(\alpha_1) + (1-\lambda)f^*(\alpha_2) \geq f^*(\lambda\alpha_1 + (1-\lambda)\alpha_2) \ . \tag{H.22}$$

Hence, by definition, $f^*(\alpha)$ is convex. ∎

Proof of Theorem H.4 *Suppose that I is left-closed and call its left endpoint a. Let us compare $f^*(\alpha) = \sup_{x\in I}\{\alpha x - f(x)\}$ (with domain I^*) and $\tilde{f}^*(\alpha) = \sup_{x\in I\setminus\{a\}}\{\alpha x - f(x)\}$ (with domain $\tilde{I}^*$). Considering $\alpha \in I^*$, from the relation $\sup_{x\in I\setminus\{a\}}\{\alpha x - f(x)\} \leq \sup_{x\in I}\{\alpha x - f(x)\}$, it follows that $\alpha \in \tilde{I}^*$ also holds. Next, considering $\alpha \in \tilde{I}^*$ and using the relation $f(a) \geq \lim_{x\searrow a} f(x)$, which is implied by the convexity of $f(x)$, we obtain the following:*

$$\sup_{x\in I\setminus\{a\}}\{\alpha x - f(x)\} \geq \lim_{x\searrow a}\{\alpha x - f(x)\} \geq \alpha a - f(a) \ . \tag{H.23}$$

The equalities $f^*(\alpha) = \tilde{f}^*(\alpha)$ and $I^* = \tilde{I}^*$ follow.
   Extension to the cases of right-closed and closed I is obvious. ∎

Proof of Theorem H.5 *Consider a function $f(x)$ and its Legendre transform $f^*(\alpha)$, with domains I and I^*, respectively. Let us assume that I^* is non-empty. From Young's inequality, (H.4), we know that $f(x) \geq x\alpha - f^*(\alpha)$ holds for every $x \in I$ and $\alpha \in I^*$, and hence we have*

$$f(x) \geq \sup_{\alpha\in I^*}\{x\,\alpha - f^*(\alpha)\} = f^{**}(x) \ . \tag{H.24}$$

This implies the relation $I^{**} \supseteq I$, where I^{**} is the domain of $f^{**}(x)$.
   Next, restricting consideration to the case of convex $f(x)$ defined on an open interval, we derive the opposite relation. From (H.4), (H.1) and (G.8), we obtain

$$\begin{aligned}
f^{**}(x) &\geq x\alpha' - f^*(\alpha') \\
&= \inf_{x'\in I}\{x\alpha' - x'\alpha' + f(x')\} \\
&\geq \inf_{x'\in I}\{x\alpha' - x'\alpha' + f(x) + \alpha(x' - x)\}
\end{aligned} \tag{H.25}$$

for all $x \in I$ and $\alpha' \in I^*$, where α is any value satisfying $f'_-(x) \leq \alpha \leq f'_+(x)$. Then, choosing α' to be equal to α, the x' dependence of the right-hand side vanishes, and the above relation reduces to

$$f^{**}(x) \geq f(x) \ . \tag{H.26}$$

Combining (H.24) and (H.26), we obtain (H.8). ∎

Before moving on to the proof of Theorem H.6, we present the following lemma.

Lemma H.7 *Consider two functions $f(x)$ and $g(x)$ defined on the same interval I that satisfy $f(x) \geq g(x)$ for all $x \in I$. The Legendre transforms of these functions, $f^*(\alpha)$ with domain I_f^* and $g^*(\alpha)$ with domain I_g^*, satisfy $I_f^* \supseteq I_g^*$ and $f^*(\alpha) \leq g^*(\alpha)$ for all $\alpha \in I_g^*$.*

Proof *Suppose that there exists α_0 for which $g^*(\alpha_0)$ is defined but $f^*(\alpha_0)$ is not defined. This implies that the function $\alpha_0 x - f(x)$ on I has no upper bound. However, because $\alpha_0 x - f(x) \leq \alpha_0 x - g(x)$ holds for all $x \in I$, clearly, $g^*(\alpha_0)$ is an upper bound of $\alpha_0 x - f(x)$. The relation $I_f^* \supseteq I_g^*$ follows.*

Then, for any $\alpha \in I_g^*$, the relation $f^*(\alpha) \leq g^*(\alpha)$ is obtained by taking the supremum over I of each side of the inequality $\alpha x - f(x) \leq \alpha x - g(x)$. $\blacksquare$

Proof of Theorem H.6 *The first assertion follows from (H.24).*

To prove the second assertion, we begin with Lemma H.7, which provides the relations $f^*(\alpha) \leq g^*(\alpha)$ for all $\alpha \in I_g^*$ and $I_f^* \supseteq I_g^*$. From these, we obtain the following:

$$g^{**}(x) = \sup_{x \in I_g^*} \{\alpha x - g^*(\alpha)\} \leq \sup_{x \in I_g^*} \{\alpha x - f^*(\alpha)\} \leq \sup_{x \in I_f^*} \{\alpha x - f^*(\alpha)\} = f^{**}(x) \,.$$

$$\text{(H.27)}$$

Then, recalling that, as demonstrated by (H.24), the domains of $g^{**}(x)$ and $f^{**}(x)$, I_g^{**} and I_f^{**}, satisfy $I_g^{**} \supseteq I$ and $I_f^{**} \supseteq I$ and that, from Theorem H.5, the relation $g^{**}(x) = g(x)$ holds for all $x \in I$, we conclude that $f^{**}(x) \geq g(x)$ holds for all $x \in I$. $\blacksquare$

Solutions to Problems

Chapter 1

1.1. From $T_1^{3/2}V_2 = T_3^{3/2}V_1$, we obtain $T_3 = (V_2/V_1)^{2/3}T_1 > T_1$.

1.2. In (a), the pressure of the gas is p_0, and hence its volume is given by $V = NRT_\mathrm{L}/p_0$. In (c), the pressure of the gas is $p_0 + mg/A$, and hence its volume is given by $V' = NRT_\mathrm{H}/(p_0 + mg/A)$. Dividing these expressions by A, we obtain the height in either case. The second half of the problem is left to the reader.

1.3. Because a full treatment of this topic would be too lengthy, here we merely present simple discussion concerning some essential points.

First, it is important to understand that definite conclusions cannot be drawn solely on the basis of theoretical considerations. Because we are inquiring about physical reality, the argument must be based on concrete experimental results. The reader should therefore consider actual experiments and carefully analyze them to understand in what sense and in which situations their results support the existence of atoms and molecules. Although in discussions of this kind, it is common to cite historical works, such as Jean Perrin's experiment on Brownian motion, we encourage the reader to also survey modern experiments and examine their implications. (For example, see Figure S.1.)

Next, it is important to understand that regarding any particular experiment or experiments as providing definite "proof" of the existence of atoms and molecules is problematic. Within science, there is no rule or method that allows us to determine conditions sufficient to judge a theory to be "proven" on the basis of experimental and/or theoretical results. For this reason, instead of "proof," we seek "preponderance of evidence." To formulate a convincing argument, it is therefore important to recognize that atoms and molecules represent an essential piece in the vast theoretical framework of natural science and that their existence is consistent with countless results obtained from experiments on systems exhibiting very broad ranges of physical phenomena.

Figure S.1 A scanning tunneling microscope (STM) image of the surface of a single graphite crystal. This image was obtained by scanning the surface with a conducting probe and measuring the tunneling current. The region depicted has sides of length $2.1\,\mathrm{nm} = 2.1 \times 10^{-9}\,\mathrm{m}$. We can clearly see a periodic pattern, which is attributed to a lattice of atoms. What can we conclude about the "existence" of atoms from this experimental result? (The image is due to Toyo Kazu Yamada of Chiba University.)

The reader may think that the argument should begin with a precise, universally valid definition of *existence* as it applies to atoms and molecules. However, in general, defining existence is a notoriously difficult problem in the philosophy of science, especially with regard to the microscopic realm. Indeed, it may not be possible to formulate precise, universally valid definitions of existence for microscopic entities.[1] For this reason, we must be satisfied with more limited characterizations, because from any given experiment, we can at most conclude the consistency or inconsistency of the experimental results with a given theory only within the context of that particular experiment.

Chapter 3

3.1. We leave the enjoyment of solving this puzzle to the reader. Let us note, however, that it is not to be solved by application of the theory of thermodynamics itself. In other words, to simply argue on the basis of Kelvin's principle that the machine could not operate as described because this would contradict the theoretical framework of thermodynamics is not a solution. The point is to confirm the impossibility asserted by Kelvin's principle through considerations outside of thermodynamics. (On the other hand, if the reader could construct a machine that operates in this manner, civilization itself—not to mention the framework of thermodynamics—would be fundamentally changed. In that case, please hurry up and get a patent!) For example, the reader should consider the density distribution of the gas and other characteristics of the system, and show where the description of the machine's operation is mistaken. The ideal situation with regard to the construction of the system can be assumed; in other words, energy losses to friction and technical problems related to the large size of the experimental apparatus, etc., can be ignored. The solution to the problem lies elsewhere.

3.2. This follows immediately from the equality $W_{\max}(T; X \to X) = 0$.

3.3. We regard N as fixed and ignore it as an independent variable. The conventional approach is to solve $\partial \tilde{p}(T; V)/\partial V = 0$ for V. However, with this approach, the computation is quite difficult. For this reason, we instead solve $\partial \tilde{p}(T; V)/\partial V = 0$ for T. This gives $T(V) = 2aN(V - bN)^2/RV^3$. Then, noting that in the limits $V \searrow bN$ and $V \nearrow \infty$ we have $T(V) \searrow 0$, and using the general relation $T(V) \geq 0$, we conclude that for any sufficiently small value of T (i.e., for $T < T_c$), the function $T(V)$ takes this value at two values of V. The maximum of $T(V)$ is realized at $V = 3bN$. Substituting this into the above expression for $T(V)$, we obtain $T_c = 8a/27bR$.

With the expression (3.37) for the pressure substituted into (3.33), integrating yields the following form for the Helmholtz free energy:

$$\widetilde{F}[T; V, N] = -NRT \log \frac{V - bN}{\{v(T) - b\}N} - \frac{aN^2}{V} + \frac{aN}{v(T)} \ .$$

Chapter 4

4.1. Consider a quasi-static adiabatic operation inducing a transition $(T; X) \xrightarrow{\text{qa}} (T'; X)$. By reversibility (see the discussion on p. 65), the time-reversed counterpart to this operation induces the transition $(T'; X) \xrightarrow{\text{qa}} (T; X)$. Clearly, unless we have $T = T'$, one of these transitions contradicts Result 4.3 (p. 67).

4.2. Fixing T, V and N, we write $p_i = p_i(T; V, N)$ and $p_a = p_a(T; V, N)$. Then, combining the two given quasi-static transitions with the transition resulting from a generalized isothermal operation that consists of merely the removal of the adiabatic walls enclosing the system (without the performance of work), we construct

---

[1] As a standard reference for this subject, see Chapter 4 of S. Okasha, *Philosophy of Science: A Very Short Introduction*, Oxford University Press, 2002. For further reading, we also recommend I. Hacking, *Representing and Intervening*, Cambridge University Press, 1983.

the following isothermal cycle:

$$(T; V, N) \xrightarrow{\text{qa}} (T + \Delta T; V + \Delta V, N) \xrightarrow{\text{i}'} (T; V + \Delta V, N) \xrightarrow{\text{qi}} (T; V, N) \ .$$

The work done by the system during this cycle is $W_{\text{cyc}} = \Delta W_{\text{a}} - \Delta W_{\text{i}} = (p_{\text{a}} - p_{\text{i}})\Delta V + O((\Delta V)^2)$. Applying Kelvin's principle ($W_{\text{cyc}} \leq 0$) to this expression yields $p_{\text{a}} \leq p_{\text{i}}$ in the case $\Delta V > 0$ and $p_{\text{a}} \geq p_{\text{i}}$ in the case $\Delta V < 0$. We thus obtain the desired identity, $p_{\text{a}} = p_{\text{i}}$.

4.3. In a quasi-static operation under which the volume of an individual ideal gas system is changed by a small amount ΔV, the work performed by the system is $\Delta W = \frac{NRT}{V}\Delta V + O((\Delta V)^2)$. Thus the quantity that we seek is given by

$$W = NR \int_{V_1}^{V_2} dV \frac{T}{V} \ .$$

In order to compute this integral, we need an expression for T in terms of V. This expression is provided by (4.45), with which the above equation can be written

$$W = NRT_1 V_1^{1/c} \int_{V_1}^{V_2} dV V^{-(1+1/c)} \ .$$

Solving, we obtain

$$W = cNRT_1 \left[1 - \left(\frac{V_1}{V_2} \right)^{1/c} \right] \ .$$

Next, replacing NRT_1 with $p_1 V_1$ and using the relation $p_1 V_1^{1+1/c} = p_2 V_2^{1+1/c}$, derived from (4.45), we can rewrite this as follows:

$$W = c \left(p_1 V_1 - p_2 V_2 \right) \ .$$

4.4. First, note that (4.45) is derived from the two expressions for the work given in (4.40) and (4.41). In the case that the system is in thermal contact with a heat bath containing the same amount of the same substance, the first of these equations still holds, while the second becomes $\Delta W = -2cNR\Delta T$. Therefore, simply replacing c by $2c$ in the expression for W in terms of V_1, V_2 and T_1 derived in Problem 4.3, we obtain

$$W = 2cNRT_1 \left[\left(1 - \frac{V_1}{V_2} \right)^{1/2c} \right] \ .$$

Proceeding analogously in the case that the heat bath consists of n systems, we have

$$W = (n + 1)cNRT_1 \left[1 - \left(\frac{V_1}{V_2} \right)^{1/(n+1)c} \right] \ .$$

Considering the $n \to \infty$ limit of this expression, let us rewrite $(n + 1)c$ as ε^{-1}. Thus, we seek the $\varepsilon \to 0$ limit of the following:

$$W = \frac{NRT_1}{\epsilon} \left[1 - \left(\frac{V_1}{V_2} \right)^{\varepsilon} \right] \ .$$

Expanding $\left(\frac{V_1}{V_2} \right)^{\varepsilon}$ about $\varepsilon = 0$, we have $\left(\frac{V_1}{V_2} \right)^{\varepsilon} = 1 + \varepsilon \log \left(\frac{V_1}{V_2} \right) + O(\varepsilon^2)$. Substituting this form into the above equation and taking the $\varepsilon \to 0$ limit, we obtain

$$W = NRT_1 \log \left(\frac{V_2}{V_1} \right) \ .$$

Chapter 5

5.1. We write the state of the compound system composed of the two individual systems in their initial states as $\{(T; X)|(T'; Y)\}$ (see Section 6.5). Then, writing the equilibrium state realized in the single, composite system resulting from the operation as $\{(\widetilde{T}; X), (\widetilde{T}; Y)\}$, we have the following transition:

$$\{(T; X)|(T'; Y)\} \xrightarrow{\text{a}} \{(\widetilde{T}; X), (\widetilde{T}; Y)\} \ .$$

In this operation, no work is performed on the mechanical world, and hence from the law of energy conservation, we have

$$U(T; X) + U(T'; Y) = U(\widetilde{T}; X) + U(\widetilde{T}; Y) \ .$$

The value of $\widetilde{T}$ is uniquely determined by this equation. Energy conservation also implies that the amount of energy in the form of heat transferred from the X system to the Y system is given by

$$Q = U(T; X) - U(\widetilde{T}; X) = U(\widetilde{T}; Y) - U(T'; Y) \ .$$

5.2. The reasonableness of (5.54) as the definition of Q is clear if one considers the flow of energy into and out of the system.

For the second half of the problem, let us consider the following isothermal cycle taking place at temperature T_2:

$$(T_2; X_2) \xrightarrow{\text{qi}} (T_2; X_3) \xrightarrow{\text{qa}} (T_1; X_1) \xrightarrow{\text{i}'} (T_2; X_2) \ .$$

The work performed by the system on the mechanical world during this cycle is

$$\begin{aligned}
W_{\text{cyc}} &= -W' + W \\
&= F[T_2; X_2] - F[T_2; X_3] + U(T_2; X_3) - U(T_1; X_1) + W \\
&= Q_{\max}(T_2; X_2 \to X_3) + Q \ .
\end{aligned}$$

Applying Kelvin's principle to these equations, we obtain $W \leq W'$ and $Q \leq Q_{\max}(T_2; X_3 \to X_2) = Q'$.

5.3. Using Kelvin's principle and the equations (3.27), (4.24) and (5.7), along with the fact that no work is performed in the third operation, we derive the result as follows:

$$\begin{aligned}
0 \geq W_{\text{cyc}} &= F[T; X_1] - F[T; X_0] + U(T; X_0) - U(T'; X_1) \\
&> F[T; X_1] - F[T; X_0] + U(T; X_0) - U(T; X_1) \\
&= -Q_{\max}(T; X_0 \to X_1) \ .
\end{aligned}$$

5.4. Let V and V' be the volumes of the gas in (a) and (c), respectively. (See the solution to Problem 1.2 for reference.) In the transition from (b) to (c), the energy increases by the amount $cNR(T_{\text{H}} - T_{\text{L}})$, and the work performed on the mechanical world is $(p_0 + mg/A)(V' - V)$. The sum of these is the amount of absorbed heat, Q_{H}. In the transition from (d) to (e), the internal energy decreases by the amount $cNR(T_{\text{H}} - T_{\text{L}})$, and the work performed on the system by the mechanical world is $p_0(V' - V)$. The sum of these is the amount of expelled heat, Q_{L}. We thus see that $W = Q_{\text{H}} - Q_{\text{L}} = mg(V' - V)/A$ is indeed the amount of work done by the system to raise the weight. Rewriting V and V' using the ideal gas equation of state and defining $\mu = mg/Ap_0$, we obtain

$$\epsilon = \frac{W}{Q_\mathrm{H}} = \left(\frac{\mu}{1+\mu}\right)\frac{T_\mathrm{H} - T_\mathrm{L}}{(c+1)T_\mathrm{H} - (c+1+\mu)T_\mathrm{L}}\ .$$

For small μ, we have $\epsilon \simeq \mu/(c+1)$, and thus the efficiency is very low. When T_H is much larger than T_L, we have $\epsilon \simeq \mu/[(1+\mu)(c+1)]$.

5.5. For a Carnot refrigerator, ω realizes the value $\omega_0 = T_\mathrm{L}/(T_\mathrm{H}-T_\mathrm{L})$, which is the largest possible value for any refrigerator (derivation omitted). Using the method described in Appendix C, it can be shown that for refrigerators, the relation $Q_\mathrm{H}/Q_\mathrm{L} \geq T_\mathrm{H}/T_\mathrm{L}$ holds universally.

Chapter 6

6.1. Let us write the temperature of the system and the volume of the gas at an arbitrary time during the operation as $\widetilde{T}$ and $\widetilde{V}$. Then, consider the amount of work performed by the system as the volume is varied by an infinitesimal amount ΔV and the temperature changes by an infinitesimal amount ΔT in response. This work can be written in two ways, $\Delta W \simeq p(\widetilde{T}; \widetilde{V}, N)\Delta V = (NR\widetilde{T}/\widetilde{V})\Delta V$ and $\Delta W = -(cNR+C_0)\Delta T = -(c + c')NR\Delta T$. Using these forms and proceeding in analogy to the derivation of the Poisson relation (4.45), it can be shown that the quantity $\widetilde{T}^{c+c'}\widetilde{V}$ is constant throughout the operation.

6.2. Recall the operation described in Problem 1.1, in which the volume of a fluid system under adiabatic conditions is increased very rapidly and then decreased slowly back to its initial value. Now, let us consider an experiment in which this operation is repeated a number of times n, and the temperature is thereby raised from T_1 to T_2. Then, we can imagine the limiting situation for such an experiment realized when we take $\varepsilon = V'-V \searrow 0$ and $n \nearrow \infty$. In this limit, we obtain an adiabatic operation in which the temperature increases from T_1 to T_2 while the volume remains fixed. (Effectively the same behavior can be realized in the case that energy is supplied slowly by an electric heating element.) This is an adiabatic operation, and hence there is no exchange of heat with the environment, but nonetheless, the entropy increases (as seen from Result 6.3 (p. 111)).

Next, consider an experiment in which a gas with initial volume V_1 undergoes adiabatic free expansion (see Sections 4.4.1 and 6.2.4) n times, with the volume increasing each time by an amount $(V_2 - V_1)/n$. Again, we can imagine the limiting situation realized when we take $n \nearrow \infty$. In this limit, we obtain an adiabatic operation in which the volume changes continuously from V_1 to V_2 with T fixed. In this case too, although no heat is exchanged with the environment, the entropy increases (as follows from (F.1) and the fact that no work is done by the system under free expansion).

6.3. Let ΔQ_i denote the heat absorbed by the system in the transition $(T_i; X_i) \overset{i'}{\longrightarrow} (T_{i+1}; X_{i+1})$, as defined in (5.54). Next, suppose that X_i' is such that the transitions $(T_i; X_i) \overset{\mathrm{qa}}{\longleftrightarrow} (T_{i+1}; X_i')$ are realizable. From the fact that $W_\mathrm{cyc} \leq 0$ holds for the isothermal cycle $(T_{i+1}; X_{i+1}) \overset{\mathrm{qi}}{\longrightarrow} (T_{i+1}; X_i') \overset{\mathrm{qa}}{\longrightarrow} (T_i; X_i) \overset{i'}{\longrightarrow} (T_{i+1}; X_{i+1})$ at T_{i+1}, we can derive the inequality $\Delta Q_i \leq T_{i+1}\Delta S_i$, where $\Delta S_i = S(T_{i+1}; X_{i+1}) - S(T_i; X_i)$. Applying the same reasoning to $(T_{i+1}; X_{i+1}) \overset{i'}{\longrightarrow} (T_i; X_i)$, we can also derive the relation $-\Delta Q_i + O(n^{-2}) \leq -T_i\Delta S_i$. We thus obtain the equality $\Delta Q_i/T_i = \Delta S_i + O(n^{-2})$. Summing these contributions over i and taking the $n \nearrow \infty$ limit, we arrive at the desired result.

6.4. Let us assume the relations $T < \widetilde{T} < T'$, where T and T' are the temperatures of the two systems before the operation, and $\widetilde{T}$ is the temperature of the composite system after the operation. From the additivity of the entropy, the change in the total entropy under the operation is $\{S(\widetilde{T}; X) + S(\widetilde{T}; Y)\} - \{S(T; X) + S(T'; Y)\}$.

With this expression, the desired result can be obtained as follows:

$$\{S(\widetilde{T};X) + S(\widetilde{T};Y)\} - \{S(T;X) + S(T';Y)\}$$

$$= \int_T^{\widetilde{T}} dT'' \frac{\partial}{\partial T''} S(T'';X) - \int_{\widetilde{T}}^{T'} dT'' \frac{\partial}{\partial T''} S(T'';Y)$$

$$= \int_T^{\widetilde{T}} dT'' \frac{1}{T''} \frac{\partial}{\partial T''} U(T'';X) - \int_{\widetilde{T}}^{T'} dT'' \frac{1}{T''} \frac{\partial}{\partial T''} U(T'';Y)$$

$$> \frac{1}{\widetilde{T}} \int_T^{\widetilde{T}} dT'' \frac{\partial}{\partial T''} U(T'';X) - \frac{1}{\widetilde{T}} \int_{\widetilde{T}}^{T'} dT'' \frac{\partial}{\partial T''} U(T'';Y)$$

$$= \frac{1}{\widetilde{T}} \{U(\widetilde{T};X) - U(T;X) - U(T';Y) + U(\widetilde{T};Y)\} = 0 ,$$

where we have used (6.15) and energy conservation.

6.5. First, establishing thermal contact between the gas and one of the solid systems and applying the quasi-static adiabatic operation constructed in Problem 6.1, we induce the transition

$$\{(T_{\mathrm{f}}; X_0), (T_{\mathrm{f}}; V_1, N)\} \xrightarrow{\text{qa}} \{(T_1; X_0), (T_1; V', N)\} ,$$

where $V' = (T_{\mathrm{f}}/T_1)^{c+c'} V_1$. Then, removing the gas from contact with the solid, we apply an operation to the gas alone through which it undergoes the transition

$$(T_1; V', N) \xrightarrow{\text{qa}} (T_{\mathrm{f}}; V'', N) ,$$

where $V'' = (T_1/T_{\mathrm{f}})^c V'$. Next, establishing thermal contact between the gas and the second solid system, we carry out another operation through which we realize the transition

$$\{(T_{\mathrm{f}}; X_0), (T_{\mathrm{f}}; V'', N)\} \xrightarrow{\text{qa}} \{(T_2; X_0), (T_2; V''', N)\} ,$$

where $V''' = (T_{\mathrm{f}}/T_2)^{c+c'} V''$. Finally, removing the gas from contact with the solid, we apply an operation to the gas causing the transition

$$(T_2; V''', N) \xrightarrow{\text{qa}} (T_{\mathrm{f}}; V_2, N) ,$$

where $V_2 = (T_2/T_{\mathrm{f}})^c V''' = (T_{\mathrm{f}}^2/T_1 T_2)^{c'} V_1$. Combining all of these operations, we obtain the quasi-static adiabatic transition

$$\{(T_{\mathrm{f}}; X_0), (T_{\mathrm{f}}; X_0), (T_{\mathrm{f}}; V_1, N)\} \xrightarrow{\text{qa}} \{(T_1; X_0), (T_2; X_0), (T_{\mathrm{f}}; V_2, N)\} .$$

The change in entropy experienced by the gas in this transition is

$$\Delta S' = N R \log(V_2/V_1) = C_0 \log(T_{\mathrm{f}}^2/T_1 T_2) ,$$

which is indeed consistent with (6.64). This increase in the entropy of the gas exactly cancels the decrease in the entropy of the two solid systems.

Chapter 7

7.1. It is only necessary to take the partial derivative of both sides of (7.8) with respect to V or N and then simplify using the appropriate relation.

7.2. The total differential of (7.8) is $dF = -pdV - Vdp + \mu dN + Nd\mu$. Combining this with (7.12), we obtain (7.62). The meaning of this equation is that for any small changes of the independent variables, the relation $S\Delta T - V\Delta p + N\Delta\mu = 0$ must hold. In particular, with T and N held fixed, dividing this equation by ΔV, we obtain the first equation in (7.61). The second equation in (7.61) can be obtained similarly.

We omit the derivation employing the Gibbs free energy.

7.3. The solution to the first half of the problem is left to the reader.

In the case of a van der Waals gas, (7.63) implies the relation $\partial^2 U/\partial V\partial T = 0$, which can also be written $\partial C_{\mathrm{v}}/\partial V = 0$.

7.4. Demonstrating that the functions in (7.64) satisfy Maxwell's equations is left to the reader.

The energy density of the electromagnetic field is given by

$$u(x, y, z, t) = \frac{\varepsilon_0}{2}|\boldsymbol{E}(x, y, z, t)|^2 + \frac{1}{2\mu_0}|\boldsymbol{B}(x, y, z, t)|^2$$

$$= \frac{\varepsilon_0}{2}E_0^2\left(\sin^2 kx\,\cos^2 \omega t + \cos^2 kx\,\sin^2 \omega t\right)\ .$$

The desired result is obtained by averaging over x and t.

The current in the $x = 0$ plane flows in the z direction with density $j(y, z, t) = (B_0/\mu_0)\sin\omega t$. Thus, there is a force per unit area of magnitude $j(y, z, t)\,B(0, y, z, t)/2 = (B_0^2/2\mu_0)\sin^2 \omega t$ in the negative x direction exerted on the conducting material. The desired result is obtained by averaging over t.

7.5. Choose arbitrary values of v and n, v_0 and n_0 (with $(v_0, n_0) \neq (0, 0)$), and define the function $g(t) = f(v_0 t, n_0 t)$. From the properties of $f(v, n)$, we know that $g(t)$ satisfies the relations $g(t) \geq g(0)$ and $g'(0) = 0$. Next, expand $g(t)$ in a Taylor series to second order:

$$g(t) - g(0) = g'(0)\,t + \frac{g''(0)}{2}t^2 + O(t^3)$$

$$= \frac{1}{2}\left[v_0^2\frac{\partial^2 f(v, n)}{\partial v^2} + 2v_0 n_0\frac{\partial^2 f(v, n)}{\partial v\,\partial n} + n_0^2\frac{\partial^2 f(v, n)}{\partial n^2}\right]_{v,n=0}t^2 + O(t^3)$$

$$= \frac{1}{2}\left(v_0, n_0\right)\mathsf{D}\begin{pmatrix}v_0\\n_0\end{pmatrix}t^2 + O(t^3)\ .$$

Because the left-hand side of this equation must be greater than or equal to 0, regarding t as small and ignoring higher-order terms, we have

$$\left(v_0, n_0\right)\mathsf{D}\begin{pmatrix}v_0\\n_0\end{pmatrix} \geq 0\ .$$

Now, note that D is a real symmetric matrix, and hence its eigenvalues are real and the components of its eigenvectors can be chosen to be real. Choosing (v_0, n_0) to be an eigenvector corresponding to either eigenvalue of D, we see that the result follows immediately from the above inequality.

7.6. Allowing the piston that had been fixed at a position θ_0 to move freely and then waiting for the system to reach equilibrium, we realize the transition $(T; \theta_0) \overset{\mathrm{i}}{\longrightarrow} (T; \theta^*)$. There is no work performed under the operation that results in this transition (i.e., releasing the constraint on the piston), and therefore from the principle of maximum work, we obtain $0 \leq W_{\max}(T; \theta_0 \to \theta^*)$. Using the definition of the

Helmholtz free energy, this can be written $F[T; \theta^*] \leq F[T; \theta_0]$. The variational principle is easily derived from this relation.

For an ideal gas (choosing $u = 0$), we have

$$
\begin{aligned}
F[T; \theta] = &-\frac{NRT}{2} \log\left\{\left(\frac{T}{T^*}\right)^c \frac{\theta}{\pi} \frac{2V}{v^* N}\right\} \\
&-\frac{NRT}{2} \log\left\{\left(\frac{T}{T^*}\right)^c \left(1 - \frac{\theta}{\pi}\right) \frac{2V}{v^* N}\right\} + mgr\sin\theta \ .
\end{aligned}
$$

With this form, we have the following:

$$
\left.\frac{\partial}{\partial\theta} F[T; \theta]\right|_{\theta=\pi/2} = 0, \qquad
\left.\frac{\partial^2}{\partial\theta^2} F[T; \theta]\right|_{\theta=\pi/2} = \frac{4NRT}{\pi^2} - mgr \ .
$$

Thus, it seems that the claim holds with $T_c(m) = \pi^2 mgr/4NR$. In fact, it can be shown by investigating the global behavior of $F[T; \theta]$ (e.g., graphically) that this claim indeed does hold.

For $T < T_c(m)$, $F[T; \theta]$ is not a convex function of θ, and for this reason, the fact that Result 7.2 does not hold is not surprising. At the heart of the problem here is that for a proper thermodynamic system, convexity follows from the extensivity of the Helmholtz free energy and the fact that it is characterized by a variational principle, while for the system considered here, the situation is more complicated. Uncovering the reason for this is left to the reader.

A metastable state can be realized by slightly raising one side of the tube. (However, this fact is difficult to prove analytically.)

7.7. The result (7.67) is readily computed from the variational principle $F[T; X^*, Y] = \min_X F[T; X, Y]$, which can be derived as in Problem 7.6. Taking the derivative of each side of (7.67) with respect to T yields

$$
\frac{\partial X^*(T; Y)}{\partial T} \left.\frac{\partial^2 F[T; X, Y]}{\partial X^2}\right|_{X=X^*(T;Y)} + \left.\frac{\partial^2 F[T; X, Y]}{\partial T \partial X}\right|_{X=X^*(T;Y)} = 0 \ .
$$

Then, applying (7.2), we obtain (7.68). (We omit discussion regarding the connection between this relation and equilibrium.) For the system depicted in Figure 8.1, increasing T causes an increase in V. As a result, the system absorbs heat, thereby acting to (slightly) decrease the temperature of the environment, in accordance with the Le Chatelier-Braun principle.

The situation is similar when Y_i is varied, as we now see. Consider a container filled with a gas that is divided into upper and lower compartments by a freely moving piston. We regard the gas in these compartments as the Y and X systems, respectively. Then, suppose that the volume of the upper compartment is increased by raising the top of the container. When this is done, the pressure in that compartment will decrease. In response, the piston will move upward. This partially counteracts the decrease in pressure experienced by the top compartment.

7.8. Let V_1 and V_2 denote the values of the volume at the points of tangency of $\widetilde{F}[V]$ and the line in Figure 7.12. The condition determining these values is

$$
\widetilde{F}'[V_1] = \widetilde{F}'[V_2] = \frac{\widetilde{F}[V_2] - \widetilde{F}[V_1]}{V_2 - V_1} \ .
$$

Let us write $\widetilde{F}'[V_1] = -p_0$. (In Figure 7.11(c), the interval of constant pressure $p = p_0$ is $[V_1, V_2]$.) Then, from the above equation, we obtain

$$0 = \widetilde{F}[V_2] - \widetilde{F}[V_1] + p_0(V_2 - V_1) = \int_{V_1}^{V_2} \{\widetilde{F}'[V] + p_0\}dV$$

$$= \int_{V_1}^{V_2} \{-\tilde{p}(V) + p_0\}dV = A_1 - A_2 \ .$$

7.9. Let us refer to the states appearing in the cycle (7.69) as A, B, C and D, in order from first to last. Because the system is characterized by phase coexistence throughout the cycle, between A and B the pressure is constant and equal to $p_{\mathrm{v}}(T + \Delta T)$, and between C and D it is constant and equal to $p_{\mathrm{v}}(T)$. In general, the change in the enthalpy, $H = U + pV$, resulting from an infinitesimal quasi-static adiabatic transition $(T; V, N) \xrightarrow{\mathrm{qa}} (T + \Delta T; V + \Delta V, N)$ is given by $\Delta H = \Delta U + p\Delta V + V\Delta p = V\Delta p$ (where we have used the energy conservation relation $\Delta U + p\Delta V = 0$). Thus, in the present case, we have $H(B) - H(C) = V_1\Delta p$ and $H(A) - H(D) = V_0\Delta p$, with $\Delta p = p_{\mathrm{v}}(T + \Delta T) - p_{\mathrm{v}}(T)$. Then, because the amount of heat transferred to the system during a quasi-static isothermal transition is equal to the change in enthalpy, we have $Q(D \to C) = H(C) - H(D) = H_{\mathrm{vap}}(T; N')$, where N' is the amount of substance that changes from liquid to gas as the state of the system changes from D to C. Similarly, we also have $Q(A \to B) = H(B) - H(A)$. Adding the changes in H over the cycle gives $Q(A \to B) = Q(D \to C) + (V_1 - V_0)\Delta p$. Combining this with the relation $Q(A \to B)/Q(D \to C) = 1 + (\Delta T/T)$, and using the identity $(V_1 - V_0)/N' = v_{\mathrm{G}}(T) - v_{\mathrm{L}}(T)$, we obtain the Clausius-Clapeyron relation.

7.10. The work performed by the system on the mechanical world is $W = -p_{\mathrm{H}}V + p_{\mathrm{L}}V'$. From energy conservation, we thus have $U(T'; V', N) = U(T; V, N) + p_{\mathrm{H}}V - p_{\mathrm{L}}V'$. Rearranging, we obtain $U(T; V, N) + p_{\mathrm{H}}V = U(T'; V', N) + p_{\mathrm{L}}V'$. The left-hand and right-hand sides are the enthalpy before and after the operation, respectively.

Chapter 8

8.1. The proof of extensivity is trivial.

If we assume that $F[T; V, N]$ is differentiable with respect to T, then for a given value of the entropy, S_0, the value of T at which $\max_T\{F[T; V, N] + TS_0\}$ is realized is that which solves the equation $S(T; V, N) = S_0$. Therefore, for given $T = T_0$, choosing S_0 to be $S(T_0; V, N)$, the value of T at which $\max_T\{F[T; V, N] + TS_0\}$ is realized is T_0 itself. We thus obtain the equality $U[S(T; V, N), V, N] = F[T; V, N] + T S(T; V, N)$. The right-hand side of this equality is identically $U(T; V, N)$.

With regard to the partial derivatives, see (F.3).

8.2. The result can be derived through direct application of (7.30) to the definition (8.36), as follows:

$$U[S, V, N] = \max_T\{F[T; V, N] + TS\}$$

$$\leq \max_T\{F[T; V_1, N_1] + F[T; V_2, N_2] + T(S_1 + S_2)\}$$

$$\leq \max_T\{F[T; V_1, N_1] + TS_1\} + \max_T\{F[T; V_2, N_2] + TS_2\}$$

$$= U[S_1, V_1, N_1] + U[S_2, V_2, N_2] \ .$$

The second half of the problem is left to the reader.

8.3. Because no work is performed in the operation, the equality $U[S_1, V_1, N_1] + U[S_2, V_2, N_2] = U[S, V, N]$ follows from energy conservation. Also, the inequality $U[S_1, V_1, N_1] + U[S_2, V_2, N_2] \geq U[S_1 + S_2, V, N]$ follows from (8.37). We thus obtain $U[S, V, N] \geq U[S_1 + S_2, V, N]$, which implies $S \geq S_1 + S_2$.

8.4. The pressure $p(T; V, N) = NRT/(V - aN)$ can be computed directly from the definition. Solving for V, we obtain $V(T, p; N) = NRT/p + aN$. Substituting this for V in the expression for $F[T; V, N]$ yields $F(T, p; N) = -NRT \log\{(T/T^*)^c RT/(v^* p)\}$, which is independent of a. Contrastingly, from the simplified definition of the Gibbs free energy, (8.12), we have $G[T, p; N] = F(T, p; N) + NRT + aNp$.

8.5. Using the definition of the Gibbs free energy given in (8.3) and the definition of a concave function (obtained from (G.1) by simply reversing the inequality), the result is derived as follows:

$$G[\lambda T_1 + (1 - \lambda)T_2, \lambda p_1 + (1 - \lambda)p_2; N]$$

$$= \min_V \{F[\lambda T_1 + (1 - \lambda)T_2; V, N] + (\lambda p_1 + (1 - \lambda)p_2)V\}$$

$$\geq \min_V \{\lambda F[T_1; V, N] + (1 - \lambda)F[T_2; V, N] + (\lambda p_1 + (1 - \lambda)p_2)V\}$$

$$\geq \lambda \min_V \{F[T_1; V, N] + p_1 V\} + (1 - \lambda) \min_V \{F[T_2; V, N] + p_2 V\}$$

$$= \lambda G[T_1, p_1, N] + (1 - \lambda)G[T_2, p_2; N] .$$

Obviously, the same steps can be carried out for multi-component fluid systems and ferromagnetic systems as well, and thus the same conclusion is reached in these cases.

8.6. We omit the derivation of the equation for the enthalpy. Taking the derivative of this equation with respect to T and p, we obtain

$$\frac{\partial H(T, p; N)}{\partial T} = C_{\mathrm{p}}(T, p; N) ,$$

$$\frac{\partial H(T, p; N)}{\partial p} = V(T, p; N) + T\frac{\partial S(T, p; N)}{\partial p} = V(T, p; N) - T\frac{\partial V(T, p; N)}{\partial T} .$$

Here we have used (8.14), (8.28), (8.16) and the Maxwell relation

$$\partial S(T, p; N)/\partial p = -\partial V(T, p; N)/\partial T ,$$

which can be derived directly from the differential form for the Gibbs free energy, (8.22). Using the above relations and the result of Problem 7.10, $H(T, p; N) = H(T - \Delta T, p - \Delta p; N)$, the desired relation can be demonstrated straightforwardly.

8.7. The derivation of the Joule-Thomson coefficient for an ideal gas is omitted.
Rearranging (3.37) and expanding to first order, we have

$$\frac{NRT}{\tilde{p}} = V\left\{\left(1 - \frac{bN}{V}\right)^{-1} - \frac{aN}{RTV}\right\}^{-1} \simeq V\left\{1 - \frac{bN}{V} + \frac{aN}{RTV}\right\} = V - bN + \frac{aN}{RT} .$$

Substituting the form of V obtained from the above for $V(T, p; N)$ in (8.41) yields

$$\mu_{\mathrm{JT}}(T, p) = (2a/RT - b)/c_{\mathrm{p}}(T, p) .$$

With the inversion temperature defined as $T_{\mathrm{inv}} = 2a/bR$, it is seen that for $T > T_{\mathrm{inv}}$, the temperature of the gas increases as a result of the operation, whereas for $T < T_{\mathrm{inv}}$, it decreases. This effect is used in actual applications to cool (and even liquefy) gases.

8.8. Taking the derivative of (8.13) with respect to T (or substituting NRT/p for V in (6.31)), we obtain $S(T, p; N) = cNR - NR\log\{(T^*/T)^{c+1}(p/p^*)\}$. Therefore, the equality $S(T, p; N) = S(T', p'; N)$ implies $p/T^{c+1} = p'/T'^{c+1}$.

8.9. To demonstrate (8.46), it is sufficient to consider (8.27), which relates $C_p(T, p; N)$ to $S(T, p; N)$, (6.24), which relates $C_V(T; V, N)$ to $S(T; V, N)$, and the Maxwell relation (7.19). The remainder is straightforward.

8.10. We directly derive the result below, leaving the determination of the necessary conditions regarding differentiability to the reader.

Let us choose Δy and Δz to satisfy $x(y, z) = x(y + \Delta y, z + \Delta z)$. To lowest order, this implies the following:

$$\frac{\partial x(y, z)}{\partial y}\Delta y + \frac{\partial x(y, z)}{\partial z}\Delta z = 0 \ .$$

Rearranging, we obtain

$$\frac{\partial x(y, z)}{\partial y}\frac{\Delta y}{\Delta z}\left(\frac{\partial x(y, z)}{\partial z}\right)^{-1} = -1 \ .$$

Then, considering the infinitesimal limit, identifying $\Delta y/\Delta z$ with $\partial y(x, z)/\partial z$, and using the relation $(\partial x(y, z)/\partial z)^{-1} = \partial z(x, y)/\partial x$, we arrive at the desired result.

8.11. Taking the difference of the two equations, we have

$$\lim_{\tau \searrow 0}\left\{G[T_{\mathrm{b}}(p + \Delta p) + \tau, p + \Delta p; N] - G[T_{\mathrm{b}}(p) + \tau, p; N]\right\}$$

$$= \lim_{\tau \nearrow 0}\left\{G[T_{\mathrm{b}}(p + \Delta p) + \tau, p + \Delta p; N] - G[T_{\mathrm{b}}(p) + \tau, p; N]\right\} \ .$$

Dividing this by Δp and taking the limit $\Delta p \searrow 0$, we obtain

$$\lim_{\tau \searrow 0}\left\{\frac{dT_{\mathrm{b}}(p)}{dp}\frac{\partial G[T_{\mathrm{b}}(p) + \tau, p; N]}{\partial T} + \left.\frac{\partial G[T, p; N]}{\partial p}\right|_{T = T_{\mathrm{b}}(p) + \tau}\right\}$$

$$= \lim_{\tau \nearrow 0}\left\{\frac{dT_{\mathrm{b}}(p)}{dp}\frac{\partial G[T_{\mathrm{b}}(p) + \tau, p; N]}{\partial T} + \left.\frac{\partial G[T, p; N]}{\partial p}\right|_{T = T_{\mathrm{b}}(p) + \tau}\right\} \ .$$

This immediately yields

$$\frac{dT_{\mathrm{b}}(p)}{dp}\frac{\partial G[T_{\mathrm{b}}(p), p; N]}{\partial T_+} + V_{\mathrm{G}}(T_{\mathrm{b}}(p), N) = \frac{dT_{\mathrm{b}}(p)}{dp}\frac{\partial G[T_{\mathrm{b}}(p), p; N]}{\partial T_-} + V_{\mathrm{L}}(T_{\mathrm{b}}(p), N) \ .$$

The result can be obtained by rearranging this expression and making use of (8.34).

Chapter 9

9.1. No work is performed in the isothermal operation under which the ordinary wall is replaced with the semipermeable wall, causing the transition

$$\{(T; V, \boldsymbol{N}), (T; V', \boldsymbol{N}')\} \overset{\mathrm{i}}{\longrightarrow} \{(T; V, \widetilde{\boldsymbol{N}}), (T; V', \widetilde{\boldsymbol{N}}')\} \ .$$

Thus, the principle of maximum work gives

$$0 \leq W_{\mathrm{max}}(T; \{(V, \boldsymbol{N}), (V', \boldsymbol{N}')\} \to \{(V, \widetilde{\boldsymbol{N}}), (V', \widetilde{\boldsymbol{N}}')\})$$

$$= (F[T; V, \boldsymbol{N}] + F[T; V', \boldsymbol{N}']) - (F[T; V, \widetilde{\boldsymbol{N}}] + F[T; V', \widetilde{\boldsymbol{N}}']) \ .$$

Note that $\boldsymbol{N}$ and $\boldsymbol{N}'$ here are identical to $\widetilde{\boldsymbol{N}}$ and $\widetilde{\boldsymbol{N}}'$, respectively, for all components other than the 1st. The variational principle and balance condition are obtained by simply rearranging this relation.

9.2. The derivations are left to the reader. The Gibbs-Duhem equation for a multi-component system is

$$S\,dT - V\,dp + \sum_{i=1}^{m} N_i\,d\mu_i = 0\,.$$

From this relation, for example, fixing T and p and varying only N_j, we can derive the equality

$$\sum_{i=1}^{m} N_i \frac{\partial \mu_i(T, p; \boldsymbol{N})}{\partial N_j} = 0\,.$$

9.3. For a two-component dilute solution, the pressure and chemical potential of the solvent as functions of T, V and N are given by

$$p(T; V, \boldsymbol{N}) \simeq p_1(T; V, N_1) + \frac{N_2 RT}{V} + \frac{N_2}{N_1}\sigma(T; \frac{V}{N_1})\,,$$

$$\mu_1(T; V, \boldsymbol{N}) \simeq \mu_1(T; V, N_1, 0) + \frac{V N_2}{N_1^2}\sigma(T; \frac{V}{N_1})\,,$$

where $\sigma(T; v) = -\partial w_2(T; v)/\partial v$. In the present case, we have $N_1 = \frac{V}{V+V'}N_1^{\text{tot}} - M$ and $N_1 = \frac{V'}{V+V'}N_1^{\text{tot}} + M$ in the left and right compartments, respectively. With these values, using the intensivity of the chemical potential, we can write $\mu_1(T; V, N_1, 0)$ as $\mu_1(T; \frac{V+V'}{N_1^{\text{tot}}}, 1 - M\frac{V+V'}{V N_1^{\text{tot}}}, 0)$ for the left compartment and $\mu_1(T; \frac{V+V'}{N_1^{\text{tot}}}, 1 + M\frac{V+V'}{V' N_1^{\text{tot}}}, 0)$ for the right compartment. Expanding these in M, we obtain the approximate forms

$$\mu_1(T; \frac{V+V'}{N_1^{\text{tot}}}, 1, 0) - M\frac{V+V'}{V N_1^{\text{tot}}}\, \frac{\partial}{\partial n}\mu_1(T; \frac{V+V'}{N_1^{\text{tot}}}, n, 0)\Big|_{n=1}$$

and

$$\mu_1(T; \frac{V+V'}{N_1^{\text{tot}}}, 1, 0) + M\frac{V+V'}{V' N_1^{\text{tot}}}\, \frac{\partial}{\partial n}\mu_1(T; \frac{V+V'}{N_1^{\text{tot}}}, n, 0)\Big|_{n=1}$$

for the left and right compartments, respectively. Using these forms for $\mu_1(T; V, N_1, 0)$ and equating the chemical potentials for Species 1 on either side so obtained, we derive the following relation:

$$M\left(\frac{1}{V} + \frac{1}{V'}\right)\frac{\partial}{\partial n}\mu_1(T; \frac{V+V'}{N_1^{\text{tot}}}, n, 0)\Big|_{n=1} = \frac{(V+V')N_2}{V N_1^{\text{tot}}}\sigma(T; \frac{V+V'}{N_1^{\text{tot}}})\,.$$

From this, we can determine the amount of Species 1 that is transferred from the left compartment to the right compartment, M. Deriving the pressures on the two sides in a similar manner, the osmotic pressure is evaluated as

$$p_{\text{osmo}} \simeq -M\left(\frac{1}{V} + \frac{1}{V'}\right)\frac{V+V'}{N_1^{\text{tot}}}\, \frac{\partial}{\partial n}p_1(T; \frac{V+V'}{N_1^{\text{tot}}}, n)\Big|_{n=1}$$

$$+ \frac{(V+V')N_2}{V N_1^{\text{tot}}}\sigma(T; \frac{V+V'}{N_1^{\text{tot}}}) + \frac{N_2 RT}{V}\,.$$

The first and second terms here can respectively be interpreted as the change in the pressure resulting from the transfer of Species 1 from left to right and the change in the pressure arising from the interaction between the two substances. Next, from the second Euler relation appearing in (7.61), we have

$$\frac{V+V'}{N_1^{\text{tot}}}\,\frac{\partial}{\partial n}p_1\!\left(T;\frac{V+V'}{N_1^{\text{tot}}},n\right)\Bigg|_{n=1} = \frac{\partial}{\partial n}\mu_1\!\left(T;\frac{V+V'}{N_1^{\text{tot}}},n,0\right)\Bigg|_{n=1}.$$

Combining this with the equation determining M, it is found that, in what seems to be a miraculous coincidence, the first and second terms mentioned above cancel.

9.4. Consider a system containing no catalyst that is in the initial state $(T;V,\boldsymbol{N}+\boldsymbol{\nu}\xi)$ for an arbitrary value of the extent of reaction, ξ. After catalyst is added, the system will spontaneously evolve toward the equilibrium composition, $\boldsymbol{N}+\boldsymbol{\nu}\tilde{\xi}$. Because no work is performed by the system on the mechanical world in this process, we have $W_{\max}(T;V,\boldsymbol{N}^{(0)}+\boldsymbol{\nu}\xi \to \boldsymbol{N}^{(0)}+\boldsymbol{\nu}\tilde{\xi}) \geq 0$. Using the definition of the Helmholtz free energy, this can be rewritten as $F[T;V,\boldsymbol{N}+\boldsymbol{\nu}\tilde{\xi}] \leq F[T;V,\boldsymbol{N}+\boldsymbol{\nu}\xi]$. The equilibrium condition $\sum_{i=1}^{m}\nu_i\,\mu_i(T;V,\widetilde{\boldsymbol{N}}) = 0$ is readily derived from this.

The remainder of the problem is left to the reader.

9.5. First we use the method employing the Gibbs free energy. The derivation is the same as that in the main text up through (9.124). However, in the present case, the A^- ions do not migrate, and for this reason, in order to maintain electric neutrality, M^+ ions must migrate in an amount that counteracts the exchange of ions between the solutions and the electrodes, as described by (9.124). As a result, $N_{L,2}^{(1)}$ and $N_{R,2}^{(1)}$ do not change. Thus, as seen from (9.126), we have $G^{(2)} = G^{(1)}$, which implies $\Delta W = 0$ and hence $\mathcal{E} = 0$.

Next we use the method employing the electrochemical potential. In the present case, the balance condition is not that in (9.134), but rather, $\mu_{M+}^{L} + F\,\varphi_3 = \mu_{M+}^{R} + F\,\varphi_4$. Thus, the membrane potential is $\varphi_3 - \varphi_4 = (\mu_{M+}^{R} - \mu_{M+}^{L})/F$. Combining this with (9.136), we obtain $\varphi_1 - \varphi_2 = 0$. This result can be interpreted as the cancellation of the membrane potential and the difference in electric potentials due to the difference in concentrations of M^+ in the two solutions.

Chapter 10

10.1. The magnitude of the magnetic field is nI. The magnitude of the emf per winding of the solenoid is $\pi r^2\, dB(t)/dt$. Multiplying this by the total number of windings, we have $\mathcal{E}(t) = \pi r^2 nL\, dB(t)/dt = nV\, dB(t)/dt$. Then, using $H(t) = n\,I(t)$, the work performed during the time interval in question can be derived as follows:

$$\Delta W_{\text{tot}} = \int \mathcal{E}(t)\,I(t)\,dt = V\int H(t)\frac{dB(t)}{dt}\,dt = VH\Delta B + O\{(\Delta H)^2\}\,.$$

For the final step, it is sufficient to substitute $\mu_0(\Delta H + \Delta m)$ for ΔB.

10.2. Straightforward computations yield the following:

$$S(T,H;N) = NN_A\left\{k_B \log\left(2\cosh\frac{\mu H}{k_B T}\right) - \frac{\mu H}{T}\tanh\frac{\mu H}{k_B T}\right\},$$

$$m(T,H;N) = N_A\mu\tanh\frac{\mu H}{k_B T},$$

$$c(T,H;N) = \frac{N_A\mu^2 H^2}{k_B T^2}\,\text{sech}^2\frac{\mu H}{k_B T}\,.$$

We omit the graphs, but as one important feature, let us note that $c(T,H;N)$ has a peak at a value $T \propto \mu H/k_B$.

10.3. Carrying out a Taylor expansion of the Gibbs free energy, we obtain

$$G[T, H; N] = -N N_A k_B T \left\{ \log 2 + \frac{1}{2} \left(\frac{\mu H}{k_B T} \right)^2 \right\} + O(\{\mu H / k_B T\}^4) \,.$$

With this form, the following expressions can be derived (to the same order) for the entropy and magnetization:

$$S(T, H; N) = N N_A k_B \left\{ \log 2 - \frac{1}{2} \left(\frac{\mu H}{k_B T} \right)^2 \right\} \,, \quad m(T, H) = \frac{N_A \mu^2 H}{k_B T} \,.$$

The Helmholtz free energy is obtained as

$$F[T; M, N] = \max_H \{ G[T, H; N] + HM \} = G[T; H^*; N] + H^* M \,,$$

where the value of H at which the maximum is realized, H^*, is determined by $M = N N_A \mu^2 H^* / k_B T$. A simple computation then yields

$$F[T; M, N] = N N_A k_B T \left\{ -\log 2 + \frac{1}{2} \left(\frac{M}{N N_A \mu} \right)^2 \right\} \,.$$

10.4. Because this is a quasi-static adiabatic operation, the equality $S(T, H; N) = S(T', H'; N)$ holds. Using the explicit expression for the entropy derived in Problem 10.3, we find $T' = (H'/H)T$. Thus, it is seen that if the magnetic field is weakened, the temperature decreases. This is one practical method for preparing a low-temperature environment that is used in actual experiments.

10.5. The first half of the problem is left to the reader.

Applying the definitions of the critical exponents (for $T < T_c$) to (10.45), we find that the relation

$$(T_c - T)^{-\alpha'} \gtrsim (\text{constant}) \times (T_c - T)^{2(\beta-1)+\gamma'}$$

holds for small $T_c - T$. The inequality that we seek follows directly.

10.6. Taking the derivative of each side of (10.29) with respect to H and setting $\tau = 0$, we have $\lambda^b m(T_c, \lambda^b H) \simeq \lambda m(T_c, H)$. Rewriting H here as H_0, which we regard as a fixed reference value, and interpreting $H = \lambda^b H_0$ as the corresponding scaled value, we obtain $m(T_c, H) \simeq \lambda^{1-b} m(T_c, H_0) \propto H^{(1-b)/b}$. The equation (10.48) follows. The scaling relations are $\gamma = \beta(\delta - 1)$, $\alpha + \beta(\delta + 1) = 2$, etc.

Bibliography

The following is a brief list of works that we found useful while writing this book.

[1] P. Atkins and J. De Paula, *Atkins' Physical Chemistry*, Oxford University Press, 2014.

We have largely followed this standard textbook with regard to notation and terminology in the context of chemical thermodynamics.

[2] H. B. Callen, *Thermodynamics and Introduction to Thermostatistics*, John Wiley & Sons, 1985.

This is a well-known textbook that employs a logical approach based on the formulation of Gibbs. That formulation, which begins by asserting the existence of entropy, is complementary to the formulation used in the present book. The concept of complete thermodynamic functions (although referred to by a different name) is given due consideration. Unfortunately, however, the convexity properties of thermodynamic functions are not fully exploited.

[3] E. Fermi, *Thermodynamics*, Dover, 1956.

This is a classic masterpiece. Although the treatment is not mathematically rigorous, it elucidates various manners of thinking important in thermodynamics.

[4] H. Flanders, *Differential Forms with Applications to the Physical Sciences*, Dover, 1989.

[5] E. A. Guggenheim, *Thermodynamics — An Advanced Treatment for Chemists and Physicists*, North-Holland, 1993.

[6] W. M. Haynes, ed., *CRC Handbook of Chemistry and Physics*, 95th ed., CRC Press, 2014.

We used this standard reference book for the numerical values of physical quantities.

[7] International Union of Pure and Applied Chemistry, *IUPAC Compendium of Chemical Terminology*, 3rd ed., 2006 (https://goldbook.iupac.org).

We relied on this resource (commonly referred to as the "Gold Book") for definitions and conventions in chemistry.

[8] A. N. Levine, *Physical Chemistry*, McGraw-Hill, 2002.

[9] E. H. Lieb and J. Yngvason, "The Physics and Mathematics of the Second Law of Thermodynamics," *Physics Reports*, vol. 310, no. 1, 1999.

This paper presents an axiomatic formulation of thermodynamics based on the concept of adiabatic transitions. The authors begin by defining and characterizing entropy through consideration of only adiabatic operations. They construct the theory of thermodynamics in a mathematically rigorous manner while also presenting a clear physical picture.

[10] K. S. Pitzer, *Thermodynamics*, McGraw-Hill, 1995.

[11] S. Sasa, *Introduction to Thermodynamics* (in Japanese), Kyoritsu, 2000.

[12] H. E. Stanley, *Introduction to Phase Transitions and Critical Phenomena*, International Series of Monographs on Physics, Oxford University Press, 1971.

[13] H. Tasaki, *Physics and Mathematics of Quantum Many-Body Systems*, Graduate Texts in Physics, Springer, 2020.

[14]   A. S. Wightman, "Convexity and the Notion of Equilibrium State in Thermodynamics and Statistical Mechanics," in R. B. Israel, *Convexity in the Theory of Lattice Gases*, Princeton University Press, 1979.

>   This appears as an introduction to Israel's book on statistical mechanics. Wightman, a mathematical physicist, gives an extremely clear account of the importance and usefulness of convexity in thermodynamics.

[15]   Y. Yamamoto, *Historical Development of the Philosophy of Thermodynamics: Heat and Entropy* (in Japanese), Gendai-Suugakusha, 1987; Chikuma, 2008.

>   The historical comments given in the present book are based mainly on this excellent book on the history of science.

Index